COASTAL AQUACULTURE ENGINEERING

A.N. Bose, S.N. Ghosh,
C.T. Yang and A. Mitra

CAMBRIDGE
UNIVERSITY PRESS

CAMBRIDGE UNIVERSITY PRESS
Cambridge, New York, Melbourne, Madrid, Cape Town, Singapore, São Paulo, Delhi

Cambridge University Press
The Edinburgh Building, Cambridge CB2 8RU, UK

Published in the United States of America by Cambridge University Press, New York

www.cambridge.org
Information on this title: www.cambridge.org/9780521427692

© 1991 A.N. Bose, S.N. Ghosh, C.T. Yang and A. Mitra

First published Routledge 1991
This edition published by Cambridge University Press 2009

A catalogue record for this publication is available from the British Library

ISBN 978-0-521-41767-9 hardback
ISBN 978-0-521-42769-2 paperback

Contents

Preface

Aquaculture is the science and technology of breeding and growing aquatic animals as well as culture of aquatic plants. Aquaculture has been in practice in China and India for more than 2000 years. However, the realisation of its role in meeting the world's food needs and of the need to develop it as an industry is more recent.

Of late, the world's evergrowing human population has necessitated harnessing of all the available natural resources particularly the aquatic ones for increased production of food to meet the current and future aquatic demands. One of the great natural reservoir of resources is our ocean. It was considered as a source of unlimited supply of food. However studies indicate that the maximum sustainable yield of marine fish through capture of natural wild stock may not exceed 100 million metric tons per year. Scientific aquaculture on the other hand, can ensure steady and regular production of fish which may be increased considerably with planned input of modern technology, similar to the case of scientific agriculture in relation to the production of land-based food crops. Aquaculture has therefore the potential of being the second largest means of food production next only to agriculture.

Aquacultural production on a viable commercial scale really lies in the realm of biotechnology, for which balanced support from biological and engineering sciences are required. A viable aquaculture system is a complex combination of various elements which include besides fish production technology, effective post harvest treatment of the product for its marketing in consideration to the highly perishable nature of the product and unsatisfied market for good quality raw or processed fish in the area of origin.

There exists considerable information, although not very perfect on biological studies of many aquatic animals for cultural purposes which helped the development of a viable base of commercial industry. This however is not true for relevant engineering disciplines for aquacultural needs particularly with regard to coastal aquaculture. The environment where coastal aquacultural operations have to be carried out are in most difficult locations from the point of view of planning, design and construction of engineering structures. Normally coastal ecosystems are varied and diverse, like rocky shores, muddy shores, sandy shores, salt marshes, lagoons, estuaries and mangroves. These are the areas where there occurs a complex underplay of tides, waves, littoral currents and sediment which primarily are responsible for modification/ changes in the coastal topography when their delicate equilibrium is disturbed. Further, the soil condition in some of these environments are mostly clay and silt and have got little or no bearing strength and special technology and means have to be taken for the design of appropriate foundation structures. In fact the coastal areas differ so widely amongst themselves that each of them demands special engineering skill for the construction of aquacultural farms.

Although there exist several other books on aquaculture, none has addressed the problem considering these important components namely the selection of site, survey, design and construction of the farm, the requirements of the culturable organisms, actual culture process and processing of aquatic animals with particular reference to coastal aquaculture. This book is an humble effort by the authors to fill up this important gap. The primary aim of the book is to present various features of coastal aquacultural operations with emphasis on theoretical considerations in the planning and design of engineering structures. The book itself is written in chapters starting with an introductory chapter on 'Fish as Food' followed by Coastal Ecology. The other topics covered include: Biology of culturable species, Coastal hydrology, Tides and waves and their computations, Fluvial processes, River protection works, Design of pond, Sedimentation problems, Basic principles of aquacultural practices, Different culture techniques, and Post harvest technology.

This book is meant not only for the aquaculture engineers but it can also be used by fishery biologists, fish processing technologists, entrepreneurs involved with coastal aquaculture and professional engineers engaged in aquacultural operations.

In the preparation of the manuscript, help has been taken from various source materials which have been duly acknowledged at the end of relevant chapters, apart from many individuals who helped in typing and drawing.

Grateful thanks are due to the family members of the authors for their constant support, sacrifice, encouragement and cooperation in the completion of the manuscript.

A.N. Bose
S.N. Ghosh
C.T. Yang
A. Mitra

CHAPTER 1

Fish as Food

1.1 Introduction

Fish has been used as a food by human beings from ancient times. It is considered to be an excellent source of nutrition. The consumption of fish has however been somewhat restricted by religious and cultural taboos as well as by ill-informed notions regarding some diseases caused by eating fish. Sometimes the method of handling and preservation of fish in vogue at a particular time determined people's reaction to consumption of fish. At one time, medical practitioners in the U.K. even considered fish to be an extremely improper diet for children (Cutting, 1955), possibly due to the high salt content of salt-preserved fish and the destruction of vitamins during salting and drying (Horisberger, 1979). The production of histamine and other toxic amides by bacterial action on some varieties of salted fish and fish sauce has been reported (Amano, 1972). Tetroden like substances present in puffer or similar fish are responsible for their toxicity. On the other hand, fish has also been recommended for its medicinal properties. Pliny the Elder described about 1000 fish or part of fish supposed to cure all sorts of disease (quoted by Horisberger, 1979). Present scientific knowledge on the nutritive value of fish as human food places it as a highly valued protein food resource available to man.

It has been generally observed that people, particularly amongst the low income groups, in populations living in areas with higher fish supply and consumption, enjoy better health. However the average per capita annual consumption of fish (17.6 kg) in developing countries is only a little more than the world average (13.1 kg) although 60 per cent of the people living in those areas obtain more than 30 per cent of their daily intake of animal protein from fish and for them fish is an indispensible item of diet. Fish contributes about 6 per cent of the world's food calorie supply. However, its share in protein food supply is about 5.4 per cent, but when the quality of protein is considered, the share goes up to 8.0 per cent. If indirect supply through domestic animal feed, of which dried fish is often a major constituent, is taken into account then fish may supply 9.0 per cent of the world's protein (Bell, 1978).

The nutritive value of fish as human food is determined by its chemical composition which compares very favourably with that of land-based animals. The principal chemical constituents of fish are: protein 15 to 24 per cent, lipids 0.1 to 2.2 per cent, mineral substances 0.8 to 2.0 per cent and water 66 to 84 per cent. Besides fish is a good source of vitamins particularly of vitamin A and vitamin D.

Fish muscle is deficient in glycogen compared to its content in livestock animals, being about 0 to 3 per cent in the former and 1.0 per cent in the latter. But certain seafoods such as mussels have a fair glycogen content, varying between 1 and 3 per cent.

1.2 Fish protein

Proteins are the most important constituents of the fish muscle from the point of view of their nutritive value. The contribution of protein-calories in the total calorie content of a protein is relatively higher in fish: about 95 per cent in the edible portion of cod, 13.9 per cent in pork meat, and 69 per cent in lean beef meat (Geiger and Borgstrom, 1962). Muscles of lean fish contain an exceptionally high level of proteins when compared to the best meats. Different types of fish contain widely varying amounts of protein, ranging from 92 to 30 per cent of the dry solids (Love *et al.*, 1959). This difference in protein content is due to the non-representation of whole fish, age, sex, seasonal variations and geographical location of the catch.

Interest in fish protein has resulted mainly from two aspects: (i) the nutritive value and the changes brought about in the physico-chemical properties of the fish proteins during processing for preservation and storage which affect the texture and taste of fish, and (ii) the commercial value of fish. The amount of protein in fish depends on a number of factors such as the variety, fat and water content, size, sexual stage, and season of catch. Some non-oily fish are exceptionally rich in protein when compared to some of the best meats.

The biological value as measured on man has been determined as for, haddock-103; cod-88; beef-104; milk-99; and casein-66. The nutritive value of a protein is determined by its amino acid composition and depends on the absolute and relative amounts of the essential amino acids present in the specific protein molecule. Most studies agree that the amino acid pattern of fish protein closely resembles that of any other protein from animal sources (Stansby, 1953; Dunn 1947), although levels of lysine and histidine are generally higher while methionine and tryptophan may be on the lower side occasionally (Rosondale, 1979). Table 1.1 shows the average amino acid content of a few fish and meat samples.

Two types of skeletal muscle tissues are found in fish,—white and red. Besides the red colour which it owes to the presence of protein—myoglobin—the red muscle differs in chemical composition and biochemical function from the white muscle. The amount of red meat varies between different species of fish but it is generally between 2 and 3 per cent of the total protein and does not exceed 10 per cent in extreme cases (Hamoir, 1955).

The protein portion of fish tissue comprises several components which react differently under conditions of processing for preservation. Knowledge of behaviour of individual components under such conditions is necessary for a better understanding of the quality changes in stored fish and fish products. The protein components of fish are roughly classified on the basis of their extraction with solutions of electrolytes at various ionic strengths into three general groups, (i) albumins comprising 16 to 22 per cent of the total protein which can be extracted with water or weak salt solutions, (ii) globulins which account for approximately 78 per cent of the protein and are soluble in a salt solution of ionic strength greater than 0.5 per cent, and (iii) stroma proteins which are not soluble even in dilute acids or alkalies and constitute about 3 per cent of the total protein except in case of elasmobranchs in which the component may be as high as 10 per cent (Dyer and Dingle, 1961). A number of protein components have been detected and in many cases isolated by electrophoresis or ultracentrifuge separation.

Table 1.1
Amino acid contents of some fish and meat samples
(as per cent of protein, N × 6.25)

Amino acid	Atlantic mackerel	Salmon	Tuna	Beef	Itan
Arginine	5.8	5.8	6.4	5.3	6.1
Histidine	3.8	2.6	3.5	5.7	3.6
Isoleucine	5.2	4.9	4.9	4.7	5.0
Leucine	7.2	7.3	7.9	7.2	7.8
Lysine	8.1	8.0	8.9	8.3	8.7
Methionine	2.7	3.0	2.5	2.8	2.7
Phenylalanine	3.5	3.7	3.8	3.5	3.8
Threonine	4.9	4.4	4.2	4.8	4.5
Tryptophan	1.0	0.9	1.0	1.0	1.0
Valine	5.4	5.6	5.4	5.1	5.2

Sources: Neilands *et al.*, 1949; and Dunn *et al.*, 1949.

The albumin fraction is derived mainly from the sacroplasm which is the aqueous fluid between fibrils in muscle fibres. A number of enzymes in the glycolytic cycle have been detected in this fraction. It also contains non-protein components, such as free amino acids, trimethylamine, ammonia and esters of organic acids.

The globulin fraction contains some clearly identified components such as myoglobin, actomyosin, myosin and tropomyosin besides action and a few unidentified ones.

Changes in proteins during processing particularly in frozen fish have been extensively studied because of the influence of these changes on the taste and flavour of fish.

Frozen fish flesh in cold storage suffers from deteriorative changes in flavour and texture. The frozen flesh tends to become dry and tough and in many cases loses the characteristic fresh fish flavour. Besides occasionally, particularly the fatty species of fish, develop a rancid flavour and brown discolouration on the surface, which is known as 'rust'. The deteriorative changes in frozen storage are not only faster in fish than in mammalian flesh but also vary in different species (Sikorski *et al.,* 1976). Although no completely satisfactory explanation is yet available, the diminution of extractable protein from the flesh has been associated with the development of toughness. The albumin fraction of the fish-protein component suffers little or no loss of extractability in frozen storage, except when severe desiccation occurs (Dyer, 1953). It is the actomyosin fraction of the protein in fish flesh which is most affected by freezing and holding in cold storage.

An important effect of freezing is the formation of drip liquid oozing out of the flesh when the frozen material is thawed. Mechanical damage of the cell wall by ice crystals in frozen flesh, allowing escape of fluid, cannot fully explain the phenomenon. The lowering of moisture holding capacity of protein is perhaps the main reason for this drip. Dyer (1951) showed that denaturation as measured by the dispersibility of protein in 5 per cent neutral NaCl was almost complete when a gel of actomyosin from cod was subjected to freezing. Actomyosin in flesh is also probably affected in the same way as an extracted fraction in gel form.

Denaturation of actomyosin, measured by the lowering of its extractibility, is dependent on the freezing rate and more so on the temperature of storage of frozen fish. Extraction of actomyosin in cod-flesh was reduced by 50 per cent with slow freezing while with quick frozen fish the reduction was of the order 25 to 30 per cent (Dyer, 1951). However, with reasonable freezing temperature, the time of storage is found to be of greater importance in the deteriorative changes in fish flesh, particularly when also examined by a taste panel (Young, 1938).

Frozen storage resulted in a decrease in the extractable actomyosin from fish flesh. The change was maximum during storage at -2 to $-6°C$. Which is the zone of maximum crystal formation and is less at lower temperatures (Reay, 1934; Dyer, 1953). The most rapid denaturation in cod actomyosin occurred at $-4°C$ and was practically complete in three to four months at temperatures between -4 and $-10°C$. At lower temperatures, denaturation was slower. During the first two to three days of storage, extractable actomyosin dropped from a value of 70 to 75 per cent down to about 65 per cent during storage at $-23°C$ and to about 55 per cent at $-12°C$.

Fish muscle is essentially a water-salt-protein system with various organic and inorganic salts in solution equivalent to 1.4 per cent of solution of NaCl (Dyer *et al.,* 1957). In the case of actomyosin, Snow (1950) demonstrated that centrifuged gel of actomyosin when subjected to freezing, underwent complete denaturation as measured by the dispersion of protein in 5 per cent NaCl.

1.3 Fish lipids

Fish oils, consisting primarily of triglycerides of fatty acids, are similar to lipids in other natural vegetable oils and animal fats. Fish oils however differ from others in their fatty acid composition. Besides palmatic, stearic and oleic acid, all fish oils contain higher fatty acids such as C_{22} acids and even some of the C_{24} series. Fish oils also have considerable amounts of polyunsaturated acids but do not contain linoleic and linolenic acids, which are present in some vegetable oils. Polyenoic acids of the C_{16} to C_{24} series and even C_{26} series are found in fish oils, the more frequent being C_{20} and C_{22} series.

The oils of fresh water fish and that of sea water fish differ in their fatty acid composition. The former contains more of palmitic acid and C_{18} unsaturated acids while C_{18}, C_{20} and C_{22} acids predominate in sea fish. The oil's in young and adult fish sometimes vary as in the case of salmon (Lovern, 1934). Oil content and composition also show seasonal and sexual variations in the same species of fish.

Besides triglycerides and free fatty acids other constituents grouped under 'unsaponifiable' matter, present in fish, have also evoked interest for their particular commercial importance. Generally the body oil of fish, as distinguished from fish-liver oil, contains very little unsaponifiable matter. But liver oils of some species of fish contain a significant quantity of unsaponifiable fractions which contain important compounds such as vitamin A that is present in large amounts in the liver oil of cod, halibut, shark etc. Liver oil rich in vitamin A has been used as such or after refinement to remove the saponifiables to serve as a vitamin A supplement in human food as well as in animal feed.

1.4 Rancidity problems in fish

Oils in the body tissue of fish are responsible for rancidity often found in preserved fish as also in wet fish stored in ice. Rancidity is indicated by the development of characteristic unpleasant odour, taste and sometimes discolouration. The two major changes to which fish are subject are (i) of an oxidative (auto-oxidation), and (ii) a hydrolytic (lipolysis) nature. These changes have undesirable effects on the quality of fish and its consumer acceptance. Of these two types, autoxidation is more important in wet and also in processed fish.

Autoxidation of lipids is initiated by the production of free radicals which is caused by the reaction of lipids with oxygen, although other chemical components in fish flesh may also initiate the formation of free radicals.

The free radicals react with oxygen to produce peroxy radicals which in turn react with unsaturated fats to give hydroperoxide and another free radical. The latter can initiate a chain reaction again (Bateman, 1954). The rate of oxidation may be affected by several factors such as by other unsaturated and high reactive compounds which are present in fish flesh (Hammond and Johnson, 1972), disposition of fatty acids within the glyceride molecule (Raghuveer and Hammond, 1967) and the presence of free fatty acids which sometimes act as a pro-oxidant (Catalano and De Felice, 1970) and sometimes as an autoxidant. Heavy metals catalyse the decomposition of hydroperoxides to produce free radicals. Protein, specially heme protein, also influences oxidation.

Although fish oils readily oxidise at an ambient temperature, wet fish, even the oily ones, do not suffer from marked oxidation of lipids, between 0°C and the ambient temperature. The explanation may be found in competition for available oxygen between the lipids oxidative enzymes and micro-organisms present on and in the fish. Prolonged storage in ice however causes oxidative deterioration of fatty fish.

Oxidation of lipids is a common occurrence during storage of frozen fish. The rate of oxidation generally decreases by a factor of 2 or 3 for 10° of lowering of storage temperature. Conditions in cold storage have a significant influence on oxidation. Dehydration of fish in cold storage, increases oxidation (Banks, A. 1952). Non-fatty fish oxidises slowly in cold storage although the extracted lipids from it oxidise readily.

One of the oxidation effects in cold storage of non-fatty fish like cod, is the development of malodour known in trade as cold storage flavour, caused by the oxidation of unsaturated carbonyls (McGill *et al.*, 1977).

Study with model systems containing phospholipids and protein has shown that oxidation of lipids is accompanied by denaturation of protein (Labuza, 1971). It is expected that oxidation of lipids in frozen fish will accelerate denaturation of proteins in fish tissue. Autoxidation of lipids in commuted dehydrated fish may proceed so rapidly as to cause sufficient rise of temperature to burn the meal.

REFERENCES

Amano, K. (1972): The influence of fermentation on the nutritive value of fish with special reference to fermented fish products of South East Asia. *Fish in Nutrition*, (eds.), Heen, E. and Kreuzer, Fishing News (Books) Ltd., London.

Banks, A. (1952): The freezing and cold storage of herring. DSIR Food Investigation Special Report No. 55, HMSO, London.

Bateman, L. (1954): Olefin Oxidation, *Q. Rev. Chem. Soc.*, 8.

Bell, F.W. (1978): *Food from the Sea*, Westview Press, Boulder, USA.

Catalano, M. and De Felice, M. (1970): Auto-oxidation of fats: Influence of free fatty acids. *Revista Ital. Sostanze Grasse*, 47.

Cutting, Ch. L. (1955): *Fish saving—A history of fish processing* from *ancient and modern times*, Leonard Hill Ltd. London.

Dunn, M.S. (1947): Amino Acids in food and analytical methods for their determination: *Food Technol.*, 1.

Dunn M.S., Camien, M.V., Eiduson, S. and Malin, R.B. (1949): The Nutritive value of canned foods I. Amino acid content of fish and meat product., *J. Nutrition* 39.

Dyer, W.J. (1953): Main problems of fish protein denaturation. *Proc. Symp. on Cured and Frozen Fish Technol. Swed. Inst.*

Dyer, W.J. (1951): Protein denaturation in frozen and stored fish. *Food Res.* 16.

Dyer, W.J., Fraser, D.L., Ellis, D.G. and MacCallum, W.A. (1957): Influence on intermittent short storage period at 15°C. *J. Fish Res. B.D. Can.* 14.

Dyer W.J. and Dingle, J.R. (1961): Fish protein with special reference to freezing. *Fish as food* Vol. *I*, Academic Press, New York.

Geiger, E. and Borgstrom, G. (1962): *Fish as Food, Vol. II,* Academic Press, New York.

Hammond, E.G. and Johnson, D.C. (1972): Factors affecting the proportion of peroxide types during cooxidation of fatty ester mixtures. *Proc. 11th. Wld. Congr. Int. Soc. Fat. Res.* Gotenberg, Sweden.

Hamoir, G. (1955): *Fish proteins: Advances in Protein Chemistry Vol.* 10.

Horisberger, M. (1979): Lessons from the past for better future utilisation of fish resources. *Nestle's Research News.*

Labuza, T.P. (1971): Kinetics of lipid oxidation in food, *CRC. Crit. Rev.* Food *Technol.* 2(3).

Love, R.M., Lovern, J.A. and Jones, N.R. (1959): The chemical composition of fish tissues: DSIR Food Invest. Spec. Rept. No. 69.

Lovern, J.A. (1934): Fat Metabolism V. The fat of the Salmon in its young fresh water stages. *Biochem. J., 28.*

McGill, A.S., Hardy, R., and Gunstone, F.D. (1977): Further analysis of the volatile components of frozen cold stored cod influence of these on flavour. *J. Sci. Food and Agri.* 28.

Neilands, J.B., Somy, R.J., Sohljell, I., Strong, F.M. and Elevehjevn, C.A. (1949): The nutritive value of canned foods II Amino acid content of fish and meat products. *J. Nutrition,* 39.

Reay. G.A. (1934): The influence of freezing temperature on Haddock's muscle, part I. *J. Soc. Chem.Ind. (London)* 52.

Raghuveer, K.G. and Hammond, E.G. (1967): The influence of glyceride structure on the rate of auto-oxidation, *J. Am. Chem. Soc.* 44.

Sikorski, K., Olley, J., and Kostuch, S. (1976): Protein changes in frozen fish, *CRC Crit. Riv. in. Food Sci. and Nutr.* 8(1).

Snow, J.M. (1950): Protein in fish muscle III. Denaturation of Myosin by freezing. *J. Fish Res. Bd. Can.* 7.

Stansby, M.E. (1953): Composition of fish: *U.S. Fish Wildlite Service Fishery Leaflet* No. 116.

Rosondale, J.L. (1979): Amino acids of flesh. *Biochem. J.* 23.

Young, O.C. (1938): The quality of fresh, frozen and stored halibut as determined by a testing panel. *Fish Res. Bd. Can. Progr. Repts Pacific Coast Station No.* 37.

CHAPTER 2

Coastal Ecology

2.1 Basic concepts

The term ecology has been derived from the Greek word 'oekologie' (*Oikos*—surrounding; *lógos*—study), meaning the relationship of different organisms with their organic and inorganic environment. The environment of each organism is always subject to varied pressures—*abiotic* as well as *biotic*, resulting in a fluctuation in its environment. The living organisms which are self-regulating and self-perpetuating have continuous interaction and inter-relationship with each other and with the non-living environment and thus are always in equilibrium with it by adapting themselves to its fluctuations. The inability of certain organisms to cope up with their respective surroundings naturally leads to their extinction from the surface of the earth. Here the term 'organism' means all that is living and the term 'environment' includes everything external to the organism constituting a physical, chemical and biological environment.

The study of the interacting, dynamic processes by which individual organisms, their populations and communities of populations respond to these diverse environmental pressures is referred as *ecology*. Odum (1959) has defined ecology as the "study of the structure and function of nature".

In ecology, the term *population* denotes groups of individuals of any one kind of organism and *community* (sometimes designated as 'biotic community') includes all the populations occupying a given area. The community and the non-living environment function together as an ecological system or *ecosystem*. The ecosystem is the basic functional unit in ecology, since it includes both organisms and abiotic environment, each influencing the properties of the other and both necessary for maintenance of life as we have it on the earth. All ecosystems, even the ultimate biosphere, are open systems—there is a continuous inflow and outflow of energy.

The environment of an organism can be of two types—*micro-environment*, the intimately local and immediate surrounding of the organism, and *macro-environment* which is the sum total of the physical and biotic conditions existing external to the organism and its micro-environment. These two environments may differ between themselves widely or slightly, for example, in a hot summer day the temperature within a burrow (micro-environment) formed by a burrowing organism on the sea beach is much less than the temperature outside the burrow (macro-environment).

The *habitat* of an organism is the place where it lives, or the place where one would go to find it. The *ecological niche,* however, includes not only the physical space occupied by an organism but also its functional role in the community, that is, the place that an animal occupies in a biotic community which expresses its functional status in relation to its food and enemies.

For any ecological study, *species* is the fundamental unit and the environment generally harbours more than one animal or plant species or species populations. *Species* are defined as groups of interbreeding natural populations that are reproductively isolated from other such groups (Mayr, 1969). Ecological studies may deal with the study of individual organisms or species in relation to their habitat *(autecology)* or the study of associations of organisms in relation to a particular area or biotope or habitat *(synecology).* While synecology is the ecology of the biotic community, *demecology,* refers to the ecology of populations. The term *ecological valency* has been used to refer to the adaptive range of both the individual and the species (Ananthakrishnan and Viswanathan, 1976).

From the ecosystem point of view, every recognisable unit of nature has two biotic components—an autotrophic component (autotrophic meaning 'self-nourishing'), able to fix light energy and manufacture food from simple inorganic substances and, secondly, a *heterotrophic* component (heterotrophic meaning 'other-nourishing'), which utilises, rearranges and decomposes the complex materials synthesised by the autotrophs. From the structural point of view, four constituents are recognised to comprise the ecosystem—(i) *abiotic substances,* basic elements and compounds of the environ, (ii) *producers,* the autotrophic organisms, largely the green plants, (iii) the *consumers,* heterotrophic organisms, mainly animals, that ingest other organisms or particulate organic matter, and (iv) the *decomposers,* heterotrophic organisms, chiefly the bacteria and fungi that break down the complex compounds of dead organisms, absorb some of the decomposition products and release simple substances useable by the producers.

The structure of an ecosystem needs to be considered from various angles to understand the interplay of structure and function. The producer-consumer arrangement is one kind of structure called *trophic* structure (trophic = food), and each 'food' level is known as a *trophic level.* The amount of living material in different trophic levels or in a component population is known as the *standing crop.* The standing crop can be expressed in terms of the number per unit area or in terms of *biomass,* that is, organism mass. Biomass can be measured as living weight or fresh weight, dry weight, ash-free dry weight, carbon weight, calories, or any other unit that may be useful for comparative purposes (Odum, 1963).

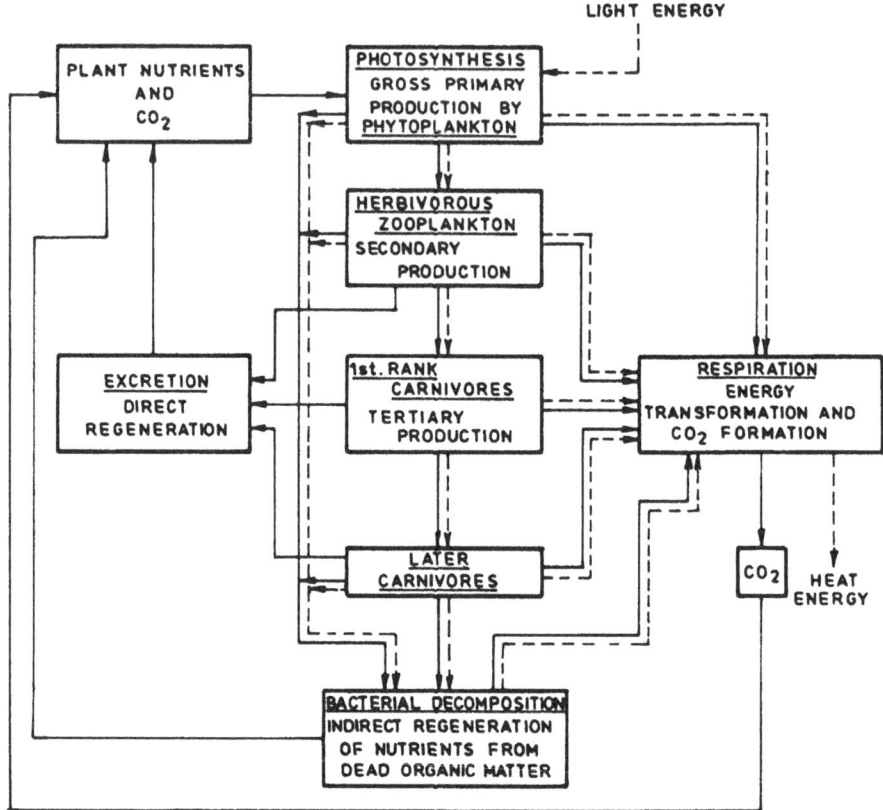

Fig. 2.1. Organic cycle in the marine environment (Based on Tait, 1981)

The transfer of food energy from the source in plants through a series of organisms (Fig. 2.1) with repeated stages of eating and being eaten is known as the *food chain*. Green plants occupy the first trophic level (the producer level), plant eaters (herbivores) the second level (the primary consumer level), carnivores that eat the herbivores the third level (secondary consumers) and sometimes a fourth level (tertiary consumers). This trophic classification is one of function, and not of the species as such; a given species population may occupy one, or more than one, trophic level according to the source of energy actually assimilated (Odum, 1963). At each transfer of energy from one organism to another, or from one trophic level to another, a large part of the energy is degraded into heat. The shorter the food chain, or the nearer the organism to the beginning of the food chain, the greater the available food energy. Losses in available biomass of about 90 per cent occurs at each step in the food chain. Thus there is only about 10 per cent as much zooplankton biomass available for harvest as there is of phytoplankton (Wheaton, 1977).

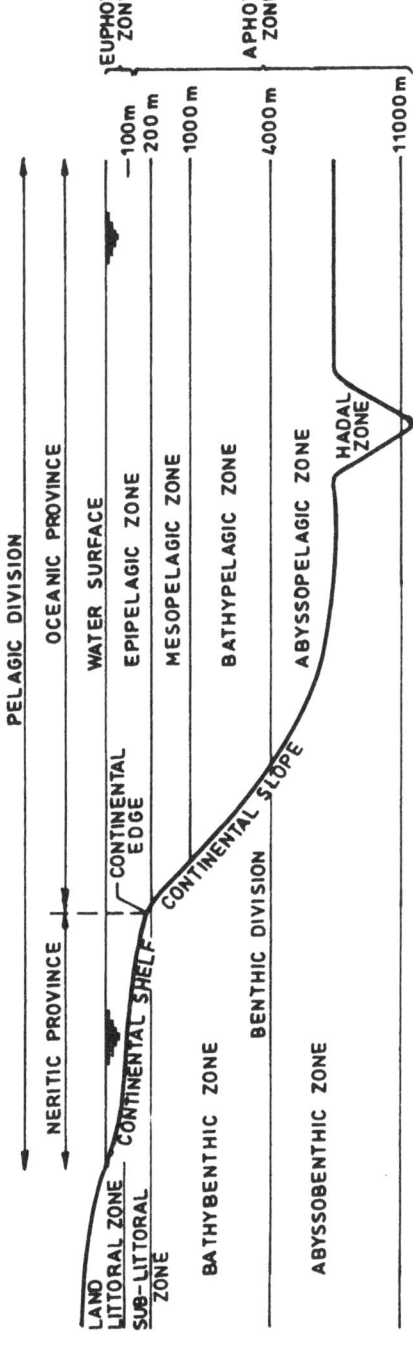

Fig. 2.2. Major subdivisions of the marine environment (Based on Tait, 1981; Boaden and Seed, 1985)

2.2 Coastal ecosystems

The marine environment is divided into different zones as shown in Fig. 2.2. The oceanic zone comprise two major realms, *benthic* (sea bed) and *pelagic* (water column), which are further subdivided according to depth. The *euphotic* (epipelagic) zone is that part of the pelagic realm which is well illuminated and extends up to a depth of 100 to 200 m depending upon the clarity of water. Below this level, where light does not penetrate, is the *aphotic* zone.

2.2.1 THE NERITIC ZONE

The neritic zone comprising the shallow (100 to 200 m depth) coastal waters overlying the continental shelf, though accounting for only about 5 per cent of the world's seas, is however highly productive and is considered important for commercial fisheries.

Two major categories of pelagic organisms are present in this region— plankton and nekton. Planktonic organisms (Fig. 2.3) have limited powers of locomotion, and are more or less passively drifted by the water currents. They may be plants *(phytoplankton)*, animals *(zooplankton)*, and bacteria *(bacterioplankton)*. Planktonic organisms which spend their entire life as planktons are known as *holoplankton* while others which become temporarily plankto-

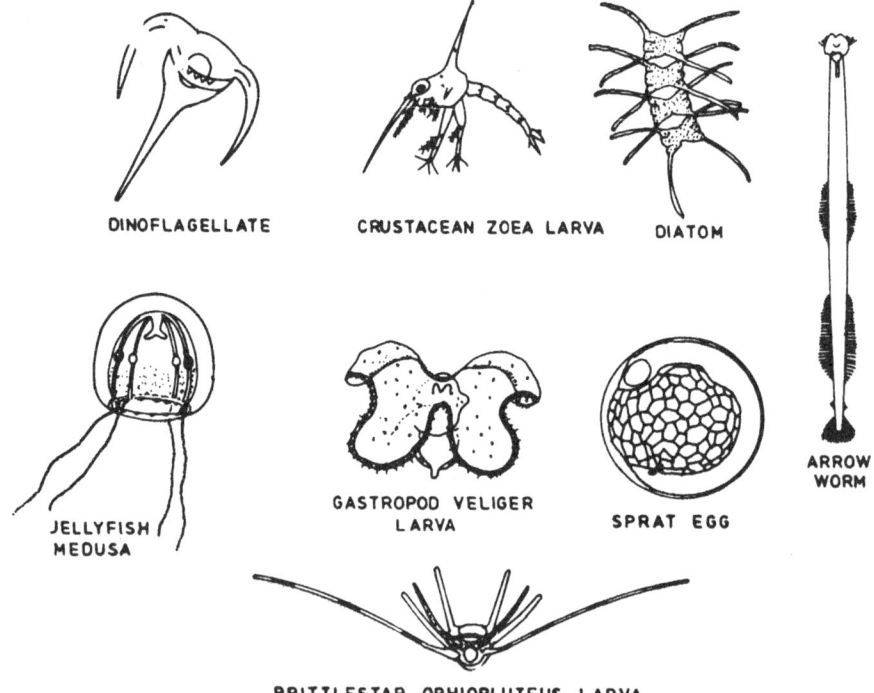

DINOFLAGELLATE   CRUSTACEAN ZOEA LARVA   DIATOM

JELLYFISH
MEDUSA

GASTROPOD VELIGER
LARVA   SPRAT EGG   ARROW
WORM

BRITTLESTAR OPHIOPLUTEUS LARVA

Fig. 2.3. Some members of plankton (not to scale) (Redrawn from Fincham, 1983)

nic at some stage of their life-cycles are known as *meroplankton*. Many small benthic organisms which are temporarily whirled up off the bottom by water movement are collectively known as *tachyplankton*.

Larger pelagic organisms, for example, fish, prawn, cephalopods and mammals, which are capable of counteracting water currents by swimming constitute the nekton. Besides plankton and nekton there is a third category, the *pleuston* comprising those organisms that thrive in the air-water interface; an example is the syphonophore *Physalia* which uses a gas filled float for the purpose (Boaden and Seed, 1985).

Though neritic and oceanic waters are adjacent, they are however populated by different plankton communities. This is largely due to the differences in the water circulation and sedimentation patterns as well as the effects of solar radiation. Coastal waters are shallow and are subjected to more fluctuations in temperature and salinity than the deeper oceanic waters. The neritic species are therefore more physiologically tolerant than the oceanic species. Also there are more meroplanktonic forms in the neritic zones while the oceanic zone contains mainly holoplanktonic species. A classification of planktons by size is given in Table 2.1.

Table 2.1
Classification of plankton by size
(Boaden and Seed, 1985)

Type	Size range	Examples[+]
Femtoplankton	0.02–0.2 μm	Viruses
Picoplankton	0.2–2.0 μm	Bacteria
Nanoplankton	2.0–20 μm	Small autotrophic flagellates
Microplankton	20–200 μm	Protozoans, diatoms, dinoflagellates
Mesoplankton	0.2–20 mm	Copepods
Macroplankton	2.0–20 cm	Krill, arrow worms
Megaplankton	0.2–2 m	Large jellyfish

[+]There are often overlaps between neighbouring categories

Phytoplankton

Phytoplanktons are the primary producers (auxotrophs) containing chlorophyll and other pigments, and can consequently utilise carbon dioxide, nutrient salts and the energy of sunlight to produce carbohydrates, proteins and fats. There are two main groups of phytoplankton—*diatoms* and *flagellates* (Fig. 2.4.).

Diatoms are non-motile unicellular algae occurring either singly (for example, *Coscinodiscus)* or as chains (for example, *Chaetoceros)*. All species characteristically have two, often elaborately ornamented, silicified valves— the frustules. Diatoms are abundant in both the Arctic and Antarctic and in

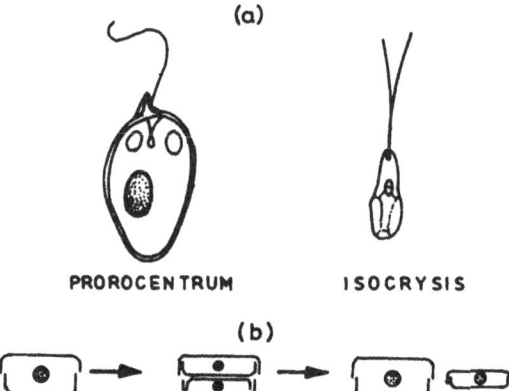

Fig. 2.4. (a) Phytoplanktonic flagellates
(b) Asexual cell division in centric diatoms (Redrawn from Fincham, 1983)

temperate and boreal coastal waters (Boaden and Seed, 1985). Reproduction is through simple asexual cell division (Fig. 2.4 b), but after a few generations, by which time the diatom 'shells' become progressively smaller, sexual reproduction takes place that restores the shell size to normal dimensions.

Flagellates are motile organisms that use their whip-like flagella to propel themselves through water. One of the important groups of flagellates, the dinoflagellates are especially abundant in the tropical and subtropical waters; in late summer and early autumn, these are found in temperate and boreal waters. Some forms (for example, *Gonyaulax*) produce toxins and during bloom conditions may cause the so-called 'red tides'. Filter feeders like some bivalve molluscs that feed upon these, consequently concentrate the toxins inside their bodies, and when, in turn, are eaten either by man, birds or fish, cause a serious neurological disorder known as paralytic shell fish poisoning through the action of these toxins.

Zooplankton

In contrast to the relatively limited taxonomic range of the phytoplankton, the zooplankton includes representatives of almost all the animal phyla, either as adults (holoplankton) or as eggs and larvae (meroplankton). Some 100,000 species out of a total of 135,000 bottom dwelling species of invertebrates have free-swimming larval stages lasting, on an average, about two to four weeks. As in case of phytoplanktons, there are single-celled forms of zooplanktons as well, though these are not capable of producing their own food (heterotrophs) (Fincham, 1983). Zooplanktons are the *primary consumers* of phytoplankton and bacteria, and provide the vital link between the primary producers and secondary consumers. Copepod crustaceans constitute an important group of zooplanktons. Euphausiids, also a group of crustaceans, are abundant in many highly productive upwelling systems. Comb jellies, arrow worms, pelagic chordates, jellyfish and siphonophores are all among the important zooplanktonic organisms (Fig. 2.3).

Planktonic adaptations

Planktons have to spend a considerable period of time in the euphotic zone either for photosynthesis or for feeding. Since they have only limited powers of swimming, if at all, they need to counteract gravitational forces to keep them from sinking. For this purpose, they have developed adaptive features, that can be broadly divided into the following two categories:

i) Reduction in body density: Some species actively exclude heavy bivalent chemical ions such as sulphate, and replace them with osmotically similar but lighter monovalent ions like chloride; others acquire gas filled floats, or oils and fats which may also serve to make the planktons more buoyant.

ii) Increase in the surface area/volume ratio: Frictional resistance being proportional to the surface area, a smaller size reduces the chance of sinking, and in addition also helps in the uptake of nutrients as well as rapid multiplication. Thus, in areas where a more efficient uptake of nutrients or rapid multiplication is necessary to offset intense grazing pressure, the smallest species of planktons are generally more abundant. Flattened shapes and elaborate spines which consequently increase surface resistance and effectively slow the rate of sinking, are found to occur in many planktonic forms. Spiny projections also serve as defensive organs.

As warm water is less viscous than cold water, tropical species, and to some extent the summer populations at higher altitudes, tend to exhibit more extreme modifications than do the species inhabiting cooler waters (Boaden and Seed, 1985).

Distribution of planktons in relation to different physical factors

Light plays an important role in the distribution of planktons, particularly phytoplanktons, which are largely restricted to the euphotic zone since photosynthesis is dependent on light. The amount of sunlight actually entering the water column after reaching the surface depends upon the angle at which it strikes the surface and the height of the sun above the horizon. Penetration· of light is greatest at the equator; in the tropics, the light intensities are optimal throughout the year, but at lower latitudes they show marked seasonal variation. Since sunlight is rapidly absorbed and also scattered by the particulate matters, planktonic cells or even water molecules, the intensity of light within the water column decreases exponentially with the increasing depth of water. This is more pronounced in the turbid coastal waters. Till the system becomes light saturated, photosynthesis also increases logarithmically with the intensity of light, to a maximal value, beyond which point, any further increase in the light intensity inhibits photosynthesis by either bleaching the photosynthetic pigments or by arresting pigment production. However, different planktonic species have different optimal light intensities for maximal photosynthesis. Though there is marked variation in photosynthesis, there is hardly any change in respiration rate due to the depth of water.

Thus, there is a point at which the energy gained through photosynthesis exactly matches that expended through respiration. This depth, at which the light intensity falls to 1 per cent of its surface value, is known as the *compensation depth* and corresponds to the lower boundary of the euphotic zone. Above this level of water, the overall production of plant material is more, whereas below this level there is a net loss of organic matter, since respiration exceeds photosynthesis. The intensity of light corresponding to the compensation depth is known as the *compensation light intensity*. There is seasonal as well as geographical variation in the compensation depth since the light conditions are variable. While in turbid estuaries it may not exceed even a few centimetres, in the clear oceanic waters it may extend to more than 250 m of depth. At times there is vigorous and extensive vertical mixing of water column caused by surface winds, as in case of inshore temperate boreal waters during winter. Such mixing causes the phytoplanktons to be carried down below the euphotic zone, due to which, frequently, the time spent by them in the aphotic zone greatly exceeds that spent in the euphotic zone and active photosynthesis may thus be insufficient. The *critical depth* occurs where the total production exactly balances the total consumption through respiration, and is always below the compensation depth. When the mixing depth is less than the critical depth, photosynthesis exceeds respiration, resulting in net production of organic matter.

Apart from dissolved carbon dioxide which is required for photosynthesis, the phytoplankton also requires other nutrients for healthy growth. Of these, the most important ones are nitrogen (as nitrate and ammonia) and phosphorus (as orthophosphate). Diatoms need silicon for the formation of frustules. Trace elements like iron, copper, or vanadium may be required in small amounts, and vitamins are also important. Carbon dioxide is present in greater amounts in sea water than other nutrients, and nitrogen is probably the most common rate-limiting process in the sea (Ryther and Dunstan, 1971). Studies on the kinetics of nutrient uptake show that oceanic phytoplanktons living in comparatively nutrient-poor waters are relatively more efficient in the uptake of nutrients than their coastal counterparts where the water is nutrient rich, though the total nutrient uptake in the latter may be more.

The euphotic zone may often be temporarily depleted of nutrients since these are utilised by the phytoplanktons, photosynthesis being a surface phenomenon. Some of the inorganic nutrients are however regenerated and recycled in this zone, though large quantities, as dead planktons or faecal matters, gradually sink to the sea bed as particulate detritus. Thus, regeneration of nutrients is more at greater depths through the decomposition of the detritus and there is a gradual draining of nutrients from the surface waters. Therefore a turbulent vertical mixing or upwelling is absolutely essential in restoring the nutrients to the euphotic zone. Some nutrients are however also regenerated directly by consumers as excretary products which can be immediately available to the phytoplankton and bacteria. The latter form the major group consuming dissolved organic material in the water column and converting it into particulate organic matter which is then available to a variety of consumers such as protozoans and copepods, and are thus an important source of food in many marine systems.

There is a marked variation in the occurrence of planktons in the temperate waters where there are variations in the light and nutrient levels in the euphotic zone. During winter months, the water column is more or less isothermal and since there is no marked density gradient (pycnocline), surface winds can cause extensive vertical mixing. Though nutrient level is high, growth of phytoplanktons is restricted due to poor light, and mixing of plant cells due to greater depths. When spring breaks, the surface waters get heated up and become lighter than the deeper layers, which impedes vertical mixing, and consequently results in the thermal stratification of the water column, which in turn effectively maintains the phytoplankton within the euphotic zone. With the increase in solar radiation, the initial high nutrient level falls by the development of the spring diatom bloom. Thus, during summer, there is a severe depletion in the nutrient levels in the surface water, and due to a pronounced thermocline preventing vertical mixing, there can be little replenishment from the deeper layers. Growth of phytoplanktons is restricted by the grazing zooplanktons, or a shortage of nutrients, and cells continually sink out of the euphotic zone. By autumn, however, the surface waters cool down sufficiently, and though there is less light available for photosynthesis, a breakdown of the thermocline and re-establishment of vertical mixing results in a rapid replenishment of nutrients to the surface waters leading to a small and short-lived but distinct phytoplankton bloom.

In the coastal waters, there is continual turbulent mixing through wind action and other natural forces. Thus, there is a constant upwelling of nutrients, which is the chief reason for the coastal waters being the sites of high productivity.

There are also geographical variations in the seasonal pattern of planktonic growth. As for example, in the tropical waters which are well illuminated and thermally stratified throughout the year, phytoplankton growth is inhibited by the low nutrient content of the surface waters. There is also no obvious relationship between the growth of phytoplankton and zooplankton. In the Arctic zone, on the other hand, the water column is never strongly stratified and nutrient level is never too low, but the phytoplankton is limited primarily by the lower light intensity; there is a single large summer phytoplankton bloom when the light is sufficient, followed by a single zooplankton peak. In the North Pacific, however, an intense grazing pressure from the zooplankton do not allow the phytoplankton to reach bloom conditions.

In addition to the seasonal variations observed in the abundance of phytoplankton, changes are also found in their species composition. This change in species composition over the year is termed *seasonal succession* (Margalef, 1962). Though more marked in the areas where temperature variations are more pronounced, nevertheless some successional changes have been observed in the polar and tropical regions as well. The actual cause of succession is still obscure, though it is felt that the changes in the water contents through the growth of organisms may be important. For example, some species may require certain metabolites (for example, vitamins) produced by species earlier in the successional sequence for their growth and survival. Thus, dinoflagellates which require more nutrients than they are able to synthesize themselves, appear much later than do diatoms which require fewer nutrients. Some of the organisms also produce certain toxic allelochemicals inhibiting the growth of other species. Apart from the type and level of nutrients available, changes in temperature, light, and salinity may also determine seasonal succession.

In the case of zooplankton, the temperate-boreal waters are initially dominated by the calanoid copepods that feed on the diatoms. Towards late spring and early summer, meroplanktonic forms are abundant while predatory species feeding extensively on herbivores become dominant in summer (Boaden and Seed, 1985).

Vertical migration of zooplankton

Like the phytoplankton, many zooplanktonic species also show diurnal vertical migration, at times more than 100 m, though the actual cause or adaptive significance of this phenomenon is still not clear in case of the latter.

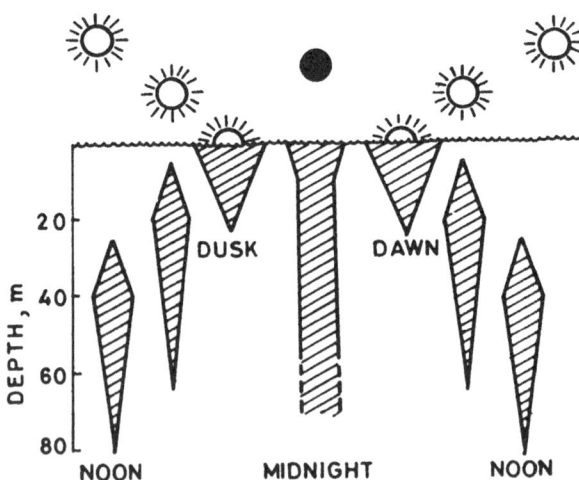

Fig. 2.5.   Diurnal vertical migration pattern of zooplankton (Redrawn from Boaden and Seed, 1985)

Though there are considerable variations in this behaviour pattern both within and between different species, the general patterns followed, show a marked similarity (Fig. 2.5). Throughout the day time, zooplankton occur deep in the water column, coming up to the surface at dusk to feed on the phytoplankton. During the night they disperse, re-assembling at the surface at dawn before descending again to their daytime depth. This migratory movement of zooplankton may be regulated by an internal biological clock since some species start their descent before dawn, and experiments show that organisms maintain their own behaviour patterns even when kept in total darkness throughout. However, the diurnal variations in light intensity have been thought to provide the stimulus for migration. Larger zooplankton have been found to swim at speeds of up to 200 m/hour (Boaden and Seed, 1985).

Trophic structure in the neritic province

Three basic types of food chains associated with oceanic (five trophic levels), shelf (three trophic levels) and upwelling regions (one or two trophic levels) can be recognised (Ryther, 1969). Differences in size of the dominant phytoplankton, the primary producers and the nektonic consumers, largely determine the length of the food chain and account for the variations. As for example, in the open seas where the water is relatively nutrient-poor, the phytoplanktons are small (nanophytoplankton and bacterioplankton) and hence, are fed upon by only small zooplanktons. Thus, energy has to be concentrated through several more trophic levels before the larger nektons (for example, fish) can enter the chain. In contrast, nutrient-rich areas contain larger phytoplanktonic forms (for example, chain diatoms) which can be consumed by the larger zooplanktons or even directly by planktivorous fish like the anchovies. Since a considerable amount of energy (about 90 per cent) is always lost between successive trophic levels (for example, as respiratory heat), shorter chains typical of nutrient-rich coastal and upwelling areas increase the yield of the top trophic levels with consequent implications on the commercial fisheries (Landry, 1977). Complex food chains are generally found to be more unstable. Environmental stability being one of the factors contributing to the length of the food chain, shorter chains are preferred in the naturally unstable areas like the coastal regions.

The ability of the photosynthetic plants to produce energy-rich organic molecules by the utilisation of solar energy and simple inorganic molecules, is termed *primary production* and forms the basis for the perpetuation of life in the seas. The total amount of energy fixed through photosynthesis is known as *gross primary production*. However, some of this is utilised by the plants themselves for respiration and other metabolic processes, and a lower amount termed as *net primary production* is only available to the consumer. Primary production in the open seas is mostly through the phytoplanktons in the éuphotic zone, below which level bacteria are the most important primary producers. Primary production in the inshore areas is enhanced substantially by macrophytic plants like the algae, mangroves and sea grasses as well.

Production is expressed as the rate of carbon fixation (measured as gm $C/m^2/year$). Secondary and tertiary production are those effected by herbivores and first-rank carnivores respectively. *Standing crop* or biomass is the amount of organic material present in any given volume of water at any given time and does not necessarily depend upon productivity. Thus, a high grazing pressure may also give rise to a low standing crop and vice-versa.

A convenient method of comparing different ecosystems is through their annual net productions. A global average of primary productivity is estimated to be about 50 gm $C/m^2/year$. However, the values vary (from 25 to 200 gm $C/m^2/year$) according to the analytical techniques used, and there are broad geographical variations also. In the open seas, the primary production is generally low. In the areas where large scale upwelling takes place, the productivity is very high (300 gm $C/m^2/year$), and some of the world's largest fisheries (for example, that of Peruvian anchovy) are situated in these types of areas. Other areas showing high productivity include frontal systems, which are characterised by water masses of different properties meeting at sharp boundaries. Due to their turbulent nature, shallow coastal waters are also nutrient-rich and consequently can yield high levels of production almost throughout the year provided suitable light conditions prevail. In the estuaries however, the high nutrient inflow from the land is offset by the high turbidity, leading to a large scale reduction in the depth of the euphotic zone (Fig. 2.6). The high level of nutrients present in the immediate coastal zone is most evident through the marked and sharp differences that are found

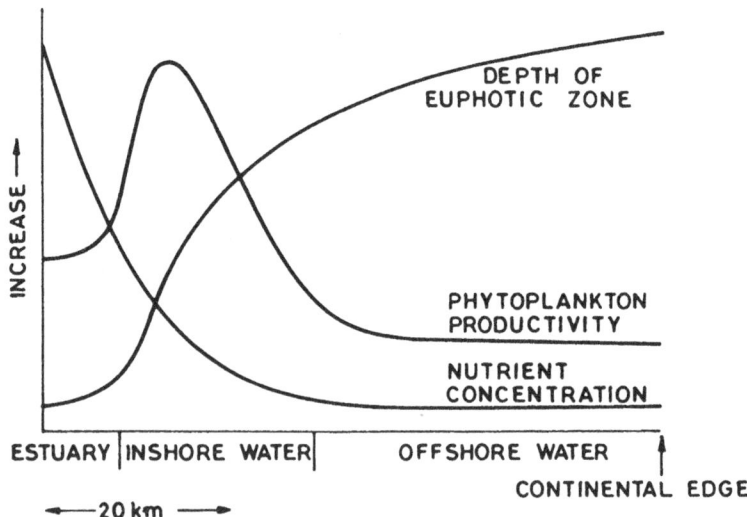

Fig. 2.6.   Variation in phytoplankton productivity, nutrient concentration and depth of the euphotic zone across the continental shelf
(Source: Livingston, 1979)

between inshore and offshore waters when compared, particularly in the tropics.

In contrast to the elaborate studies on primary production, production by zooplanktons or secondary production is not so well studied. However, secondary production is generally found to follow primary production quite closely, being high (27.5 gm C/m^2/year) in the coastal and upwelling areas, and much less (4.5 gm C/m^2/year) in the tropical oceanic waters, though very high levels of primary production may sometimes lead to anoxic conditions as found in the eutrophic lakes, sharply reducing the secondary production. Also the bacteria and the particulate detritus may provide alternate energy sources to the different pelagic consumers (Boaden and Seed, 1985).

2.2.2 ROCKY SHORES

Those intertidal shores that are composed of hard materials are termed *rocky shores*. Rocky shores can comprise smooth uniform slopes or irregular masses of rocks, boulder beaches or from wide, gently sloping platforms to steep, overhanging cliffs etc. This type of shores are most densely populated with macro-organisms and exhibit the most diverse plant and animal life.

Adaptations of rocky shore populations

Substratum is the chief factor determining the distribution pattern and adaptations of all benthic organisms. Rocks provide quite a stable surface to which organisms may attach themselves or even bore into. The texture of the surface, which depends upon the rock type, in turn, determines attachment or boring success. Organisms that live on the open rock surfaces (epifauna) are more exposed to the outside environment than are the infauna. Of the epifauna, mobile species specially need adaptations to prevent dislodgement by waves and current action. Most species show one or more of the following adaptive features to maintain their positions in the face of strong *water movement*:

 i) Strong attachment devices like the algal holdfasts, cementation by oysters and barnacles, byssus threads of mussels, tube-feet of starfish and sea urchins, the adhesive foot of the gastrapods, and the modified sucker-like fins of many fish.
 ii) Devices for boring into the rock surface.
 iii) Changes in orientation to minimise shear stress.
 iv) Use of the rock crevices.
 v) Formation of dense aggregates to expose only a small surface area.
 vi) An increased flexibility as in case of the algal stripes.
 vii) Irregular surface contours to check turbulence and drag, like the ridged or crinkly fronds of many kelps (Boaden and Seed, 1985).

Organisms that inhabit the intertidal rocky shores have in addition, adaptations to prevent *desiccation* (Fig. 2.7). Thus, the upper shore barnacle *Chthamalus montagui* has tightly fitting opercular plates and a nonporous

BEADLET ANEMONE   EDIBLE WINKLE
ACTINA EQUINA   LITTORINA LITTOREA

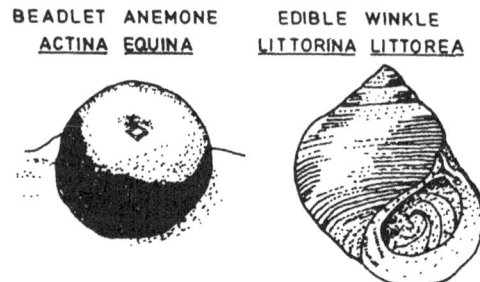

Fig.2.7. Some adaptations to withstand desiccation. Beadlet anemone with tentacles retracted, body covered with mucus and water stored inside the body. Edible winkle with entrance to shell sealed
(Source: Fincham, 1983)

shell to prevent water loss. Snails have either a horny or calcareous operculum, for the same purpose. Limpets, like *Patella,* return to a 'homescar' where their shells exactly fit the contours of the rocks to form an effective seal against desiccation. Anemones, like *Actinia,* or the hydroid *Clava squamata* produce mucus that can prevent water loss. Other forms of high intertidal zones like the algal genera *Porphyra, Fucus,* and *Enteromorpha* that do not possess any such device, are however adapted to withstand a severe loss of water from their tissues, even up to a 60 to 90 per cent loss, and still recover. After a long exposure at low tide, these species are found to be dry and brittle, but quickly start all normal metabolic processes when the tide returns. Among some intertidal secondary species, chitons can tolerate up to 75 per cent water loss, and different species of limpets from 30 to 70 per cent.

Adaptations to counteract high *temperature* include pale, heat reflecting shells of some tropical molluscs, ridges, spines and other forms of shell sculpturing of others, which all serve to increase the heat radiating surface. Some intertidal organisms can also cool themselves by evaporating water, which also prevents excessive desiccation. For this purpose, some extra water is stored in the mantle cavity, as in case of intertidal barnacles and limpets. Mobile organisms, like many crabs and snails, migrate with the tide or seek refuge in crevices or weed beds during low tide to avoid extreme temperature. Extreme low temperatures are also damaging, causing the body fluids to become increasingly concentrated. However, many intertidal organisms like *Mytilus, Fucus* etc. can survive extended periods of sub-zero ($-20°C$) temperatures.

Since high light intensity can be damaging to the plants, they also exhibit some adaptive features to protect themselves. As for example, in some algae, like *Ulva* or *Codium,* the photosynthetic pigments are housed in chloroplasts which can retreat from the cell surface at times of high light intensity; other species possess certain masking substances, like polyphenol granules, which can protect the chlorophyll molecules against photolysis;

even others, like many high shore lichens, are brightly pigmented to screen out excess light.

A good number of rocky-shore dwellers are sessile. Though serving to protect them against wave action, this however limits the methods of *food acquisition,* and a majority of rock dwellers are therefore suspension feeders. The different adaptive features for such feeding behaviour include the possession of gills (bivalves), setae (barnacles), tentacular crowns (tube worms, bryozoans) and pharyngeal baskets (ascidians). Some organisms can also utilise dissolved organic matter in the water passing through their feeding structures. Other feeding adaptations include the file-like radula of gastropods and chitons to scrape the rock, or the complex series of teeth (Aristotle's lantern) of sea urchins. These animals are all herbivores.

There are also many anti-predator devices found among these organisms. Examples of structural adaptations are spines of urchins, or thickened shells with narrow openings of many gastropods. Mobile prey have developed escape behaviours and can even respond to water-soluble chemicals exuded by the predator. Chemical defenses also include secretion of toxins and acids (Boaden and Seed, 1985).

Zonation

In response to the periodic inundation of the tidal rocky shores, there is a clear-cut zonation parallel to the water line in the habitat of the shore dwellers. Thus, organisms that cannot tolerate exposure to the air, live on the lower parts of the shore while those that can withstand desiccation, dwell on the upper reaches. Earlier these vertical zones were classified based on the tidal levels, in which the terms 'littoral' and 'intertidal' were synonymous (Fig. 2.8). Later, however, Stephenson and Stephenson (1972) introduced a system of classification using biological rather than the physical criteria, and though species composition varied geographically, the universal occurrence of these biological zonations provided a broad framework which fitted all rocky shores and into which local features or variations could be accommodated. Lewis (1964) recognised three biologically defined zones,: (i) the littoral fringe, (ii) the eulittoral, and (iii) the sublittoral. The term 'supralittoral' describes the lowest belt of terrestrial vegetation and is dominated by orange and grey-green lichens. Thus, here the littoral zone is quite distinct from the intertidal zone of the earlier classification, and as the determinants of the vertical zonations are considered, the true significance of using biological criteria becomes more evident.

The littoral fringe is the transitional zone between the land and the sea, and is relatively arid. Consequently few species live here, the most common ones being the small littorinid snails, black crustose lichens (*Verrucaria*) and small blue-green prokaryotes. Inhabitants of the eulittoral zone are mostly barnacles, mussels and other sessile forms like oysters. Towards the lower limits of this zone a narrow band of red algae can often be found. The sublittoral zone, which is only exposed during spring tides is a particularly rich

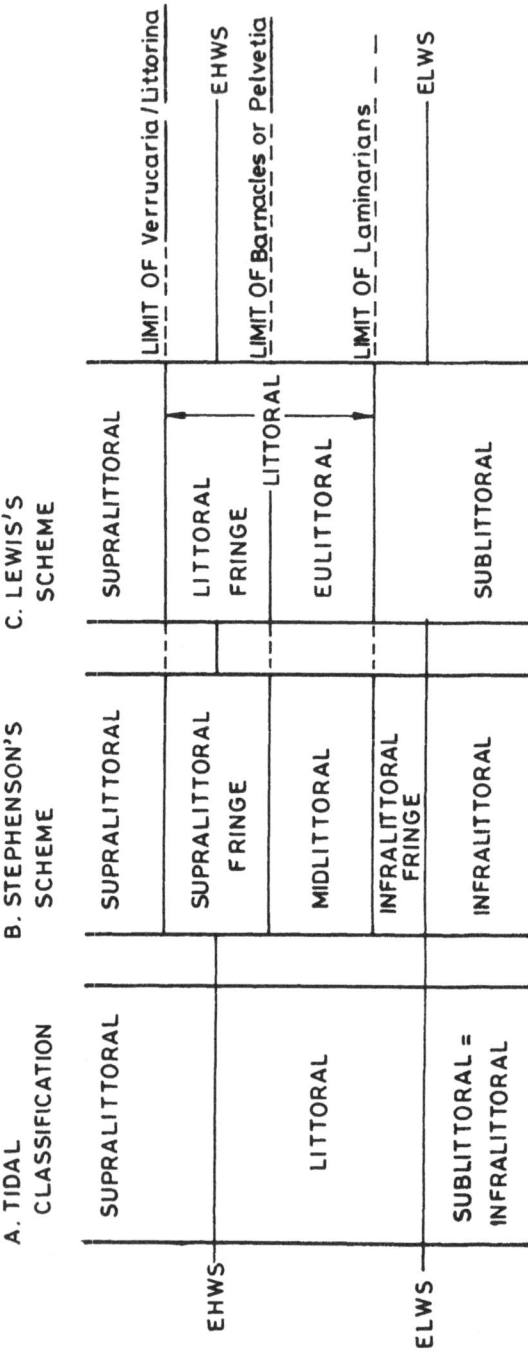

Fig. 2.8. Schemes used to classify shore zones (Source: Boaden and Seed, 1985)

zone with respect to flora and fauna. The organisms here are dominated by kelps in almost all the regions (cold temperate, boreal and antiboreal). In tropical regions luxuriant growths of corals are found here and in warm temperate regions, ascidians or turfs of red algae (Boaden and Seed, 1985).

Trophic structure

Rocky shores offer quite an unique habitat for the flora, since they provide a stable surface for attachment (algae possess holdfasts in place of roots) and shallow well illuminated water that is often turbulent and rich in nutrients. Macro-algal populations are dominated by kelps of the genera *Laminaria, Macrocystis,* and *Ecklonia,* and fucoids of the genera *Fucus* and *Ascophyllum.* These occur mainly in the cooler waters but extend up to subtropical seas, particularly at places where the cooler bottom layers are stirred up. Compared to the world phytoplankton production, the total production here is smaller, but on an area-for-area basis, the coastal algae are usually more productive. Net primary productivity in kelp beds can be around 1,000 gm C/m^2/year—a figure comparable to other highly productive ecosystems like the tropical rain forests and coral reefs. The giant kelp *Macrocystis pyrifera* is one of the fastest growing plants (up to 45 cm/day) often growing up to a length of 60 m. Gas bladders situated at the base of the fronds keep these giant forms in the surface water. The fronds of kelps are renewable belts of tissue and as new tissue are added basally the older tissue erodes distally, gradually replacing the whole frond. Fronds may be replaced several times a year when growth is rapid. This constant replacement of tissue generates large quantities of particulate organic matter that are attacked by the micro-organisms and enter the detrital food chain.

In addition, kelps exude large amounts of dissolved organic matter, which on flocculation by bacteria also enters the detrital food chain. The major herbivores in the kelp beds are sea urchins, that usually feed on detritus, detached drift weeds, or on newly established plants. But if they are allowed to reach unlimited numbers without predatory control, as had happened along parts of the Californian coast through a virtual extermination of the sea otter, or in Nova Scotia as a result of large scale lobster fishing, they can also be responsible for the mass destruction of kelps. Apart from the sea urchins and some gastropods, there are only a few kelp feeders and most of their production eventually adds to the detritus. This in turn supports a diverse population of suspension feeders like sponges, mussels, holothurians and ascidians; these again are consumed by predators like starfish, lobsters and fish. The uprooted plants that are cast onto the shore are degraded by amphipods and isopods. The whole complex trophic relationship is helped on by strong wind and wave action. Thus upwelling generated by offshore winds provides the supply of nutrients to the autotrophs. Strong wave action cuts off tips of fronds, breaks off fragments of weeds or even whole plants, creating space for the growth of new plants, eventually providing a constant supply of detritus. By keeping the plants in constant motion, the wave action also minimises grazing pressure.

The major factors influencing plant growth being the availability of nutrients and the degree of water movement and light, macro-algal growth may be nitrogen-limited at certain times of the year. As a result, some species have become adapted to photosynthesise at low temperature and light levels so as to utilise the high nutrient content during autumn and winter months. Though vigorous water movement over the pond surface constantly renews the supply of CO_2 and nutrients, facilitates their diffusion by reducing boundary layer resistance and removes metabolic wastes, very turbulent conditions however increase stress conditions in the plants which consequently utilise more resources in producing tissues to strengthen themselves. Seaweeds have been found to contain photosynthetic pigments that are most suited to their particular depth zone. With the increase in depth, algae may increase the relative proportions of the component pigment types as well as the total pigment concentration so as to enable the photosynthetic mechanism to become saturated at lower light intensities. Intertidal algae can also acclimatise by changing into 'sun' or 'shade' forms, as required, merely by changing the pigment concentration only. Light requirements are also known to vary according to the developmental stage and may also be one of the factors determining lower zonal boundaries (Boaden and Seed, 1985).

Other habitats in the rocky shores

Tide pools
One of the characteristic features of most rocky shores are the tide pools. Tide pools may be of different sizes, depths and can occur at any level in the intertidal zone. But since low level pools are almost identical to the surrounding sea, mid and infralittoral pools have only been considered here.

Though tidal pools provide a refuge against the periodic desiccation the intertidal forms are exposed to, the finite size of tide pools that restricts the volume of water is responsible for the wide fluctuations in physical factors that are less pronounced in the sea, the vast quantity of water in the latter moderating the effects of changes in the physical conditions. Consequently, flora and fauna of tidal pools are only those that are able to withstand large variations in physical factors.

There are three main factors that are subjected to variations in tide pools. The first is *temperature*. During the hot summer or cold winter, these water bodies may experience a greater change in temperature than the nearby ocean. Another problem is, while the water may change temperature gradually over a long period when exposed to the air, it may suddenly change at the return of tide with the sea water at a different temperature. The organisms here, therefore, have to be adapted to withstand sudden fluctuations of temperature. The second factor to vary in a tide pool is *salinity*. During exposure to air at low tides, they may get heated up and lose water by evaporation, thus increasing salinity. In some instances, in the hot tropical summer, the salinity increase may be as high as to precipitate out some salt. The opposite condition results in cases when heavy rains occur at low tide, flooding the pool with fresh water and suddenly lowering the salinity. As before, when the tide returns and the pool is flooded with sea water, there is again an abrupt return to the normal salinity conditions. Thus tide pool inhabitants have to be adapted to withstand wider ranges of salinity as well than do either typical marine or intertidal forms. The third physical factor that is subjected to changes in the tidal pools is *dissolved oxygen*. Since the amount of oxygen that can be held in sea water is inversely proportional to the temperature and the salinity, it naturally follows that the pools that get heated and lose water by evaporation on exposure to air will also lose oxygen. Under normal conditions it may not be serious enough to cause oxygen stress, but if the pool is crowded with organisms, such a situation may arise. In some tropical tidal pools, the oxygen level has been recorded to have fallen to about 18 per cent of saturation level.

Algal epifauna

Macro-algae, which are typical inhabitants of many sheltered rocky coasts, provide a suitable and large surface, in the form of fronds, for the attachment of many suspension feeding invertebrates, or shelter and food for many mobile species including fish. Many kelps possess complex three-dimensional holdfasts that can support both epifaunal and infaunal species.

Crevices

Rock crevices offer a damp and protected micro-habitat to organisms that are usually not encountered elsewhere on the shore. Factors that influence the zonation of organisms within the crevices include humidity, intensity of light, water circulation, or the amount and nature of accumulated sediment, all of which are in turn dependent on the shore level and the size of the crevice. Terrestrial air-breathing forms such as insects, myriapods, or mites may frequently dwell here, utilising the small pockets of air trapped within the crevice. At low tides, crevices provide refuge for many transient species such as crabs, and isopods like *Ligia*.

Tidal rapids

Tidal rapids are areas that are protected from direct wave action, but which receive clear, fast flowing water and occur in channels between adjacent land masses, specially where the entrance to a bay or sea-lough is narrow by comparison to a larger area beyond. Tidal rapids are especially rich in flora and fauna, the flora being dominated by large algae, like *Laminaria*, the fauna by attached suspension feeders.

Boulders

The upper surfaces of intertidal boulders that are also well illuminated, show organisms similar to those on adjacent rocky beds. The underboulder life, however, is protected from extreme desiccation and wave action and is dominated by the shade-loving (sciaphilic) forms like sponges, bryozoans, ascidians, hydroids and tube worms. The actual species composition varies according to the degree of tidal exposure. When the boulders are embedded in anoxic sediment (for example, in extreme shelter), colonisation of the undersurface naturally ceases (Boaden and Seed, 1985).

2.2.3 SANDY SHORES

Sand constitutes many types and sizes of particles, like, coarse sand 2.0 to 0.5 mm; medium sand 0.5 to 0.25 mm; fine sand, 0.25 to 0.062 mm; silt and clay deposited from many sources; silica particles mixed with silt, clay and organic debris as in case of grey, muddy beaches; fragments of shells, diatoms, calcareous algae, foraminifera, coral (in lower latitudes) etc. Yellow beaches consist almost entirely of coarse silicious sand (Tait, 1981). Sandy beaches are the most common type of beaches occurring throughout the world. Here the forces of nature make a majority of the organisms inhabiting them burrowers.

Physical conditions

The most important physical factor that affects life on sandy shores is the *wave action* and its effect on the *size of particles*. In the case of light wave action, the particles are fine, but where it is strong the particles remain coarse and form deposits that are called gravel or shingle rather than sand. The capacity to *retain water* and suitability for burrowing by the animals depend upon the particle size, which thus is responsible for the distribution of organisms. Through its capillary action, fine sand can retain water in its interstices long after the tide retreats. In contrast, water drains away quickly through gravel and coarse sand. Consequently fine sand beaches offer an area more protected against desiccation than a coarse gravel beach, and are also more amenable to burrowing (Fig. 2.9) (Nybakken, 1982).

Another important physical feature of sandy beaches is wave-induced *substrate movement*. Since the particles in sandy or gravel beaches are not stable enough, with every passing wave, they are churned up and redeposited. Light wave action disturbs the substratum only to a small depth and shifts the fine particles only minimally. Strong and heavy waves, on the other hand, disturb the substratum to a greater depth and carry the fine particles in suspension through a greater distance away from the beach allowing only particles heavy enough to be deposited on the beach, which is the reason why coarse and gravel beaches are found in places where the wave action is strong and fine sandy beaches only where the wave action is gentle. Any change in wave action thus disturbs and alters the profile of the beach as well as the grain size. Such changed profiles are a common seasonal occurrence in many temperate zone beaches, where a gentle slope of fine sand occurs during

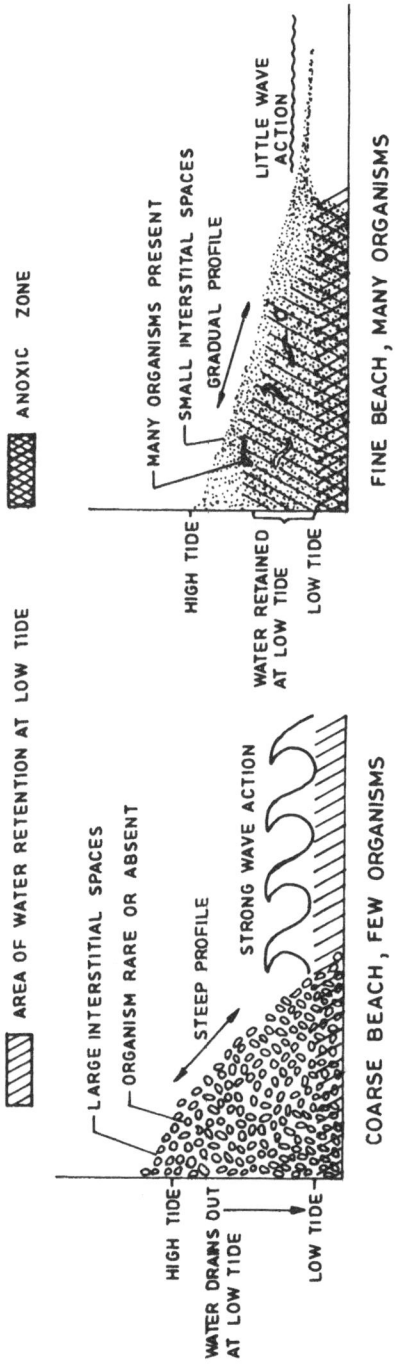

Fig. 2.9. Comparison of the physical conditions between fine-grained and coarse-grained beaches (Source: Nybakken, 1982)

summer, which is replaced during winter storms by a steep, coarse beach (Nybakken, 1982).

Sandy beaches appear to be largely barren, since the constant movement of the surface layers by wave action allows only a few large organisms to permanently occupy the surface. However sandy beaches also offer some advantages to marine organisms with respect to other physical factors. Sand, being an excellent buffer against large temperature and salinity changes below a few centimetres of depth, the temperature and salinity remain very near to that of the surrounding sea water. This is in part due to the insulating properties of sand itself and also due in part to the water being held in the interstices of the deeper layer where it retains its basic character. Any fresh water that may flow or fall over the upper layer is restricted there since the saline water in the interstices has a higher density. Also, sand being opaque to light, remains an effective barrier against direct sunlight, and *desiccation* is not a problem as long as the particles are fine enough to hold some water by capillary action at times of low tide. Thus, sandy beaches provide a sheltered life to the burrowing forms. However the limiting factor in the deeper layers of fine sand beaches is a somewhat lower *oxygen content*. Though the oxygen content of the water bathing the beach is constantly kept at saturation level due to the turbulence generated by wave action, the oxygen in the water of the interstices gets used up by the burrowing organisms and has to be replenished when the tide returns. The rate of exchange of oxygen is more rapid in coarse sediment (Nybakken, 1982).

Organisms present in sandy shores

Sandy beaches are dominated by three invertebrate groups, (i) polychaete worms, (ii) bivalve molluscs, and (iii) crustaceans. The groups most noticeable by their absence are large plants and sessile animals like barnacles and mussels, due to the lack of a stable surface for attachment.

Zonation

A three zone division that quite closely follows Stephenson's universal scheme can also be recognised in most sandy beaches. The highest part of these beaches that correspond to the supralittoral fringe are inhabited by the talitrid amphipod crustaceans like the beach hoppers in the temperate zone, and the fast moving ghost crabs of the genus *Ocypode* in the tropical zone. Both these groups are scavengers, excavating burrows. The broader midlittoral zone exhibits variable life. A common group encountered here are isopods of the family Cirolanidae. When the area is awash with waves, various crabs and animals of similar feeding habits may also come into this zone. The infralittoral fringe or the lower-most part contains the highest number of species, which include the large surf clams like *Tivela* or *Spisula*, sand dollars of the genus *Dendraster*, various polychaete worms, crustaceans and the large carnivorous snails like *Natica* and *Polinices* (Nybakken, 1982).

Adaptations

The basic features of sandy beaches make it essential that the organisms inhabiting these areas be burrowers. There are two routes by which the burrowing forms are adapted to tolerate the unstable and constantly moving substratum. The first is by being able to burrow deeply enough so as to be at a depth below that of the sediment affected by passing waves, as done by many large clams like the *Tivela stultorum*. These animals also generally have heavy shells to keep them down in the substratum, though a very severe storm may generate waves large enough to pull them out and throw them onto the beach. The second route of adaptation is the ability to burrow again very quickly after being pulled up by a passing wave. This is a more common adaptive feature, found in many annelid worms, all clams and crustaceans. The burrowing molluscs have very smooth shells that reduce resistance while burrowing into the sand. Similarly shore echinoderms like the sand dollars have much reduced spines to facilitate burrowing.

One special type of adaptation is seen in small sand dollars (*Dendraster excentricus*) which tend to accumulate iron compounds in a special area of their digestive tracts, serving as a 'weight belt' to keep them down in the face of wave action. Another type of special adaptation is the presence of various types of screens fitted on the intake siphons of sandy beach clams that prevent the entry of suspended sand particles while allowing a free passage to the incoming water necessary for respiration and food (Nybakken, 1982).

Food and feeding habits

Since large multicellular plants are lacking in sandy beaches, there is virtually no primary productivity here, and though diatoms are present, the opaque nature of sand restricts them to the surface layers only. As a result, sandy beaches do not have macroscopic herbivores, and the animals living here must depend upon either phytoplankton and organic debris carried in the sea water over the beach for food, or consume other beach animals. However there are few true carnivores among beach animals since it would involve active movement across the substratum in search of prey, which is difficult in the face of wave action. Thus, the majority of beach dwellers are either suspension or detritus feeders (Fig. 2.10).

2.2.4 MUDDY SHORES

There can be no sharp demarcation between sandy and muddy shores. As sandy shores become more protected from wave action, they become more and more fine grained, collect more organic matter, and become more 'muddy'. Sandy and muddy beaches are thus the two extremities of a continuous spectrum, and consequently the flora and fauna follow a change from those of a typical open sand beach to those of a typical muddy shore (Nybakken, 1982).

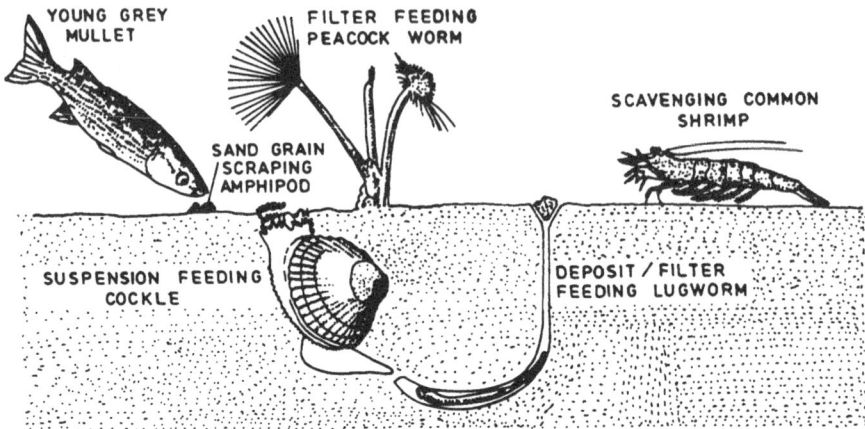

Fig. 2.10. Food web interrelationships between different organisms of sandy beaches (not all drawn to the same scale)
(Source: Fincham, 1983)

Physical features

Typical muddy shores are developed in areas completely protected from open ocean wave activity and where there is a source of fine grained sediment particles. They are thus located in different partially enclosed bays, lagoons, harbours, and more notably estuaries. Since water movement is at a minimum, most muddy shores are much flatter than sandy beaches and are often called 'mud flats'. The finer particle size coupled with their flat angle of repose enables these areas to retain water much longer. This again, together with a poor interchange with the tidal sea water above and a high internal load of bacteria, lead to a complete depletion of oxygen within the sediments at a depth of a few centimetres from the surface. This anaerobic condition below the surface is one of the major characteristic features of mud flats. However, accumulation of high amounts of organic material also ensures a potentially rich source of nutrients for the resident organisms, even though the heavy shower of small organic particles often tend to clog their respiratory surfaces.

Organisms present

Diatoms are most abundant on bare mud flats and give a brownish colour to the surface at low tide. Macro-algae are also found on muddy shores. The lowest tidal levels may show growths of different sea grasses. Such vegetation thus leads to a substantial level of primary productivity in these regions. Mud flats also contain a large number of bacteria that feed on the abundant organic matter. Bacteria are abundant in the anaerobic layers also, where the dominant forms are the chemosynthetic or sulphur bacteria capable of utilising energy obtained by oxidation of the reduced sulphur compounds (for example, H_2S). Mud flats are therefore unique among different coastal environments, in that they have two separate layers in which primary productivity occurs, the surface layer where the green plants carry on photosynthesis, and a deeper layer where autotrophic bacteria conduct chemosynthesis.

The most abundant macrofauna encountered on muddy shores are mainly various polychaete worms, bivalve molluscs, and crustaceans.

Zonation

No detailed study has been made on the zonation of muddy shores. The supralittoral zone is usually inhabited by different crabs, many burrowing into the substrate. The midlittoral area is very extensive and houses most of the common species of clams and polychaetes. There is no well defined boundary with the infralittoral fringe, and similar organisms extend up to there.

Adaptations

As in the case of sandy shores, the surfaces of mud flats also seem bare. Most of the inhabitants are burrowers, and some live in permanent tubes inside the substratum. Unlike sandy beaches, however, mud flats exhibit holes of various shapes and sizes on their surface betraying the presence of organisms below. Here also, one of the primary adaptations is the ability to burrow or to form permanent tubes. However, since wave action is negligible on mud flats, there is no necessity for the development of either rapid burrowing habits or heavy bodies to retain their positions within the substratum.

Another major adaptive feature concerns the ability to counteract the anaerobic conditions prevalent in the substratum. Since most multicellular organisms cannot survive without oxygen, one of the most common adaptations is to develop some way to bring the overlying surface water with its supply of oxygen down to them. However, some organisms are even adapted to live at oxygen tensions lower than that for similar forms inhabiting open sand beaches. In this case the organisms often develop special carriers of oxygen (like haemoglobin) that are effective at much lower concentrations of oxygen than are similar pigments in other organisms.

Trophic relationship

Due to the large amount of organic matter present in the muddy shores and increased productivity through both bacteria and plants, these areas support the growth of a large population of macrofauna. The dominant feeding types on mud flats are the deposit feeders, and suspension feeders, the former being more abundant due to the larger amount of organic material and the huge population of bacteria in the sediments. Deposit feeders mainly comprise the polychaetes and bivalves, while the suspension feeders include various clams, crustaceans and also polychaetes. Despite the relative abundance of plant material on muddy shores, there are only few herbivores, and most of the vegetation enters the food chain through the deposit feeding food web after being broken down into smaller pieces. The carnivores of muddy shores are the fish which feed when the tide comes in, and birds when the tide goes out.

The trophic structure of mud flats is thus essentially built up from two bases, a detritus-bacteria base, and a second one based on the diatoms.

The detritus base web is derived from the plants and other organic material including bacteria living upon the detrital particles. These particles can either be taken in directly by the large invertebrates or be consumed by bacteria. The bacteria are in turn consumed by different nematode worms, the latter occupying a position similar to the herbivores of other food chains. Bacteria can also be consumed by the deposit feeders as they ingest organic particles upon which the bacteria commonly occur. These deposit feeders and the nematodes are then consumed by different carnivores, including predatory invertebrates and fish.

The second food web depends on the microscopic diatoms which forms the autotrophic base. The diatoms are consumed by several polychaetes, molluscs and crustaceans, which in turn are preyed upon by large birds and fish (Nybakken, 1982).

2.2.5 ESTUARIES

Estuaries can be defined as semi-enclosed bodies of coastal water that retain a free connection with the open sea, and within which sea-water is measurably mixed with fresh water of terrestrial origin (Pritchard, 1967). Though estuaries are the interface between open marine and a fully fresh water environment, the estuarine water is not a simple dilution of sea water. Many changes that vary according to the local conditions are involved. For example, some of the minor constituents of sea water, like Fe, Si, PO_4 etc. are often present in much higher concentrations in the river waters, which are therefore reflected in the estuaries. Similarly, some rivers are relatively rich in Mn, Cu or Zn.

Types of estuaries

Based upon the geomorphological characters and their modes of formation, estuaries can be classified into the following types (Boaden and Seed, 1985):

i) *Drowned river valleys* (or *coastal plain estuaries*): These are the most common types of estuaries and have been formed by a relative rise in the sea level compared to the land, resulting in the release of ice-hold water at the end of the last glaciation. Examples: Chesapeake Bay; R. Thames.

ii) *Tectonic estuaries:* Members of this small class of estuaries originate through the invasion by the sea of fault-lines extending up to the coast due to both land subsidence and a rise in sea level. Example: San Fransisco Bay.

iii) *Fjords* (or *drowned glacial valleys*): These are glacially overdeepened valleys now invaded by sea water. One important characteristic found almost universally in fjords is the presence of silt at their mouths, which can rise within 200 m from the surface though their depths may be over 1250 m. The silt restricts the interchange of water between the fjord and the adjacent sea so far as to effectively isolate the water beneath the sill level even making it occasionally stagnant. Fjords can be commonly encountered along the coasts of Norway, Chile and British Columbia.

iv) *Bar-built estuaries* (or *semi-enclosed bays or lagoons*): These are shallow estuaries occurring chiefly in the low-lying areas and are formed by partial isolation of the river from the sea by sand or shingle bars running parallel to the shore line. Examples: Pamlico Sound in North Carolina; Dutch Waddenzee.

Characteristics of the estuarine environment

Salinity and water movement

According to their salinity, brackish water can be classified into several zones (Table 2.2). Basically, the water movement in estuaries is two-fold: fresh water flowing seaward at the surface, which is replaced by denser saline sea water flowing up the estuary at a depth. These two opposing layers of water show a varying degree of mixing depending upon the local topography, tidal velocity, the relative volumes of fresh water and saline water, and the friction generated at their interface.

In *highly stratified or salt wedge estuary* (Fig. 2.11), the saline sea water enters the river bed in the form of a salt wedge. Fresh water, which is of less density, flows seawards over the salt wedge in a progressively thinner layer. Frictional forces at the interface also lift up some sea water into the fresh water layer, and to compensate this loss there is a slow upstream flow within the salt wedge. The surface layer becomes increasingly more saline as it moves towards the sea, but the lower layer remains more or less similar to that of the open sea in the face of the virtual absence of downward mixing of the fresh water. The limited mixing also results in a marked discontinuity in

Table 2.2
Venice system for the classification of brackish waters

Zone	Salinity (parts per thousand NaCl)
Hyperhaline	>40
Euhaline	40–30
Mixohaline	30–0.5
Mixo-euhaline	>30 but <adjacent sea
—polyhaline	30–18
—mesohaline	18–5
—oligohaline	5–0.5
Limnetic (fresh water)	<0.5

From: McLusky (1981).

the salinity or halocline between the two circulating water layers. This type of estuary occurs where the river flow is larger compared to the tidal flow, and where the width/depth ratio of the estuary is also comparatively small, exemplified by the Mississippi estuary where the depth is 30 m at places.

A relative increase in the tidal flow over the river flow results in two-way vertical mixing abolishing the salt wedge and the establishment of a *moderately stratified or partially mixed estuary* exemplified by the Chesapeake Bay (Fig. 2.11). Here the salinity increases seawards throughout the water column, but the salinity of deeper layers is always somewhat higher than that of the surface layer despite the two-way mixing and there is a persistent weak halocline.

In places of especially vigorous tidal mixing concurrent with a weak river flow, all vertical stratification breaks down and a *vertically homogeneous estuary* results (Fig. 2.11) where there is no halocline with only a marked

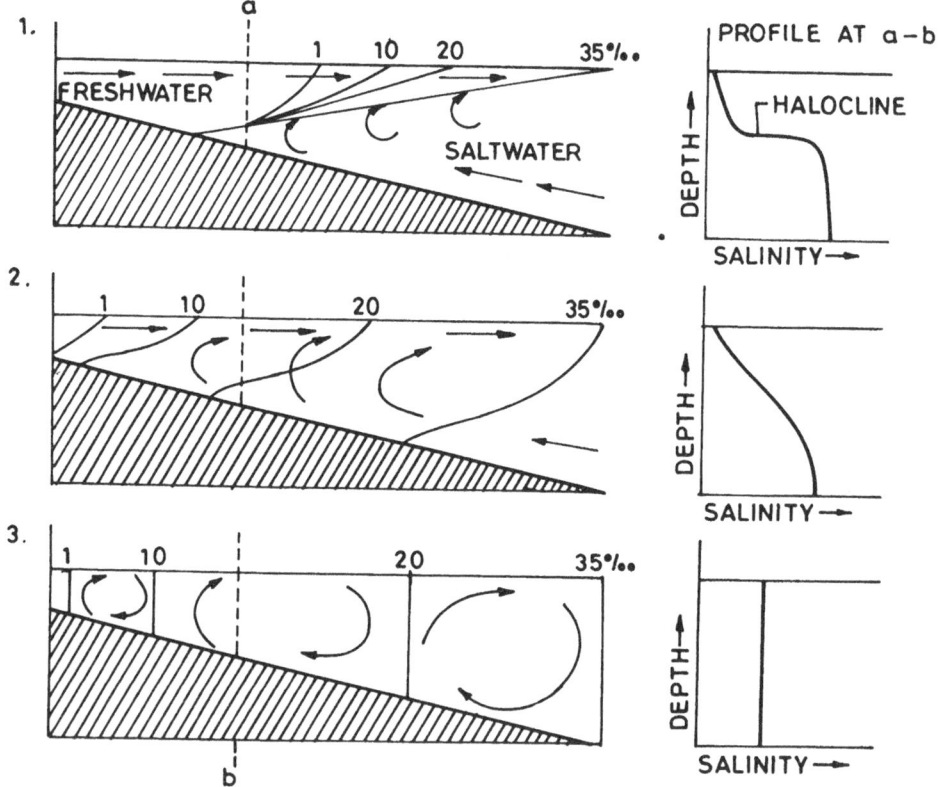

Fig. 2.11. Distribution of isohalines in
    1. a salt-wedge estuary;
    2. a partially mixed estuary;
    3. a vertically homogeneous estuary
    (Redrawn from Boaden and Seed, 1985)

salinity gradient existing along the length of the estuary as exemplified by the R. Thames estuary.

However, in addition to the above water circulation patterns coriolis effect can also be observed, which deflects the water to one side of the estuary. The opposite banks of an estuary may therefore exhibit markedly different salinity levels, which in extreme cases in the wide estuaries may result in an almost vertical fresh water/sea water interface.

Depending upon the volume of fresh water input, there are also seasonal variations in the position of the mixing zone between the two types of water masses within the estuary, in addition to the differences related to the effects of tidal ebb and flow. In estuaries with a specially large input of fresh water, as in the case of the R. Amazon, the mixing zone is located in the sea. Again, in very hot climates, the fresh water input may be negligible, and the evaporation from the surface is also high, so that the surface water becomes hypersaline and dense causing it to sink and move out of the estuary as a bottom current. This is called a *negative estuary* (Boaden and Seed, 1985).

Sedimentation

Sedimentation is dependent on water movement, topography, and the amount and type of available suspended matters. Coarser materials carried downstream by the river are mainly deposited in the upper reaches of the estuary while fine particles or silt of the size range of 4 to 63 μm are carried further down before being deposited. Clumping of particles or flocculation occurs at the halocline where there is mixing of water. Flocculation and the fall velocities of the floccules depend upon salinity, temperature, and the amount of suspended organic and inorganic matter. Illite and kaolinite flocculate in salinities less than 4 ppt, while montmorillonite flocculates over the whole range of salinity.

Floccules usually tend to sink after formation, though they may be carried into the outflowing fresh water as well, whence they deflocculate, thus setting up a flocculation-deflocculation cycle. Larger floccules sink and either adhere to the substratum or flow as a layer in the form of liquid mud where the concentration of sinking floccules is high enough (around 10 gm silt/litre of water) though some may be resuspended by current action at ebb tide. In most estuaries, there is an overall accumulation of mud on an average of about 2 mm per year. Sediments are derived in part from the sea as well as through tidal action as in the Marsey in northwest England. However, sedimentation is largely related to the seasonal input of fresh water. Storms and floods can also dump large amounts of sediments at times. Most estuaries are found to be muddy, though in places the rivers drain hard igneous rock or where the currents are sufficiently strong to prevent the deposition of finer particles, sandy estuaries may also develop.

Estuaries also contain deposits of large amounts of detritus or particulate organic matters that are derived in part from outside sources like salt marshes or mangrove swamps (allochthonous) and partly from within the estuary itself (autochthonous).

Light penetration

The considerable amount of suspended particles present in the estuaries render it a turbidity that is much higher than that of the adjacent open sea. Nearly all the light is absorbed in the first 1 to 2 m. The high productivity frequently associated with the estuarine ecosystem is thus largely due to the constant supply of organic material to power the detrital food chain.

Wave action and tides

In estuaries that narrow gradually and at a distance from the mouth, there is limited fetch in most directions and only small waves are generated, encouraging a further deposition of mud. However, the narrowing inland of many estuaries can funnel the tidal flow and actually increase tidal height. A typical example is the R. Severn flowing into the Bristol Channel. At its mouth where the width is about 65 km, the average tidal range is around 6 m, whereas 130 km further inland where the width is 8 km, the tidal range has increased to about 10 m. This funnel also produces a huge wall of water or 'bore' 2.5 m high that sweeps up the river at 25 km/hr at every high tide.

Oxygen

The oxygen level is generally low, both in the water and within the sediments. However, there may be local pockets of supersaturation during the day particularly in *Zostera* beds due to intense photosynthetic activity; but here also oxygen deficiency occurs at night. Microbial decomposition of the heavy load of organic material in the estuaries sets up a high oxygen demand leading to a further depletion of oxygen even in the interstitial water within the fine sediments deposited already. In the fjords, there may be severe depletion of oxygen in the water below sill level.

Other factors

The variable input of fresh water that is also subject to seasonal changes in temperature, together with the large surface area/volume ratios of estuaries are dually responsible for the frequently encountered widely variable *temperature* ranges of the estuarine waters.

Since the ionic content of seawater and fresh water differ, the *chemical composition* of estuarine water varies according to the water budget of the particular estuary (Barnes, 1984).

Nature of the fauna and flora

The fluctuating and stressful conditions prevalent in the estuaries are reflected in the relatively poor diversity of species.. A lack of diversity of habitat may also contribute to the general paucity of species (since most estuaries are uniformly muddy), as also do their comparatively recent origin.

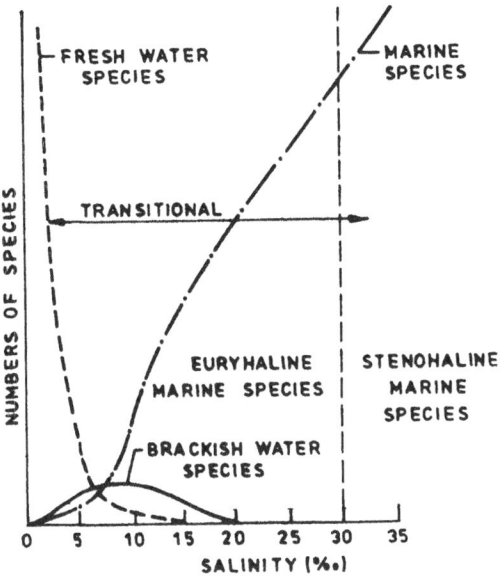

Fig. 2.12. Numbers of species in each of the three major components of the fauna of estuaries
and their distribution with salinity
(Source: McLusky, 1971)

Groups of animals like the cephalopods and echinoderms are completely
absent. However, some brackish water species like *Macoma baltica* can occur
here, presumably due to a reduced interspecific competition and an
abundance of food. The stratified nature of estuaries allows the benthic
organisms to move further ashore than the pelagic forms. The midupper
shore is awash with more or less fully saline water at high tide and is exposed
to the air at low tide, thus enabling the marine forms to reach a higher
intertidal zone. Organisms further down the shore line are exposed to much
greater fluctuations of salinity, since they are covered by fresh water at low
tide. The epifauna likewise experiences much larger salinity variations than
does the infauna which is under the buffering influence of interstitial water.

All three major aquatic faunal groups namely marine, fresh water and
brackish water, are encountered in the estuarine water (Fig. 2.12). Of these,
the marine component is the largest one and includes both stenohaline and
euryhaline species. The stenohaline forms that can only tolerate a narrow
range of salinity are restricted to the lower reaches of the estuary; euryhaline
species, tolerating much wider fluctuations, are found on the other hand,
further up the estuary. The fresh water species with a salinity tolerance rarely
over 5 ppt occur chiefly at the head of an estuary. Areas with a salinity
range of 5 to 8 ppt mark the *critical salinity range* for many species and
contain the least number of species. The small group of brackish water
organisms comprising *Hydrobia, Nereis diversicolor, Macoma* etc. occur in
the middle reaches of the estuary where salinity fluctuations are at their

highest. Essentially of marine origin, these organisms have lost their seaward distribution probably due to predation and competition with other marine species.

Apart from the resident fauna that is dominated by the polychaetes, bivalves and crustaceans in the cooler waters and gastropods and crabs in the tropical and subtropical areas, estuaries also exhibit migratory species including fish, birds and decapod crustaceans which use it as a feeding and/or a nursery ground. Other migrants also pass through estuaries *en route* to breeding grounds either in rivers or in the sea.

Estuarine vegetation is also limited. Muddy sediments being generally unsuitable for the attachment of macro-algae, only a few species that are tolerant of reduced salinity like *Enteromorpha,* or *Ulva* may occur seasonally. Growth of phytoplanktons is also limited due to the high turbidity leading to poor light conditions. Sea grasses may occur in the lower reaches of estuaries. However, an abundance of emergent rooted plants is one of the chief features and forms an important component of estuarine macro-flora. The high organic content of estuarine mud is however conducive to the rich growth of micro-organisms and meiofaunal species (Boaden and Seed, 1985).

Adaptations of estuarine organisms

The foremost adaptation required for an estuarine life is for *osmotic* and *ionic regulation* due to an often very quick changing salinity gradient present in the estuaries. Mechanisms of such adaptations may be varied. However some estuarine species, the osmoconformers, are unable to regulate either the cell volume or its ionic composition, their survival being dependent solely on their tolerance to wider fluctuations in the concentration of their body fluids. Most marine species however show some degree of adaptation. Thus, in some species, even those incapable of regulating their extracellular fluids, some sort of intracellular regulation is effected by changing the concentration of dissolved amino acids in the face of changes in the external salinity, so that intra and extracellular fluids remain more or less isosmotic and a constant cell volume is maintained. Many higher crustaceans are strong osmoregulators, while most bivalves are osmoconformers, closing their shell if the salinity falls too far below the acceptable level. Osmoregulatory abilities appear to be dependent on temperature, and may be one of the reasons why tropical regions have more estuarine species, and the stenohaline forms extend further upstream in these regions than they do in the higher latitudes, or why some species migrate into deeper waters during the colder season.

Some estuarine organisms avoid wider salinity changes by either burrowing into the sediment, or by migrating. The shore crab *Carcinus maenas* thus moves up and down the estuary with the tide so as to remain more or less at the same salinity level. Many crustaceans like the blue crab *Callinectes* move out of the estuaries for breeding since their eggs are more vulnerable to low salinities (Fig. 2.13). Conversely, some other organisms like the mullet or flounder use estuaries as their breeding grounds. The circulation of water in

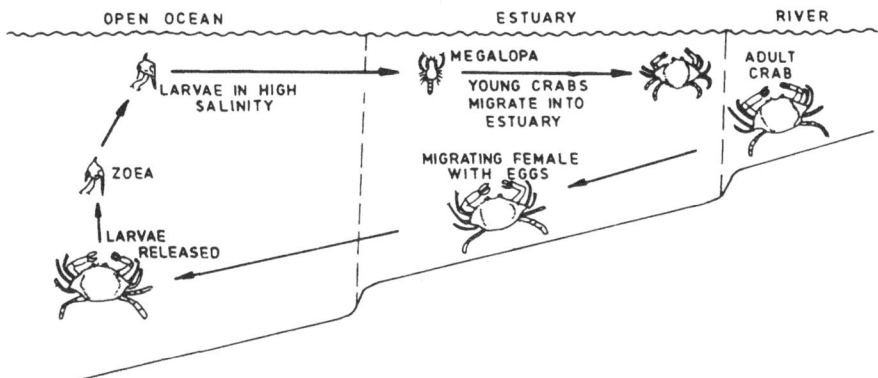

Fig. 2.13. The life-cycle of the blue crab, *Callinectes sapidus*
(Source: Nybakken, 1982)

the estuaries can be responsible for the loss of a large number of planktonic larvae. This may give rise to different adaptive features as is the case of spionid polychaetes which are adapted to remain within the estuary by selectively swimming into the water column in flood tide and remaining at or near the bottom at ebb tide. Many estuarine species have even eliminated the planktonic stages from their life-cycles. For example, the larvae of *Nereis diversicolor* are benthic, eggs of the sand prawn *Callianassa kraussi* hatch directly into miniature adults, etc. In some other species, as for example in the mud snail *Nassarius kraussianus*, the plankton stage is very brief.

Some morphological and anatomical peculiarities have often been observed in the typically estuarine organisms, much of which are related to a life in muddy sediments. Thus, fresh water forms that inhabit brackish water may be partially sterile, and the original marine forms show lower rates of reproduction. This might suggest that many estuarine populations have diverged enough to form genetically distinct races as opposed to the conspecific populations on the adjacent open coastal areas (Boaden and Seed, 1985).

Productivity and trophic structure

In estuaries, the primary production is effected by growths of phytoplankton, benthic algae and rooted plants, though phytoplanktons are clearly at a disadvantage in most estuaries due to high turbidity and a net transport of surface waters to the open sea. In addition, macrophytic plants compete with the phytoplanktons for nutrients, and among them some mangroves may even release tannins that suppress phytoplankton growth. However, phytoplankton growths may be quite substantial at places, and in particular in larger estuaries where the interchange of water with the open sea is restricted. In general, the zooplanktons of the estuarine waters are not very efficient feeders of phytoplankton, so that only about 50 per cent of the net phytoplankton production may ultimately be available either as living cells or as detritus, to the benthic forms. The seaweeds generally cannot tolerate low salinities and are therefore confined more or less to the lower reaches of the estuaries with extensive connections to the open sea. Diatoms that live either on or within the sediment are generally an important source of food. However, a large proportion of the primary production in most estuaries comes from the rooted plants growing on the intertidal mud banks and on the margins of the estuary. Most of this production is however utilised through the detrital pathway since very few herbivores directly feed on this vegetation. This autochthonous detritus and the organic matters brought in by the rivers provide the principal food source fueling the entire ecosystem. The nutrients released from the detritus and the animal faeces are carried back up stream by water movement in the bottom layer and are again available for plant growth.

Estuaries are abundant with detritus feeders which are mainly constituted by the benthic invertebrates and include both deposit and suspension feeding forms. Of these, the latter are rarer presumably because their filtering devices may be more easily clogged by the sediment, though some species including the commercially exploited *Mytilus edulis* may be locally abundant. However, the degree of assimilation of the detritus feeders is not very high and the sediments are thus continually being reworked (this is known as bioturbation). The detritus feeders are consumed by various predators of which the more important groups are crabs, birds and fish, and which have a marked effect on the community structure of estuarine ecosystems.

The major interactions in a typical estuarine food web have been shown in Fig. 2.14. Efficient utilisation of the available nutrients and their extensive recycling through the biologically active bottom sediment helped on by the water movements make the estuaries highly productive. A seasonal abundance of food in estuaries attracts migrant birds serving as their feeding grounds as they do different types of fish and many invertebrates that use estuaries as a nursery ground.

Other brackish water habitats

Estuarine realms: These occur beyond the mouths of the rivers where a discharge of large volumes of fresh water results in areas of coastal water with distinctly brackish surface layer. These are bounded by the coastline on the land side and by a salinity front towards the sea (Boaden and Seed, 1985). Example: waters off the eastern coastline of North America.

Brackish seas: These are large semi-enclosed water bodies with salinity usually lower than that of many estuaries. Unlike estuaries, the water of brackish seas are permanently brackish, and consequently the osmotic problems faced by the resident organisms are somewhat different here, though the diverse species present here also include originally marine and fresh water forms (Boaden and Seed, 1985). Example: Caspian sea.

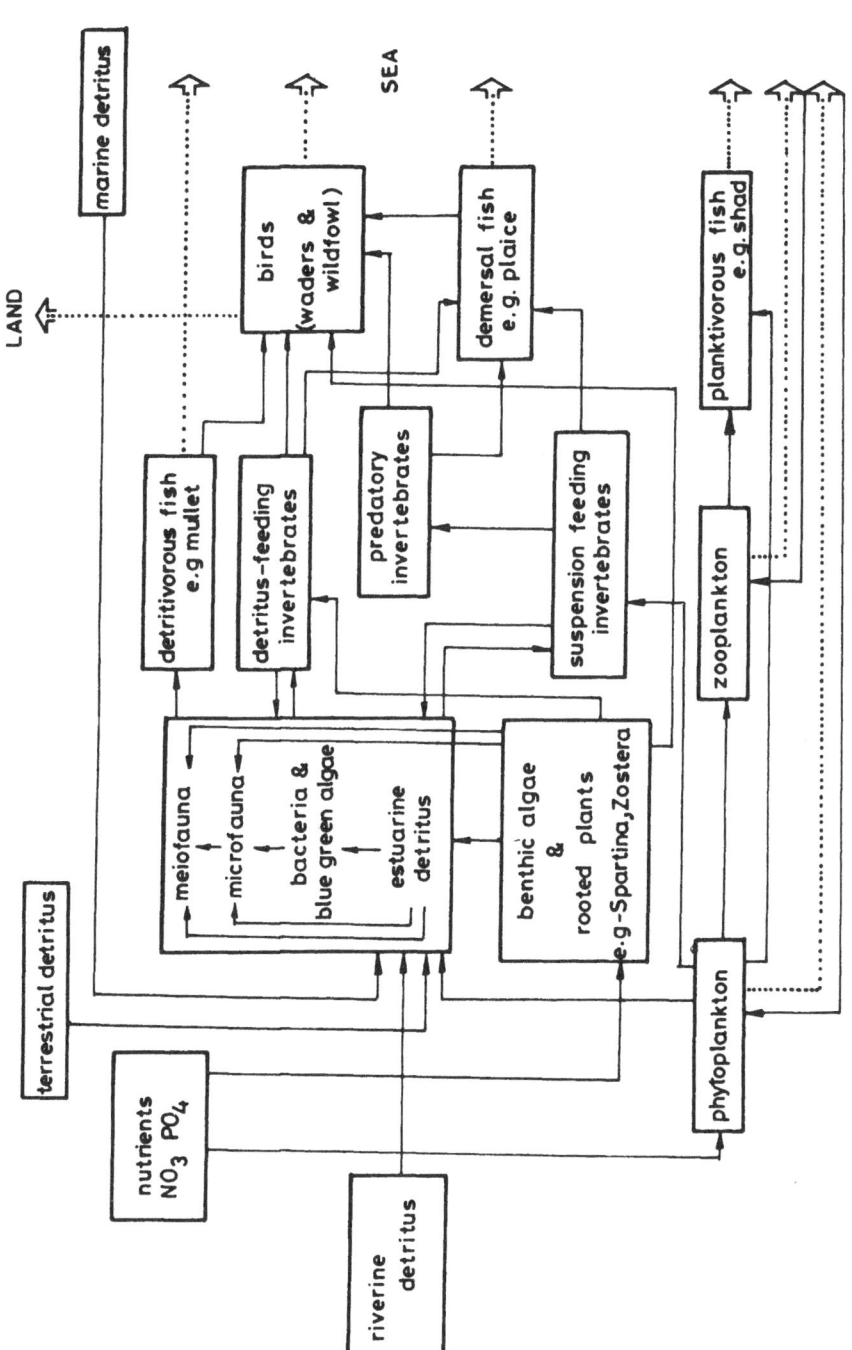

Fig. 2.14. Schematic representation of estuarine food web. Dotted lines indicate 'losses' from the estuarine system (After: Barnes, 1984)

Brine seeps: These may occur subtidally in places where hypersaline water emerges on the sea floor and have a characteristic biota of anaerobic micro-organisms with a few associated meiofaunal species (Boaden and Seed, 1985). Example: East Flowergarden seep in Texas.

2.2.6 SALT MARSHES

Salt marshes occur in the areas where the salinity range is between 5 and 38 ppt. They have been defined as "natural or semi-natural halophytic grassland and dwarf brushwood on the alluvial sediments bordering saline water bodies whose water level fluctuates either tidally or non-tidally" (Beeftink, 1977). Even though they can occur both on fairly open as well as on enclosed coastlines, some measure of shelter is generally necessary since their formation and maintenance requires accumulation of sediment. The dominant organisms of salt marshes are originally terrestrial rooted plants that cannot grow in the absence of a sediment stable enough to permit rooting and growth. The establishment of salt marshes also requires the presence of finer sand particles, with some admixtured silt, that are able to hold back water. These areas also tend to be flat and favour a growth of protists, prokaryotes and filamentous algae as a surface layer. However, deposition and accretion concentrated at some places raises them somewhat from the surrounding area thus confining surface drainage to the latter and consequently a system of small channels is formed. During this process angiosperms (for example *Salicornia and Spartina*) begin to colonise on the raised surfaces either by germination or by transport of vegetation.

Environmental characteristics

Salt marshes are characterised by widely fluctuating environmental conditions. Due to the interaction between the river flow and marine water, the salinity varies as it does in estuaries, but the intertidal nature of salt marshes makes these fluctuations much wider and more sudden (as 20 to 40 ppt in a single·tidal cycle) since a sudden rainstorm at low tide may reduce surface salinities to a near zero level, while the return of tide may wash the area with full strength sea water.

Similarly the temperature also varies widely. At low tide, the marsh and the substratum are exposed to the extreme air temperature of a terrestrial environment with a return to the original levels at high tide. The mud temperatures have been known to vary by as much as 10°C within a single day.

The substratum of salt marshes is typically muddy and similar in nature to the estuarine sediments with a high salt content resulting from a perfusion of sea water concomitant with a high rate of evaporation.

Flora and fauna

The bacterial population is high both in the sediment (up to $10^9/cm^3$) as well as in the overlying water ($10^6/cm^3$). Protists are also common. However, the rigorous environmental conditions do not allow a very large or diverse population in these areas, and the organisms that are found to occur here are taxonomically similar over a vast geographical area. Thus the dominant plant species occurring world-wide in salt marshes are grasses of the genera *Spartina, Salicornia* and *Juncus.* Marine animals found in salt marshes include various crabs (*Uca, Hemigrapsus*), mussels (*Modiolus*), some species of snails (*Littorina, Cerithidea, Melampus*) and smaller crustaceans like the amphipods. Terrestrial animals, that may also be permanent inhabitants of marshes, are represented by various types of insects. In addition, many duck and geese species, rats, moles, rabbits and hares also graze on the salt marshes and may also have a controlling effect on the vegetation. At low tide the terrestrial species dominate, while with the return of the high tide marine and estuarine animals enter the area (Nybakken, 1982).

Zonation

A basic zonation pattern of species that may be found to vary according to the tidal regime, drainage, slope, climate and some other environmental factors, can usually be recognised in a salt marsh that extends from the tidal creek to the terrestrial vegetation.

The lowest zone is that of the creek bank, and is usually bereft of any macrophytic vegetation, containing mainly infaunal estuarine and marine animals and many mud snails. The next higher zone is a border zone between the creek bank and the main marsh land comprising the creekside marsh. Then comes the main part of the marsh which can be further subdivided. In the east and Gulf coasts of the North American continent the recognisable zones are a lower tall *Spartina* zone, an upper short *Juncus* or *Spartina* zone with a final marsh zone of *Salicornia* and some other plants. The main marsh land of the west coasts of North America can be subdivided into a narrow *Spartina* lower zone, followed by a broad *Salicornia* zone covering most of the mid-intertidal zone with the highest intertidal area covered by the diverse population comprising commonly *Jaumea, Distichlis,* and *Limonium* (Nybakken, 1982).

Productivity

Primary production is carried out by emergent marsh plants and various micro-algae growing on the surface of these plants and the mud. Productivity is higher in the salt marshes than in the estuarine waters and mud flats. Complete data on the productivity is available for the east coast marshes of North America (Teal, 1962). The major primary producer is the different species of *Spartina.* The average annual net productivity of Georgia marshes has been shown to be 1600 gm/m^2/year, decreasing towards the north to a value of 325 gm/m^2/year in New Jersey. On the west coast on the other hand, the net production is reported to be 50 to 1500 gm/m^2/year in the San Francisco marshes. Due to the occurrence of only a few herbivores, most of this is however not directly consumed, but utilised after the dead plant materials have been attacked and broken down by the bacteria in the mud surface or the overlying water. The indigestible cellulose being broken down into digestible carbohydrates by these micro-organisms, this process actually increases the nutrient value of the vegetation. The detritus thus ultimately forms the basis of the food chain for the different marsh animals. Some nutrients may even be carried into the estuary by tidal flow. The high nutrient content of salt marshes make them effective as nursery grounds for many crustaceans and fish.

2.2.7 MANGROVES

Mangroves are a group of flowering halophytic shrubs and trees growing up to 30 m of height, belonging to several unrelated families occurring around sheltered tropical shores. The term 'mangroves' is also used to describe this whole ecological community or the association of such plants, though the term 'mangal' is sometimes used to describe the same. Mangroves occur along protected sedimentary shores, particularly in the tidal lagoons, embayments and estuaries. Capable of tolerating full saline conditions, mangroves, are never totally isolated from the sea, though they may grow far inland. Compared to other tropical and temperate vegetation, mangroves are however a relatively less diverse group presumably because the rigours of life at the interface of the land and sea is not conducive to much diversity of forms. However, the dense evergreen growth of mangroves shelters a large number of terrestrial animals, while their complex root systems, together with the thick silty sediments in which they grow, harbour a variety of marine invertebrates. Filamentous algae are abundant on the substratum as well, and utilise the fine sediments while the leafy algae directly attach to the mangrove roots. In addition, numerous fish and crustaceans move in and out of the mangrove vegetation with the tidal flow.

Though mangroves thrive most in warm moist coastal belts as found in the tropics, some species like the *Avicennia marina* extend up to Japan and Bermuda in the north, and Australia and northern New Zealand in the south. The limiting factor for the distribution of mangroves seems to be low temperature. Though tolerance to lower temperature varies from species to species, most mangroves do not survive below a temperature of 4°C, with the optimal conditions at around 20°C and seasonal variations < 5°C. Seed dispersal in many species may be over very long distances and are responsible for the wide distribution of such species. Two main groups of mangroves have been recognised: those from the Indo-Pacific regions, and those of the Atlantic, there being very little overlapping of genera between the two regions. Of the two, the Indo-Pacific region contains the more number of genera and species.

Zonation

Mangrove vegetation occurs as distinct zones of a single or groups of species growing over the marine-terrestrial environmental gradient with each zone being dominated by a single species. Such clear-cut zonation of mangroves has been attributed to the variations in such physico-chemical and biological factors as the salinity, tidal regime, climate, geomorphology, interaction between species, stochastic factors, the influence of man etc.

Adaptation of mangroves

Mangroves provide a typical example of convergent evolution among taxonomically unrelated groups of plants inhabiting the same ecological niche. Adaptations are mainly for combating salinity stress and growth in water logged anoxic sediments. For this, all mangroves have developed a system of shallow laterally spreading cable roots. Two main types of roots are found above ground—the pneumatophores, and the prop roots (Fig. 2.15). Being negatively geotropic in nature, pneumatophores extend upwards through the soil and their extensive development enables the mangrove to grow under extremely anoxic conditions. Prop roots develop from tree trunks well above ground, curving downwards towards the ground, and as the name suggests, provide a firm anchorage in, and allowing proper exploitation of the particularly soft sediments. Both these types of roots contain numerous small pores called lenticels that are permeable to air but not to water and are connected with an elaborate system of aerenchyma tissue which are responsible for rapid transport of gas to the underground root system. The oxygen trapped thereby is utilised in times of oxygen starvation when the roots are submerged. It has been shown that an experimental coating of the lenticels with grease rapidly depletes the oxygen within the roots.

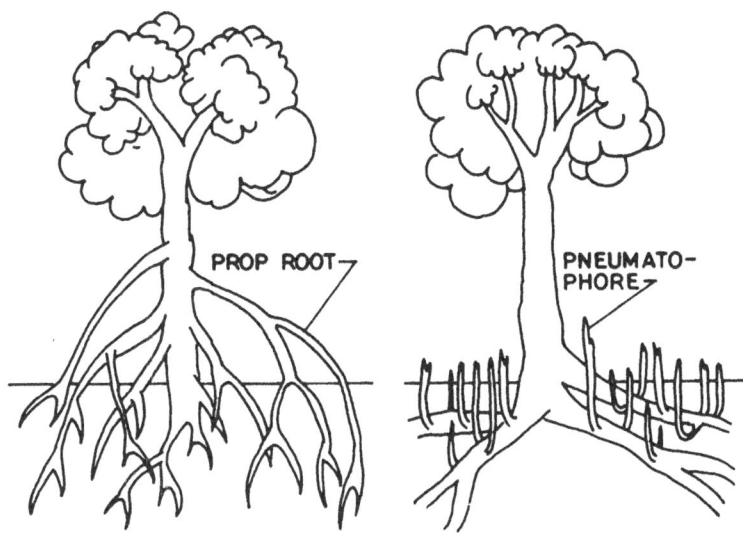

Fig. 2.15. Prop roots of the red mangrove *(Rhizophora)* and pneumatophotephores of the black mangrove *(Avicennia)*
(After Boaden and Seed, 1985)

Another set of adaptations is for facing the difficulty in absorbing water in the face of an osmotic pressure gradient existing between the saline sediment water and the vascular system of the plants. The adaptation in response to this condition is either by accumulating high concentrations of salt that is subsequently eliminated by special salt-secreting glands present in the leaves, or to exclude salt from the roots.

Many mangroves like *Rhizophora, Bruguiera,* or *Ceriops,* are viviparous and develop unusual types of seeds that germinate while still attached to the tree, each propagate producing a long torpedo shaped fruit, the hypocotyl, which on release drops vertically down and gets established in the muddy substratum (Fig. 2.16) (Rabinowitz, 1978). Sometimes these may even be transported over long distances through ocean currents. Some other mangroves like the *Avicennia* are however not viviparous, but the ability of their seedlings to respire anaerobically allows them to withstand extended periods of tidal inundation.

Mangrove fauna

Mangrove fauna consists of both marine as well as terrestrial components. Marine animals are dominated by different molluscs, crustaceans and fish. Many sedentary invertebrates and littorinid snails live on the mangrove

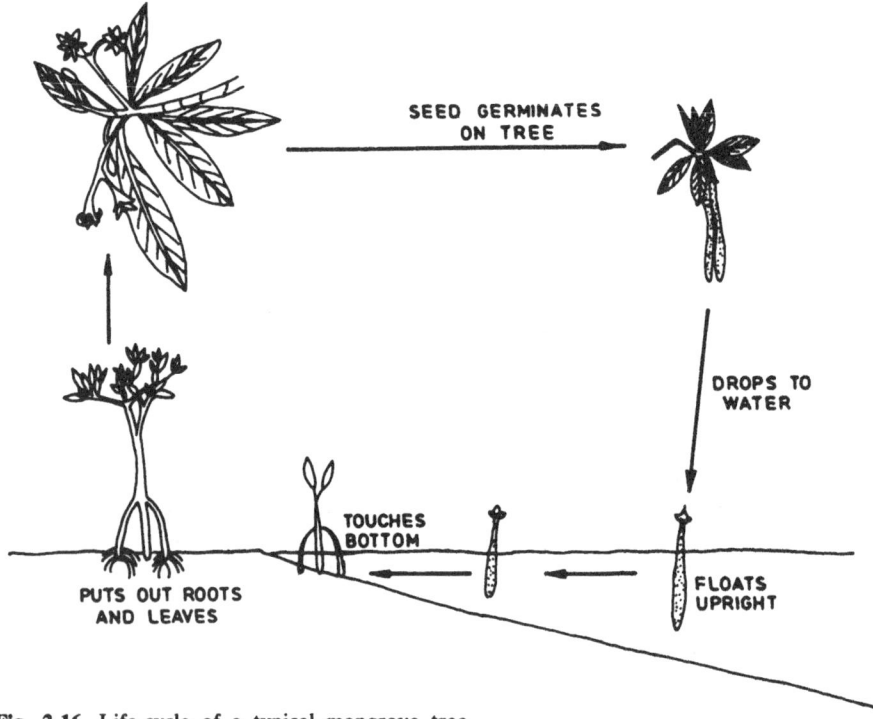

Fig. 2.16. Life-cycle of a typical mangrove tree
(After Boaden and Seed, 1985)

trunks and their extensive root systems. The mangrove snail *Cerithidea* feeds on the mud surface at the time of low tide, and climbs up the tree trunk with the return of high tide to avoid predators. Tropical land crabs, fiddler crabs and ghost crabs have mouth parts adapted for straining organic detritus from the mud and are successful burrowers. Some of these mangrove species also have highly vascularised and lung-like gill chambers and are consequently partially adapted to breathe air. Vascularised sacs present in the mouth and gill chambers of mud skippers and gobioid fish allow them also to breathe air and consequently these are found to climb over the roots and 'walk' perched on their modified pectoral fins, or 'skip' by using the fins and tails. They have eyes situated on the top of the head for better vision in air. Seeds of mullets and penaeid prawns also come along with the tides into mangrove areas. Snakes and crocodiles are also among the common mangrove fauna (Boaden and Seed, 1985).

Productivity and trophic interrelationships

Being highly productive, mangroves may yield a net primary production of 350 to 500 gm C/m^2/year/. A major portion of this production comes to the water as plant debris which is broken down by the microbes and ultimately provides food for the detritivorous organisms that may be either grinders, deposit feeders or filter feeders. These species devour the micro-organisms present on the decaying plant surfaces and give out finely macerated debris as their faecal matters which are again reworked on by the micro-organisms and the cycle is repeated. Small carnivores like the minnows, and the higher carnivores like the carnivorous fish, fish-eating birds, crocodiles etc. enter the food chain at a later stage. On the whole, the food web in the mangroves is controlled by the tidal flushing, riverine input, accumulation and decomposition of plant detritus, faunal behaviour, and some other physical and biological factors that determine the rates of import and export of both organic and inorganic material (Boaden and Seed, 1985).

REFERENCES

Ananthakrishnan, T.N. and Viswanathan, T.R. (1976): *General Animal Ecology*. The Macmillan Company of India Limited.

Barnes, R.S.K. (1984): *Estuarine Biology*, Edward Arnold, London.

Beeftink, W.G. (1977): Salt marshes, In: *The Coastline*, (ed.), R.S.K. Barnes, Wiley, Chichester, 93–121.

Boaden, P.J.S. and Seed, R. (1985): *An Introduction to Coastal Ecology*, Blackie.

Fincham, A.A. (1983): *Basic Marine Biology*. British Museum (Natural History), Cambridge University Press.

Landry, M.R. (1977): A review of important concepts in the trophic organization of pelagic ecosystems. *Helgol. Wiss. Meeres*, **30**, 8–17.

Lewis, J.R. (1964): *The Ecology of Rocky Shores*. English Universities Press, London.

Livingston, R.J. (1979): *Ecological Processes in Coastal and Marine Systems*. Plenum, New York.

Margalef, R. (1969): Succession in marine populations. *Adv. Front. Pl. Sci.* **2**, 137–188.

Mayr, E. (1969): *Principles of Systematic Zoology*, McGraw-Hill. Inc.

McLusky, D. (1971): *Ecology of Estuaries*. Heinemann Books Ltd.

McLusky, D.S. (1981): *The Estuarine Ecosystem.* Blackie.

Nybakken, J.W. (1982): *Marine Biology: An Ecological Approach.* Harper and Row Publishers, New York.

Odum, E.P. (1959): *Fundamentals of Ecology.* Saunders, Philadelphia.

Odum, E.P. (1963): *Ecology.* Amerind Publishing Co. Pvt. Ltd., New Delhi.

Pritchard, D.W. (1967): Observations of circulation in coastal plain estuaries. In *Estuaries* (ed.), G.H. Lauff., American Association for the Advancement of Science, Washington D.C., 37–44.

Rabinowitz, D. (1978): Mortality and initial propagule size in mangrove seedlings in Panama. *J. Ecol.* **66**, 45–52.

Ryther, J.H. (1969): Photosynthesis and fish production in the sea. *Science,* **166**, 72–76.

Ryther, J.H. and Dunstan, W.M. (1971): Nitrogen, phosphorus and eutrophication in the coastal marine environment, *Science,* **171**, 1008–1012.

Stephenson, T.A. and Stephenson, A. (1972): *Life Between Tide Marks on Rocky Shores.* Freeman, San Francisco.

Tait, R.V. (1981): *Elements of Marine Ecology. An Introductory Course. Butterworths.*

Teal, J.M. (1962): Energy flow in the salt marsh ecosystem of Georgia. *Ecology,* **43**, 614–624.

Wheaton, F.W. (1977): *Aquacultural Engineering. John Wiely and Sons.*

CHAPTER 3

Biology of Culturable Species

The following organisms are at present being cultured in coastal waters at different parts of the world:
  i) Fin fish (mullets, milk fish, yellow-tail, eel, pearlspot, red sea bream, grouper, tilapia, plaice, salmon, pompano etc.)
 ii) Shell fish
     a) Molluscs (edible oysters, pearl oysters, clams, mussels, scallops, abalones, conch, squid)
     b) Crustaceans (shrimps, lobsters, crabs)
iii) Turtles
 iv) Other invertebrates (sponges, bait worms)
  v) Seaweeds

3.1 Fin fish

Fin fish are the true fish. They are cold blooded animals with a vertebral column and a dorsal nerve cord, have paired as well as unpaired fins supported by fin rays, breathe mainly by means of gills, their skins are usually covered by scales and they have a streamlined body with a post-anal tail. The main morphological features of a fin fish are shown in Fig. 3.1.

3.1.1 MULLETS

The culture of mullets has a long history that dates back to when the Romans and Egyptians cultivated mullets several centuries ago. Mullet culture has now been developed in many countries around the world.

Mullets are available almost throughout the tropical and subtropical waters and are available in plenty along the coasts of the southern Atlantic and the Gulf states of the U.S.A. The most common specie currently farmed is the grey mullet or striped mullet, *Mugil cephalus*. It is frequently cultivated along with other brackish-water varieties of fish and prawn in India and Israel. The exclusive farming of mullets in marine water is confined to France and Hawaii.

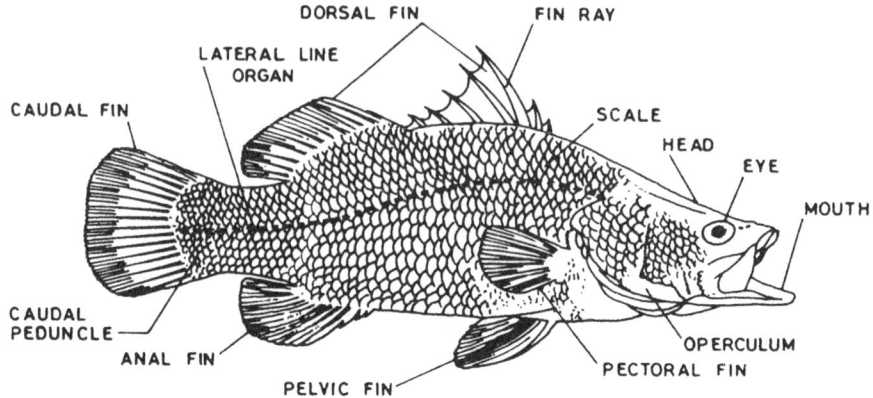

Fig. 3.1. Morphological features of a fin fish *(Lates calcarifer)*

Mullet fry feed principally on plankton and are believed to prefer diatoms and epiphytic Cyanophyceae. The fry of one predominantly fresh-water species, *Rhinomugil corsula* of India, Pakistan and Burma, prefer copepods and small insects. Adults of all species are primarily benthic feeders, consuming algae and vegetable detritus, with an incidental intake of small animals. Decayed higher plants are readily accepted when available. Mullets feed in a head-down position, moving their heads from side to side so violently that at times their entire bodies vibrate. They appear to be surrounded by a cloud of mud during feeding, for they suck up mud from the bottom sieving it in their mouths to remove the food and then rejecting the mud particles. They also feed on surface scum which contains diatoms. Mullets generally form a school when feeding near the surface. Their intestines, over three to six times the length of the fish show the vegetable nature of their diet (Iversen, 1976).

Mullets occur in the sea, in estuaries, and some even migrate into fresh water. Several males and a single female spawn at sea during the cold months, laying pelagic eggs which hatch within two days. The minute (26 mm in length), heavily pigmented fry move into estuaries and coastal tide pools in late winter or early spring, to remain there until moving offshore the following fall or winter (Fig. 3.2). Mullets apparently prefer warm water. Though most mullets are found at a salinity of 30 ppt or below, they may thrive in areas of very high salinity, for example, Laguna Madre in Texas, where they occur in salinities that are seldom below 50 to 60 ppt and may be as high as 80 ppt. They have also been obtained from the Bulgarian coastal lagoons where salinity is as high as 83 ppt.

M. cephalus reaches lengths of 50 to 55 cm and weighs 1.2 to 2.0 kg in four to six years. Most of the other species are slightly smaller, although *Liza tade* of the Indo Pacific region reaches a maximum length of about 70 cm.

OFFSHORE

LARVAE

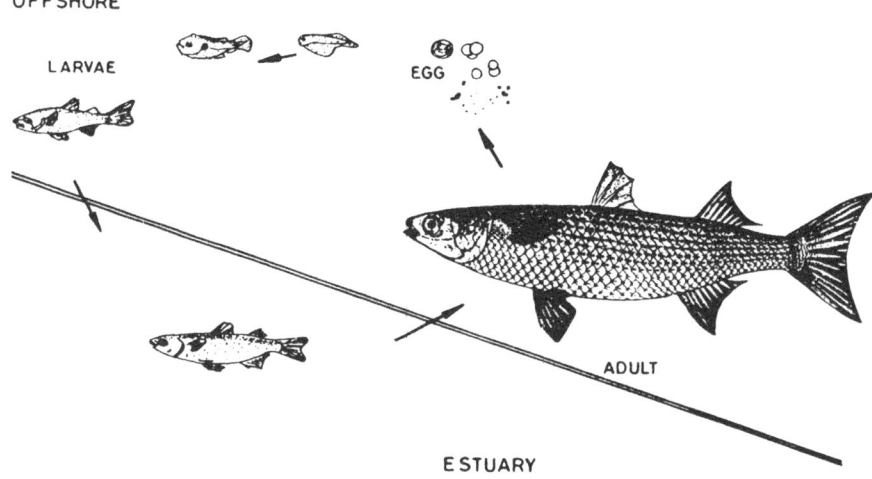

EGG

ADULT

ESTUARY

Fig. 3.2. Lifecycle of mullet (Based on Iversen, 1976)

3.1.2 MILK FISH

A popular fish, milk fish (*Chanos chanos*) (Fig. 3.3) is found in tropical and sub-tropical waters bordering the Pacific and Indian Oceans. It is not found in the Atlantic and a part of the Pacific. Its body is sturdy with a large deeply forked tail and large eyes. It has a small toothless mouth and long intestines, often thirteen times the body length, both adaptations to its herbivorous habit. A large powerful fish, the milk fish weighs up to 20 kg. Adult milk fish spawn close to the coasts. The fry and fingerlings spend the early part of their lives near the shore. The young ones which enter fresh water (rivers) do not attain large sizes and sexual maturity, and have to return to the sea to spawn. In some areas milk fish spawn throughout the year, though in the Philippines they spawn from March to July. Females release 1.5 to 7 million buoyant eggs during a single spawning. Fertilisation is external and occurs in the sea. When the fry are about 1.3 cm long they drift towards the shore, by which time they have exhausted the stored food in their yolk sacs. Here, they eat an assortment of blue-green algae and associated organisms (bacteria, protozoans) till they reach the fingerling stage (15.6 cm),

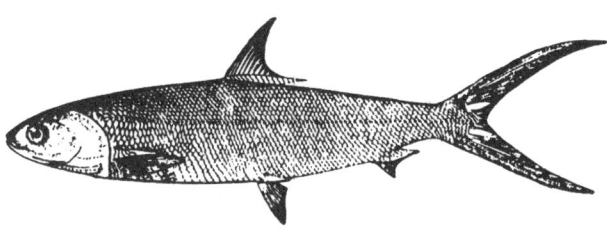

Fig. 3.3. Milk fish *(Chanos chanos)* (After Chen, 1976)

when they change their diet to filamentous green algae. Their growth is very rapid, attaining about 0.8 kg in the first year and 2.7 kg by the fourth year (Iversen, 1976).

Milk fish have a wide range of salinity tolerance. However, if put directly into fresh water, their mortality rate is quite high. Fry of 1.3 cm seem to be more tolerant than fingerlings of 5 to 7.6 cm length. However, in the Philippines, fingerlings have been known to tolerate abrupt changes from fresh water to a salinity of 20 to 30 ppt. Temperature below 15°C induces sluggishness and paralysis and may even cause death. Milk fish have been found in water as warm as 40°C.

3.1.3 YELLOW-TAIL

Yellow-tail (*Seriola quinqueradiata*) is a popular fish cultivated widely in Japan, where it accounts for 98.6 per cent of the total production of sea farming. Three techniques, namely floating net cages, coastal ponds, and net enclosures have been employed in culturing this fish.

The maximum size attained is 96 cm in length and 13 kg in body weight. The dorsal surface of the laterally compressed fusiform body is dark blue in colour and the ventral surface is silvery white with a yellowish line present laterally on either side of the body. The larval form is 3 to 8 cm in length and yellowish-brown in colour with six to 11 reddish-brown lateral bands on the body. The dorsal fin has five spines and 29 to 36 soft rays and the anal fin has one branched spine and 17 to 22 soft rays. There are 210 to 220 scales on the lateral line (Fig. 3.4).

Eggs are pelagic and 1.15 to 1.44 mm in diameter with an oil globule, 0.24 to 0.37 mm in diameter. A reticulate pattern is observed on the surface of a fully mature egg. They hatch out around 51 hours after spawning at a water temperature between 18.0 and 24.5°C, when the larvae measure 3.5 mm in total length. Larvae are most abundant in waters of 19 to 21°C temperature and 19.1 to 19.3 ppt chlorinity. They attach to the drifting seaweeds and grow from 1.5 to 15 cm in length, after which they detach and begin a free-swimming life. Fish less than 4 cm in length feed mainly on copepods such as *Calanus*, while after growing, the important foods are young forms of anchovy and other small fish. Once they grow to 15 cm, they feed on sardines,

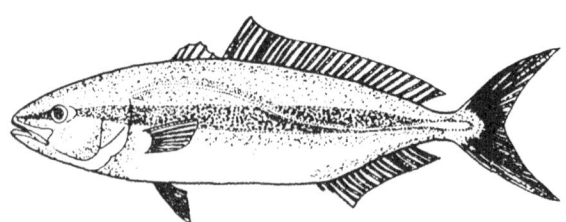

Fig. 3.4. Yellowtail *(Seriola guingueradiata)* (After Kafuku and Ikenoue, 1983)

mackerel, squid etc. Yellow-tails live up to seven years (Kafuku and Ikenoue, 1983).

3.1.4 EEL

Japan is the only country which has been culturing eels for a long time, and here *Anguilla japonica* is farmed commercially. Another species of eel, *Anguilla anguilla* grows in Europe, Israel, United Arab Republic, Soviet Union, West Germany etc.

Anguilla japonica

The maximum size attained is 60 cm in males and more than 75 cm with 1 kg body weight in case of females, while the commercial size is 120 gm to 200 gm in body weight. Larval eels (elvers) are transparent and while ascending the river, the body colour darkens to black after two to four weeks. The dorsal side of an adult eel is black and the ventral side is whitish-silver. However, body colour varies according to the quality of food, water and other environmental conditions. Fully mature eels descend from the river to the sea for spawning. They then exhibit a nuptial colour of metallic lustre on the lateral parts and faint purple colour on the abdomen.

The body is elongated and without a pelvic fin. Dorsal, caudal and anal fins are continuous and do not have spines. Scales are very small and are embedded in the skin (Fig. 3.5).

The eel generally spawns in the middle layer of a 400 to 500 m depth in the Pacific Ocean between 20°N, and 28°N, and 121°E and 128°E, where the water temperature is between 16 and 17°C and the salinity is about 35 ppt. Simultaneously 7 to 12 million eggs are spawned. These are pelagic and almost 1 mm in diameter. The fertilised eggs hatch within two to three days. The fry gradually move up to the surface layer of the sea and repeating the diurnal vertical movement, they are dispersed in many directions by the

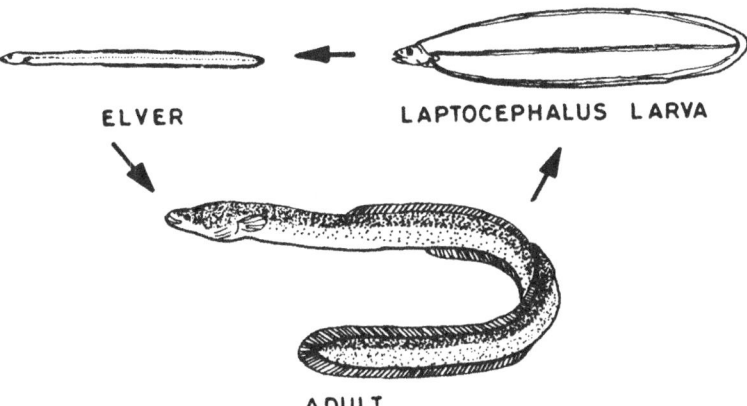

ELVER   LAPTOCEPHALUS LARVA

ADULT

Fig. 3.5. Lifecycle of eel (Anguilla japonica) (After Kafuku and Ikenoue, 1983)

surface ocean current. Larvae of this stage are willow-leaf shaped and are called Leptocephalus larvae. These are brought to shallow coastal areas by the ocean current, settle to the sea bottom and develop into the elver stage (Fig. 3.5). Elvers migrate upstream into the river from November to April when the water temperature is between 8 and 10°C. As eels are nocturnal, elvers burrow in river beds during the day. Eventually they migrate, and reach a middle course in the river, lake, marsh or estuary and settle there. Being nocturnal in habit, feeding is observed at night only. Young elvers feed mainly on detritus. As the temperature rises, their feeding activity increases and they feed on many kinds of small animals. Adult eels feed on a wide vareity of animals including earthworms, shrimps, crabs fish, frogs etc. The male eel attains sexual maturity at three to four years age and the female at four to six years age. Mature eels that descend to the sea for spawning do not feed. Although eels which migrate to the sea seem to die after spawning, there are examples of eels surviving 20 to 35 years in a pond without migration to the sea (Kafuku and Ikenoue, 1983).

3.1.5 PEARLSPOT

Pearlspots (*Etroplus suratensis*) are found in India, Pakistan and Sri Lanka. The name is derived from the pearl-like spots found on the majority of the scales on the dorsal side of the body. They can be cultured in both brackish as well as in fresh water. During the second year they become sexually mature and breed naturally in ponds. For spawning, the female cleans algae and other growth from a small area on submerged objects and attaches 1300 to 6000 eggs there, whereupon the male fertilises them. The female guards the young which hatch in about seven days. The young fish feed almost exclusively on plants and the older fish on fish larvae, aquatic insects, algae and decaying plant remnants. The pearlspot attains a length of 10 to 12 cm by the end of the first year and reaches a maximum length of about 30 cm in adulthood (Iversen, 1976).

3.1.6 PERCH

The perch (*Lates calcarifer*) (Fig. 3.1) is widespread in India, Thailand, Vietnam, Philippines, Indonesia, Cambodia, Malaya and Australia. It has a projecting jaw and a serrated gill cover and reaches a length of about 170 cm. Perch does not breed in ponds. At the end of the first year it can weigh about 500 gms and attains a length of approximately 38 cm. It is carnivorous in habit and feeds mainly on fish and crustaceans in the open sea (Iversen, 1976).

3.1.7 TARPON

Tarpon (*Megalaps cyprinoides*) occur around the Indian and Pacific Ocean, and is reared in India. They breed in the sea and are cultured in ponds. The young feed mainly on crustaceans, diatoms and algae. Being predatory in nature, 50 per cent of the diet of adults consists of fish. In some brackish-water ponds they may attain a length of about 40 cm by the end of one year and thereafter may reach a length of more than 1 am (Iversen, 1976).

3.1.8 OTHER FIN FISH

Red sea bream (*Pagrus major*), Grouper (*Epinephelus* sp.), Rainbow trout (*Salmo gairdnerii*), Salmon (*Salmo salar, Oncorhynchus* sp.), puffers (*Fugu rubripes, F. vermicularis*), red porgy (*Chrysophrys major*), black porgy (*Mylio macrocephalus*), Pampano (*Trichinotus carolinus*), Plaice (*Pleuronectes platessa*), Sole (*Solea solea*), *Therapon jarbua*, *Scatophagus argus*, Sea bass (*Morone labrax*), Tilapia (*Oreochromis mossambicus*) etc. are some other fin fish cultured in the coastal areas. These are either of regional importance or at the initial stage of cultivation.

3.2 Shell fish

Shell fish are not true fish but are invertebrates. They do not have fins supported by fin rays. They have either external or internal shell to support their bodies.

3.2.1 MOLLUSCS

They are members of the phylum Mollusca which includes invertebrates like oysters, clams, squids, snails etc. They are distinguished by the presence of a muscular foot, a calcareous shell secreted by the underlying integument called as mantle, and a feeding organ, the radula. A few molluscs of commercial importance are shown in Fig. 3.6.

i) Bivalves

Bivalves are the molluscs belonging to the class Bivalvia, also called Pelecypoda or Lamellibranchia. They are laterally compressed and possess a shell with two valves, hinged dorsally, that completely encloses the body. The foot, like the remainder of the body, is also laterally compressed and known as a hatchet foot. The head is greatly reduced in size. The mantle cavity is very spacious and the gills are usually very large.

Edible oysters

The oyster is a sedentary mollusc with two hard shells hinged together. The opening and closing of the shell is operated by a strong muscle that is responsible for breathing and feeding (Fig. 3.7). The various oyster species farmed throughout the world fall into two categories: either flat or cup shaped oysters. The flat oysters, so named because both their shells are flat, belong to the genus *Ostrea*, of which there are two commercial species. In cup-shaped oysters the lower shell is rounded or cup-shaped. They belong to the genus *Crassostrea*, of which there are seven commercial species. Both have worldwide distribution in the warmer temperate waters. In addition to the indigenous species, other varieties are often transplanted to replace or supplement the existing stocks.

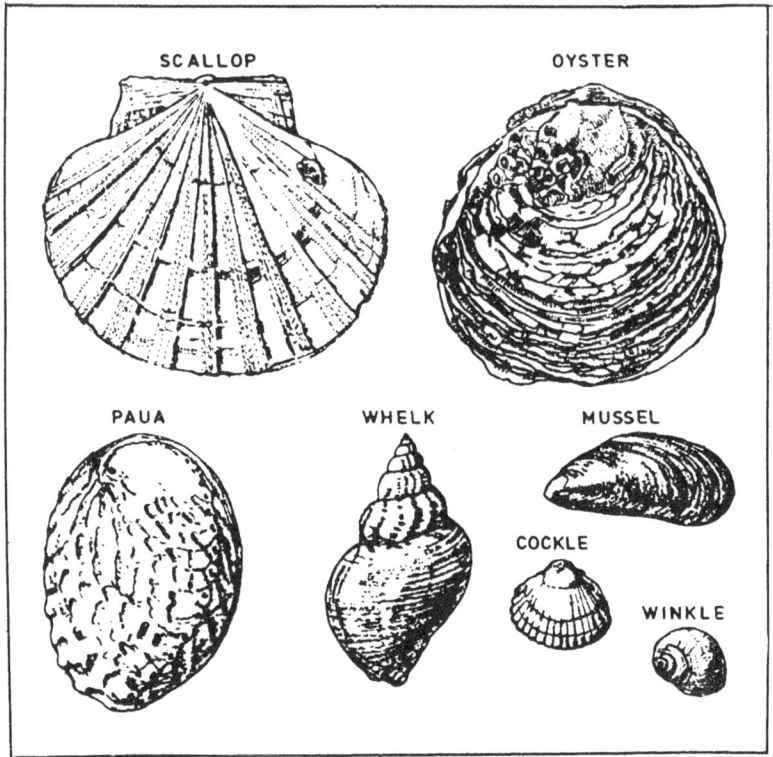

Fig. 3.6. A few molluscs of commercial importance (After Fincham, 1983)

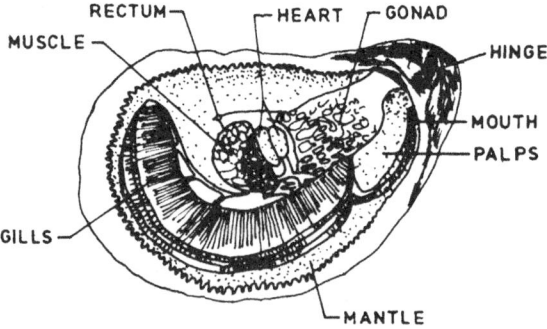

Fig. 3.7. Anatomy of an oyster (After Iversen, 1976)

The European flat oyster, *O. edulis,* is farmed mainly in Britain, France, Norway and Spain. The other flat oyster, *O. lurida,* or Olympia oyster, used to be farmed on the Pacific coast of the U.S.A. from southern Alaska to lower California, but due to their small sizes, these native flat oysters have now been replaced in many areas by the imported Japanese oyster, *C. gigas*

(Milne, 1972). This species is being successfully farmed in Japan since the eighteenth century and the seed finds a ready export market in countries like Britain, Canada, the Philippines and the U.S.A. The oyster, *C. virginica* is farmed extensively throughout the U.S.A. and Canada and is known as the American oyster. Specialised hatcheries and farming techniques for this purpose have been developed in Long Island Sound. Another important cup-like species. *C. angulata,* (Portugese oyster)' is found on the coasts of Europe, and is mainly farmed in the Arcachon region of France and Britain. Other cultivable oyster species include the Sydney rock oyster, *C. commercialis,* in Australia, the rock oyster, *C. glomerata,* in New Zealand, and the slipper oyster, *C. eradelie,* in the Philippines. Commercial farming of the mangrove oyster, *C. rhizophorae,* in Cuba and Venezuela seems promising as well. Russia and some parts of Africa have also started oyster hatcheries.

Biology of American oyster *(C. virginica)*

The spawning season of this species is from mid-July to early October in the Long Island Sound, mid-June to mid-October in the Chesapeake Bay, May to October in South Carolina and April to November in the Gulf of Mexico. Under natural conditions, females probably spawn many times in a single breeding season. Spawning starts when the water temperature reaches 20°C. Larvae develop well at 17.5 to 32.2°C.

A single ripe American oyster female may release 14 to 114 million buoyant eggs during one spawning. Fertilisation is external. Fertilised eggs hatch out into free-swimming 'veliger' larvae within a few hours. Veligers are delicate, weak swimmers carried by the current for about two to three weeks, after which they attach themselves by a small but well-developed foot to a relatively clean and preferably hard surface. After attachment, they transform into the 'spat' stage, when shell formation begins. Like many other bivalves, the American oyster may also change its sex, a young male being converted into a female in the later part of its life, sometimes again changing back to male in between spawning seasons. American oysters grow in the intertidal zone at a depth of more than 30 m. However, spawning and spat settling are more successful in the estuaries.

Feeding takes place by pumping water between the two valves, filtering the micro-organisms present in the water by specially adapted hair-like structures on the gills which also serve to move the food into the mouth. The rate of feeding is reduced or even stopped altogether by closing the valves if the water temperature falls appreciably. Similarly, the valves are closed if oysters are left exposed to the air by the outgoing tide in the intertidal zone, or even if the water contains certain pollutants. Oysters can also extract oxygen from the water present within their shells. Frequent or prolonged gaps in feeding however reduce their growth rates. Though tolerant to wide salinity fluctuations, they prefer salinity above 32 ppt; larval development occurs between 16.5 and 22.5 ppt, though larvae have been known to survive even at 7.5 ppt.

To compensate for a high mortality rate under unfavourable environmental conditions, competitors, predators and diseases, the females release an extremely large quantity of eggs. The American oyster grows to a length of 9 cm in four to five years in long Island Sound under suitable conditions, whereas in the Gulf of Mexico it reaches the same length within two years. The growth rate is lowered in colder waters and during winter hibernation. Oysters growing in tight clusters assume long slender forms, but if grown individually, acquire a circular outline.

Biology of Japanese oyster *(C. gigas)*

The Japanese oyster may acquire a maximum size of 35 cm in shell height, 8 cm in shell length and 10 cm in shell width, though its commercial size is usually 8 cm in shell height, 5 cm in shell length and 3 cm in shell width. The shell surface is rough with numerous scaly leaves, and its colour is yellowish-white with brownish-violet radial bands, while the flesh is yellowish to greenish-white and the margin of the mantle is black. The shell is basically oval in shape but shows wide variation according to the environmental conditions. This oyster attaches itself to rock or any other hard surface with its left shell which contains the soft body and is concave in shape. The right shell is almost flat. It can tolerate temperatures of 15 to 30°C. Optimum salinity for larval development is 23 to 28 ppt and spat sets best at 15 to 18 ppt. The spawning begins in summer when the sea water temperature rises to at least 23°C. In the warmer southern part of Japan where water temperature is higher than 23°C, the spawning season lasts longer and spawning takes place several times a year, whereas in the colder northern region, spawning takes place only once in one or two years. Females release eggs of 60 μm diameter and the fertilisation is external. Eggs develop cilia on their body surfaces with the help of which they start a rotating movement approximately five to six hours after fertilisation. The larva passes through the trocophore and veliger stages before developing into the D-shaped form of 70 to 90 μm in size when it comes up to the surface. The D-shaped larvae develop into the umbo size larvae two to three days after fertilisation. Two to three weeks after fertilisation, oyster larvae attain their full size of 270 μm. At this stage, their planktonic life ends, after which they attach to rocks or other surfaces in the intertidal zone of rocky shores with a cementing substance they excrete themselves. The oyster later spends a sessile life feeding on microplankton and detritus filtered by its gill filaments (Kafuku and Ikenoue, 1983) (Fig. 3.8).

Pearl oyster *(Pinctada fucata)*

The maximum size attained by pearl oysters is 10 cm in shell height and shell length, while the commercial size is 5.5 to 6.0 cm in shell height and shell length, with 30 gm total body weight. The outer side of the shell is coloured dark brown or dark reddish-brown with six deep violet radial bands, while the inner side has a bright pearly lustre. Its hinge line is straight. The umbo is

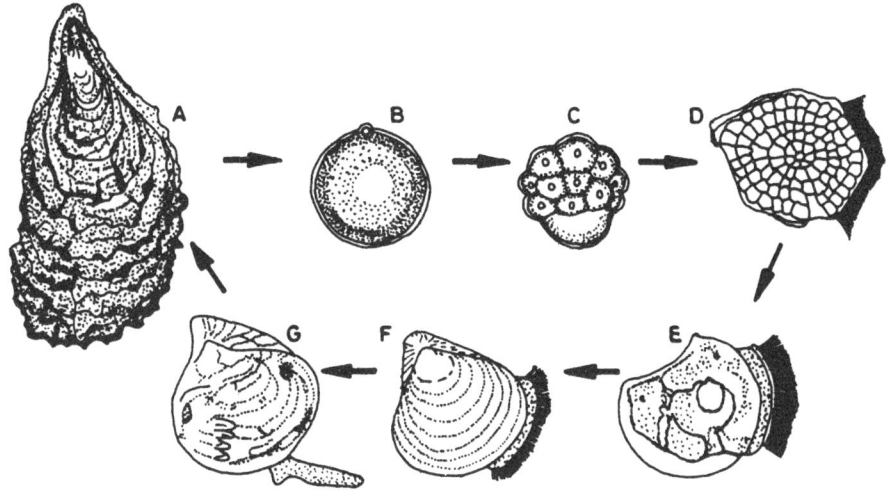

Fig. 3.8. Lifecycle of edible oyster (*Crassostrea gigas*)
A—Adult, B—Fertilised egg, C—Cleavage of egg, D—Veliger stage, E—D-shaped larva stage, F—Umbo stage, G—spat (fully grown larva (Based on Kafuku and Ikenoue, 1983)

located in the anterior part of the shell and the lower part of the shell is rounded. The shell surface is covered with thin scaly leaves. The left shell is more concave than the right one.

The Japanese pearl oyster is a temperate zone species thriving on coarse sand, gravel or rock bottoms of inner bays with clean water and moderate sea conditions. The preferred depth is from 0 to 10 m. The optimum temperature and salinity are 20 to 25°C and 27.2 to 33.7 ppt (1.020 to 1.025 sp gr) respectively. It is a dioecious species spawning from late spring to early summer, the peak spawning being observed when the sea water temperature is about 25°C. Fertilised eggs develop into D-shaped larvae, 70 μm in diameter within 20 hours after fertilisation, which grow to the umbo stage larvae and finally develop to full grown larvae of (220 μm) shell length in two to three weeks, after which they end their planktonic life and enter a sessile life (Fig. 3.9). Their sessile life is spent attached to rocks or gravel with their strong byssuses (tuft of strong threads). Sometimes, they even shift to other places by cutting the byssuses. They hibernate when the water temperature falls below 13°C in winter (Kafuku and Ikenoue, 1983).

Clams

Sea farming of clams is widely practised in the U.S.A. The two species cultivated are the hard clam, *Mercenaria mercenaria* (popular names: quahog, quahaug, hard-shelled clam, little neck clam etc.), indigenous along the Atlantic coast from Maine to Florida, and the soft clam, *Mya arenaria*

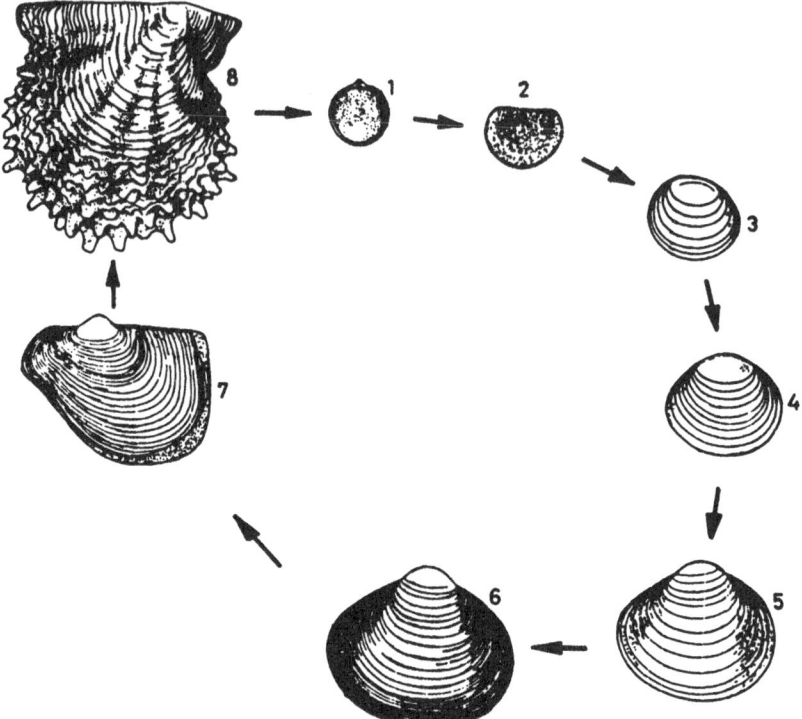

Fig. 3.9. Lifecycle of pearl oyster *(Pinctada fucata)*
  1—Just fertilised egg (41 × 53 m)
  2—D-type larva (straight hinge stage, 63 × 80 m)
  3—Umbo stage (S.L. 115 m)
  4—Umbo stage (S.L. 135 m)
  5—Umbo stage (S.L. 235 m)
  6—Full grown stage (S.L. 240 m)
  7—Young (S.L. 1.1 mm)
  8—Adult
  (Based on Kafuku and Ikenoue, 1983)

(popular names: soft-shelled clam, long necked clam etc.), found from Labrador to North Carolina.

In Japan, at least nine species namely, haigai *(Anadara granosa)*, Sarubo *(Anadara subcrenata)*, tairagi *(Atrina japonica)*, torigai *(Fulvia mutica)*, hokkigai *(Mactra sachalinensis)*, bakagai *(Mactra sulcataria)*, hamaguri *(Meretrix lusorina)*, agamaki *(Sinovacula constricta)*, and asari *(Tapes japonica)* are cultured (Bardach *et al.*, 1972).

Hard clam *(Mercenaria mercenaria)*
    This clam can occur in waters 15 m deep, though the largest numbers are found just below the low tide level, living equally well in both sandy or muddy

bottoms. Fertilisation is external. The young larvae hatch as free-swimming veligers with a spinning motion, but are generally carried away from the place of birth by the water current. This free-floating stage has an important bearing on the ultimate survival rate of this clam. Since the adults are sessile and have to depend upon the water current for their food and dissolved oxygen and also to remove metabolic wastes, survival would not be possible if a large number were congregated at one spot. Thus, it is precisely this congregation that is avoided by the early dispersal of the larvae. After drifting for one or two weeks, the swimming hairs of the veligers disappear, and a sessile existence begins at the bottom. In time a strong muscular foot and the gills and siphons form, and the now pin-head sized clam attaches itself to sand grains, plants or any suitable material with the help of the byssus. But it can still move short distances on its muscular foot. On reaching a size of about 5 mm the clam digs a burrow with its foot, often making several trial digs before settling down as a burrower for the rest of its life which may be as long as 25 years. The byssus forming glands regress after the clam burrows successfully. The clam may reach a length of about 14 cm towards the end of its life; and, given suitable conditions to grow as in New England, it takes the hard clam about five years to reach half its maximum size.

Soft clam

The morphology and life history of the soft clam (*Mya arenaria*) is similar to that of the hard clam. Like the hard clam, it is rarely found on tidal flats, inlets and bays; it thrives in the more exposed areas of the coastline, surviving best in gravel to mud bottoms where it buries itself, leaving only its long siphons protruding, out of the substratum. Harvesting soft clams thus involves digging to remove them from their burrows. Since it is dependent on plankton carried by the water up to its siphons, its growth and survival requires an adequate water current, as well as water that is more or less free from silt since silt may clog the opening causing mortality as induced by hurricanes along the Atlantic coast or tropical storms. The accompanying rain in case of the latter also drastically reduces the salinity of water.

Mussels

Although there is a greater demand for mussels in Europe, their culture is not as profitable as that of oysters, though they are often easier to farm with less attention being required. The most common edible mussel is *Mytilus edulis*. Spain is now the world's leading producer of mussels, France, Holland and Italy being the other main European countries with large-scale mussel farms. The success with mussel culture in Spain has led to its introduction in Germany, Ireland, Norway and Scotland to investigate the commercial prospects of *M. edulis* farming in these countries. The other mussel commercially exploited in Europe is *M. galloprovincialis*, the mediterranean mussel widely farmed in Italy. Mussels are also cultivated in the Philippines, where the green mussel, *M. smaragdinus*, is farmed extensively. Several other species of mussel are also undergoing experimental cultivation techniques to determine their commercial prospects. These are *M. edulis planulaties* in Australia, *Perna canaliculus* in New Zealand and *P. perna* in Venezuela.

The common mussel, *M. edulis*, has a thin shell coloured dark purple to black. It remains attached to rocks and collectors by means of its byssus that remain throughout its life, and feeds on plankton, small diatoms, protozoans and detritus, floating in the water, that are drawn up into the mouth through the currents set up by its gills. It grows about 2.5 cm per year for the first three years. A single female spawns 5 to 12 million eggs annually. The ciliated veliger larva is formed about four hours after fertilisation, floats briefly before attaching to its resting place by means of its byssus. A large number of mussels accumulate in places where the environmental conditions are favourable. All the organs are formed after 10 weeks (Field, 1923).

Scallops

Sea scallops (*Patinopecten yessoensis*) are being raised by the long-line method in Japan's bays, the bay scallops (*Argopecten irradians*) are being considered for mariculture at the Virginia Institute of Marine Sciences, U.S.A., and research is underway in Russia on *Mizuhopecten yessoensis* and *Spisula sachalinensis* to find suitable commercial techniques.

The commercial size of *P. yessoensis* is 10 cm in shell length though it may reach a maximum shell length of 21 cm. The left shell shows individual colour differences with a variety of shades from bluish-violet to reddish-brown. The right shell is however always white. The shell is rigid and almost spherical in shape, the right shell being more concave outside than the left shell. About 25 low radial ribs are present on both shells. It is a cold water species occurring at 10 to 30 m depths in sea beds of sand with mud or gravel, in the northern parts of Japan. The spawning season is earlier in the south than in the north, mid-March to April in Mutsu Bay, May in Funka Bay and May to June in Lake Saroma. During the spawning season, males discharge sperm into the sea as often as 60 times, which stimulates the discharge of eggs by females (a single female releasing as much as 10 million in number) in successive spawning acts. The sperm or eggs are released both from the ventral margin, and the anterior and posterior parts of the dorsal margins on a vigorous opening and closing of their shell. A fertilised egg develops to blastula, gastrula, trochophore, veliger (D-shaped, $72 \times 58 \mu$m), veliger ($104 \times 87 \mu$m), umbo ($118 - 104 \mu$m), umbo ($200 \times 180 \mu$m) and full grown larva, 40 hours, two days, four days, five to seven days, eight to ten days, 15 to 17 days, 30 to 35 days and 40 days after fertilisation respectively (Yamamoto, 1964). Fully grown larval shells (spat) terminate planktonic life and attach themselves to seaweeds with byssuses. After five months, when they reach a shell length of 0.7 to 1.0 cm they cut their byssuses and sink to the sea bottom to begin a benthic life. Scallops settle with the left shells facing up in shallow depressions dug on the sandy sea bottom. Scallops stop growing, wherein the resting marks appear on their shells, when the water temperature rises above 20°C.

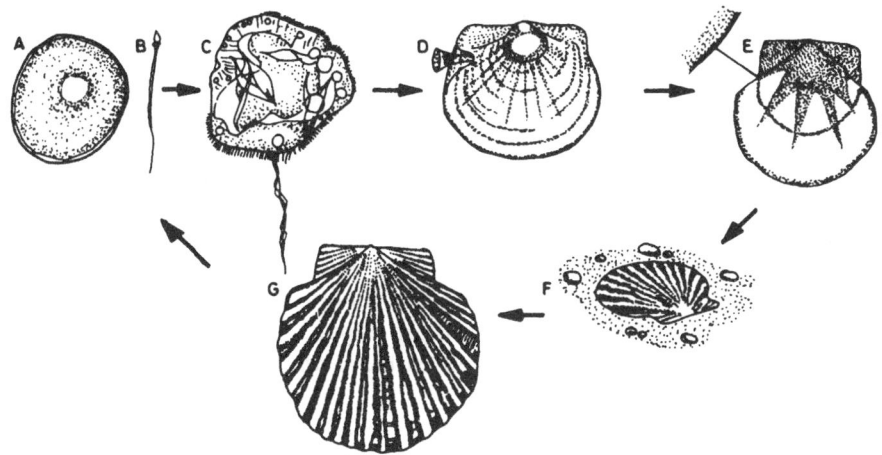

Fig. 3.10. Lifecycle of scallop (*Patinopecten yessoensis*)
A—Egg, B—Spermatozoa, C—Trochophore, D—Spat (attaches to sea weed or other objects by means of byssuses), E—Spat (moves to sea bottom by cutting byssuses), F—Young scallop (digs a shallow hole in the sea bed) and G—Adult scallop (Based on Kafuku and Ikenoue, 1983)

They grow up to a shell length of 10 cm by 20 months, after which they begin spawning (Fig. 3.10). Scallops swim by continuously opening and closing their shells. They extend short tentacles from the margin of the shells during benthic life and any accidental contact with each other makes them immediately swim away to maintain a distance between them. Scallops feed on phytoplankton and detritus and grow to about 13 cm in shell length and 325 gm in body weight after four years. The longevity of a scallop is eight years. (Kafuku and Ikenoue, 1983).

ii) Gastropods

Gastropods belong to the class Gastropoda under phylum Mollusca. The essential characters of gastropods are the development of a head which bears a pair of tentacles with an eye at each tentacle base, the conversion of the shell from a shield to a protective retreat and the body having undergone torsion or twisting. Many edible marine gastropods are considered delicacies. In Florida and the Bahamas, various conches are highly prized as food. Limpets and periwinkles, despite their small size, are collected for human consumption. But the most valued marine gastropods in this respect are the abalones.

Abalones

There are about 100 species of abalones distributed worldwide, of which only about 10 large ones, those confined to temperate waters, that are usually commercially exploited. Important abalone fisheries occur only in Japan, China, South Africa, New Zealand, Southern Australia, Mexico and the

Pacific coast of the U.S.A. Abalones occur mainly in low tidal zones to about
20 m depths in rocky shores of the open sea along with brown algae. Being
slow moving they are easy prey, which accounts for their dwindling
population. It is therefore somewhat surprising then, considering the scarcity
of abalones and the high price they command, that their culture has been
undertaken only in Japan and California (Bardach *et al.*, 1972).

At least eight species of abalone are found along Japanese shores, but
the ones most important commercially are *Haliotis discus, H. diversicolor
aquatilis, H. gigantea* and *H. sieboldi.* One sub-species. *H. discuss hannai,*
which is morphologically, ecologically and genetically distinct from *H. discus*
is distributed along the coasts of northern Japan. Of these five, the greatest
success in rearing has been achieved with *H. discus.* In California the red
abalone (*H. rufescens*) is cultured.

Maximum size reached by the abalone is 20 cm in shell length and about
1 kg in body weight; its commercial size however being 12 cm in shell length
and 550 gm in body weight. The shell is ear shaped and reddish-brown in
colour with ridges and growth rings on the surface. Black and yellow stripes
are present on the dorsal surface of the soft body and greenish-purple stripes
on the pedal sole.

H. discus spawn from October to December while the spawning season
of *H.D. hannai* is from July to October, though the periods may vary
locally according to differences in water temperature. *Haliotis* spp. are
dioecious and the sexual maturity is determined by the olive-green colouration of the ovary and ivory-white colouration of the testis (Fig. 3.11).
Abalones exhibit a tendency to congregate together during the spawning
season. Spawning takes place after dusk with the male releasing sperm into
the sea, followed by the female releasing the eggs, stimulated by the presence
of sperm in the water. The eggs are demersal, bright green in colour and 0.2
to 0.25 mm in diameter. The fertilised eggs of *H. discus* hatch out to a
trochophore stage after about 15 hours at 20°C. Under the same conditions,
the incubation time of *H.d. hannai* is about 13 hours. The trochophore stage
is succeeded by the veliger stage in about 20 hours. The larva develops an
epipodial foot by the end of the veliger stage and settles down at the bottom

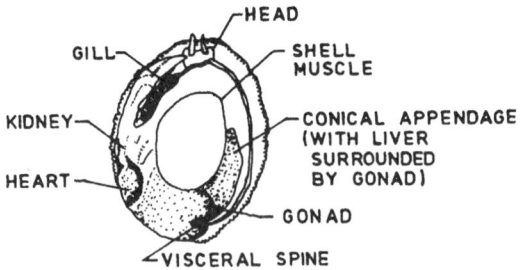

Fig. 3.11. Dorsal view of abalone with main organs and gonad (After Kafuku and Ikenoue, 1983)

(Fig. 3.12). Newly settled abalones are found in small crevices of rocks or under stones. For their growth, abalones select shelters in deeper zones.

Conches

The conch with a high sea farming potential is the queen conch (*Strombus gigas*), a large snail attaining a total weight (with shell) of more than 2.7 kg. Characterised by a large red shell lip, the queen conch is found throughout the Caribbean sea and Atlantic ocean east of Florida from shallow waters to a depth of about 60 m. In addition to consuming it fresh, natives also preserve it by sun-drying.

The conch lays egg masses containing about 750,000 individual eggs in 50 to 75 mm long strings on sandy parts of the sea bottom. After hatching there

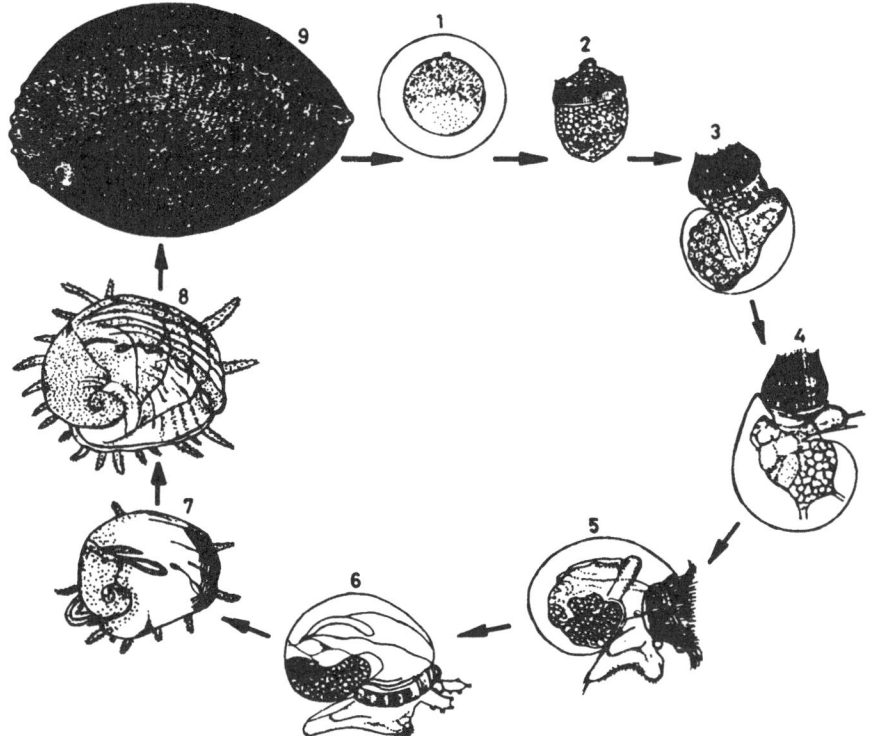

Fig. 3.12. Lifecycle of Japanese disc abalone

1—Fertilised egg (egg diameter 0.23 mm and yolk diameter 0.18 mm), 2—Trochophore larva in swimming stage (19 hours after hatching), 3—Early veliger stage before torsion (29 hours, S.L. 0.26 mm), 4—Swimming veliger having stigmas and cephalic tentacles (2.5 days, S.L. 0.29 mm), 5—Full grown veliger in early creeping stage (5 days, S.L. 0.29 mm), 6—Creeping larva begins to secrete peristomal shell (13 days), 7—Young having respiratory pores (140 days, S.L. 3.0 mm), 8—More developed young shell (160 days, SL. 3.7 mm), 9—Adult. *Haliotis (Nordotis), discus*

is a short drifting larval stage. The young conch with a thin shell is at its most vulnerable and falls easy prey to spiny lobsters, hermit crabs, octopuses, logger-head turtles and at least 10 species of fish. At this stage the conch generally buries itself into the sand during the day and emerges at night to feed on algae and sea grass. Conches grow about 5 cm per year on an average.

iii) Cephalopods

These are molluscs under the class Cephalopoda and includes nautilus, cuttle fish, squid and octopus. The class as a whole is adapted for a swimming existence. The most striking feature of this class is the head being projected into a circle or crown of large prehensile tentacles or arms.

Squid

Squid (Fig. 3.13) and octopus are greatly appreciated as food animals in the Far East and the mediterranean countries of Europe. Commercial suppliers of edible squid presently rely entirely on capture fisheries as a source, but *Sepioteuthis lessoniana* and the related but less commercially valuable cuttle fishes *Euprymna berryi*, *Sepia esculenta*, *Sepia subaculeata*, and *Sepiella maindroni* have also been experimentally cultured, with considerable success in Japan since 1960 or even before. *Sepioteuthis sepioidea* is being successfully cultured in Miami. The propagation and culture of octopus has been attempted in several laboratories and has even met with some preliminary success, particularly in Japan, but the technique is not yet sufficiently advanced.

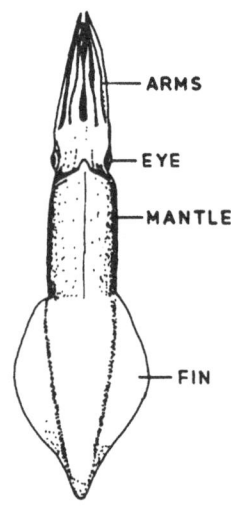

Fig. 3.13. Dorsal view of the squid, *Loligo* in swimming position (After Barnes, 1980)

3.2.2 CRUSTACEANS

Crustaceans are invertebrates belonging to the phylum Arthropoda and subphylum crustacea, with the unique characteristic of jointed appendages, and includes animals like shrimps, lobsters, crabs, cray fish, barnacles as well as a large number of small animals that occupy important positions in the aquatic food chain. Though extremely diverse in both structure and habit, crustaceans characteristically possess two pairs of antennae on the head, which also bears one pair of mandibles and two pairs of maxillae. The trunk specialisation varies greatly, but a carapace that covers all or part of the body is common. Appendages are typically biramous (Fig. 3.14). Crustaceans are primarily aquatic in habit, and most of them are marine.

i) Shrimps

Of the crustaceans suitable for sea farming, the shrimp (Fig. 3.15) is the most important one and is extensively farmed all over the world. The various shrimp species farmed throughout the world fall into three genera, *Penaeus, Metapenaeus* and *Parapenaeopsis.*

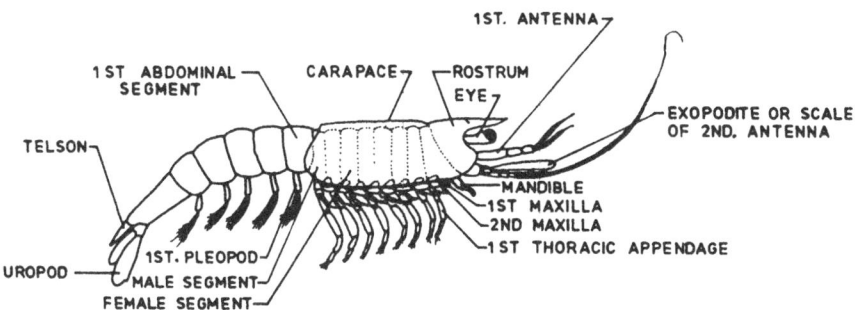

Fig. 3.14. Morphological features of a shrimp (a crustacean) (after Barnes, 1980)

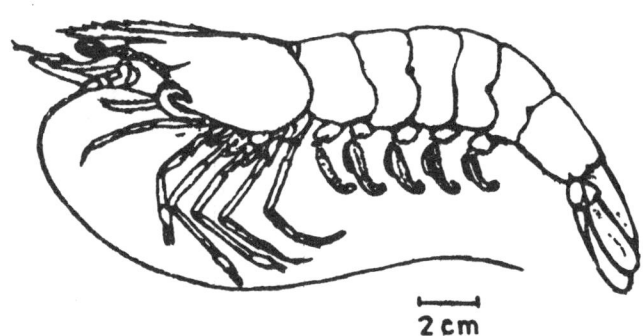

Fig. 3.15. *Penaeus aztecus*

Singapore has the largest number of indigenous commercially exploited shrimps. The main species harvested from the ponds in Singapore are *Penaeus indicus, Penaeus merguiensis, Penaeus monodon, Penaeus semisulcatus, Metapenaeus brevicornis, Metapenaeus burkenroadi, Metapenaeus ensis* and *Metapenaeus mastersii*.

Considerable attention has been paid in Japan to the commercial farming of the Japanese shrimp, *Penaeus japonicus*.

In the U.S.A., the shrimp species mainly cultured are the pink shrimp, *Penaeus duorarum*, the brown shrimp, *Penaeus aztecus* and the white shrimp, *Penaeus setiferus*.

In Indonesia, the shrimp species commercially farmed include *Penaeus merguiensis, Penaeus monodon, M. brevicornis* and *M. monoceros*.

In Australia, the king prawn *Penaeus plebejus*, the greasyback *M. bennettae*, and the school prawn *M. macleayi*, are cultured.

In the Philippines, culture of *Penaeus monodon, Penaeus indicus, Penaeus merguiensis* and *M. ensis* is very popular.

In different coasts of India mainly the seed of *Penaeus monodon, P. indicus, P. merquiensis, P. semisulcatus, M. monoceros, M. brevicornis, M. dobsoni, Parapenaeopsis styliferus* and *Parapenaeopsis sculptilis* are available and these are cultured also.

In contrast to the non-penaeid prawns the pleurae present on either side of the exoskeleton of the second abdominal segment in the penaeid prawns overlap the third segment only and the first three thoracic legs (pereopods) are chelate. Penaeid prawns are bisexual, the females being generally larger than the males, the two sexes being easily recognised through a difference in the external morphology. The female possesses a sperm-storing organ (thelycum) on the ventral side of the head (cephalothorax) between the fourth and fifth walking legs; the oviducts open at the base of the third pair of walking legs. The ovary is present on the dorsal side extending throughout its entire length. The posterior portion of the developing or mature ovary within the abdominal segments can be clearly visualised through the cuticle in a live specimen held against a source of light. In a fully ripe female ready to spawn, the ovary shows a lateral expansion in the first abdominal segment. The male sexual organ or petasma (Fig. 3.16) is a modified part of the first pair of swimming legs (pleopods) and is attached to the ventral side of the first abdominal segment. The two sperm ducts open at the base of the fifth pair of walking legs. The posterior-most portion of the sperm duct is enlarged to form the terminal ampoule in which the spermatophore or the sperm packet is stored and is visible in mature males as a white mass at the base of the fifth pair of walking legs. Sperms are non-motile and are transferred in spermatophores to the thelycum of the female at the time of mating. Mating takes place soon after the moulting of a female, that is, when it is still in a soft condition. The transferred spermatophores show up in impregnated females as a whitish mass below the translucent cuticle of the thelycum. At the time of impregnation the ovary is still immature, so that there is an interval between

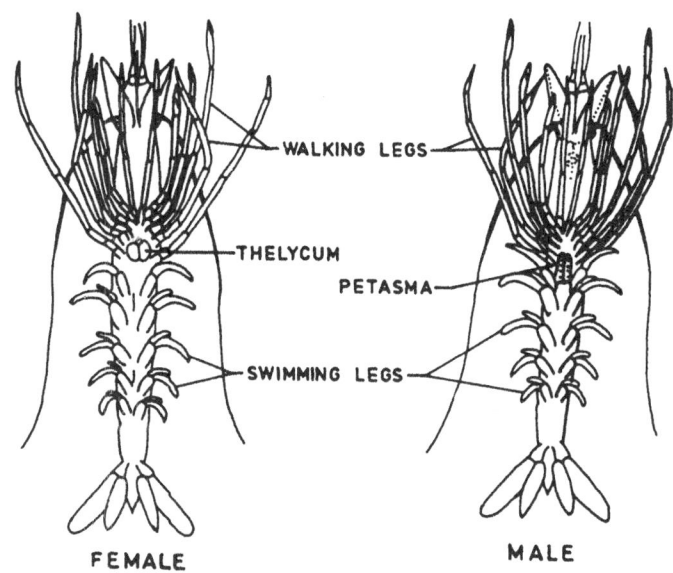

Fig. 3.16. *Penaeus indicus* (Ventral view) (After Silas *et al.*, 1985)

mating and spawning. The spermatophores can be stored inside the thelycum for the duration of the intermoult period and are discarded along with the moulted cuticle at the time of moulting. The sperms remain viable throughout this period and can be used for fertilising successive batches of eggs if the female happens to spawn more than once during the intermoult period. Spawning invariably takes place during the night, and the fertilisation is external. When the eggs are shed, the female simultaneously releases the sperms also from the thelycum. The fertilised eggs sink to the bottom remaining close to the substratum and can be easily stirred up if the water is agitated.

Kuruma prawn *(Penaeus japonicus)*
    This can attain the maximum size of 27 cm in body length and 130 gm in body weight, the commercial size being above 12 to 13 cm in body length and 20 to 25 gm in body weight. Live prawn of 12 cm in body length and 20 gm in body weight fetches the highest price in the Japanese market. The carapace and abdominal segments typically show brown bands while the tip of the telson contains narrow bands of blue, yellow and red colours. The shell surface is smooth. The rostrum is with eight to ten dorsal and one or two ventral spines. A female of 17 cm body length spawns about 700,000 eggs at a time. The diameter of an egg is 0.26 to 0.28 mm. The eggs hatch out into a nauplius larva about 14 hours after fertilisation at a water temperature of about 27°C. The nauplius obtains nourishment from the yolk sac and moults six times within 36 to 37 hrs during which its length increases from 0.3 to 0.5 mm. The nauplius stage is followed by the zoeal stage, when the larva

begins to feed on small phytoplankton. The zoea moults three times in four to five days, increases from 0.5 to 2.5 mm in body length and passes into the mysis stage. The mysis metamorphoses into post-larva after three moultings within five days, growing from 2.5 to 5.0 mm in body length. Fifteen days after the metamorphosis, the prawn is carried by ocean and tidal currents from deeper waters of high salinity to a shallow nursery in a bay or inland sea of lower salinity and settles on the sea bed, thus marking the end of its planktonic life and beginning a sedentary life. It now enters the juvenile stage and feeds on small benthos, detritus and algae. The growing stage is reached when the prawn increases in body length to 2.5 cm and acquires a nocturnal habit burrowing into the sand during the day, emerging only after sunset to feed. As it grows, the prawn gradually migrates from the shallow nursery to deeper waters. At 10 cm body length, it passes into the adult stage and inhabits the sea at a depth of 5 to 6 m, and at 20 cm body length moves to a depth of 8 to 9 m. With the drop in water temperature during winter, larger sized prawns move to a depth of 30 m or more.

Both the male and female gain sexual maturity when one year old, with the male 12 cm and the female about 15 cm in body length. The life span of the prawn is two years for a male and three years for a female.

Tiger prawn *(Penaeus monodon),* Sugpo, grass prawn

It is the largest number of the penaeid group with individuals from offshore catches reportedly reaching 500 to 600 gm in body weight. It can survive through a wide range of temperature and salinity conditions.

Adult tiger prawns are bottom dwellers in the offshore areas at depths of 20 to 70 m. The life cycle starts with the spawning or release of approximately 500,000 eggs by a gravid or ripe female. The larvae hatch out 12 to 15 hours after spawning. They pass through the nauplius, zoea and mysis stages and after 10 to 12 days, metamorphose into post-larvae. The young post-larvae or fry move towards the shore and are encountered in coastal waters about two weeks after entering the post-larval stage. They continue their shoreward migration entering mangroves and other brackish-water areas which serve as their nurseries or feeding grounds, growing to larger juveniles, post-juveniles and subadults (Fig. 3.17). At this stage, *P. monodon* is mainly carnivorous, feeding on slow-moving microscopic animals, small crabs, shrimps, molluscs, marine worms and also detritus. Although the first mating occurs in the estuaries, it is only during or after migration back to the offshore areas that a full ripening of the ovaries takes place. The first spawning occurs at 10 months of age after arrival in the offshore areas, and is followed by a further two or three consecutive spawnings within a single season. The adults remain in the ocean and live up to the age of three, four or even five years unless they die earlier of predation or disease.

Some *Metapenaeus* species, however, reportedly undergo spawning and the larval stages in brackish-water areas, completing the cycle without returning to the open sea.

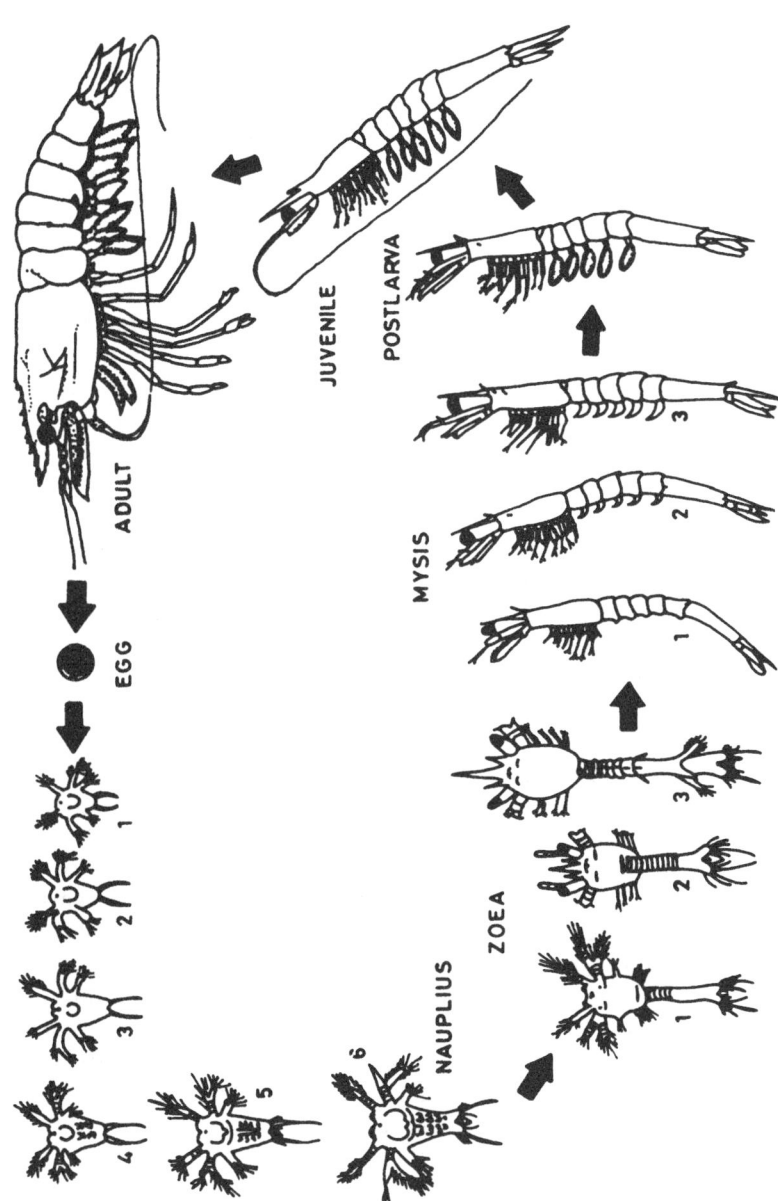

Fig. 3.17. Lifecycle of *Penaeus monodon*

ii) Lobsters

Lobster is one of the most highly prized of all seafoods. Two important species, American lobster (*Homarus americanus*) and its European counterpart, *Homarus vulgaris*, occur in the North Atlantic. However, the major proportion of the much sought after lobster tail comes from one or another of the spiny lobsters, which are found virtually throughout the world but are more abundant in tropical water. All species of spiny lobsters are distinguished from *Homarus* by the lack of the formidable claws or chelipeds. The claws and tail of *Homarus* are both eaten but only the tail of the spiny lobster is used.

Homarus

The general biology of the two species of *Homarus* is virtually identical. The mating, which takes place in summer, follows the general pattern of the decapod crustaceans and depends upon the moulting of the female. Though copulation may successfully occur up to 12 days after moulting by the female, the chances are highest within 48 hours. The success of mating seems to depend at least partially on the relative size of the male and female, and mating either proceeds with difficulty or may even cease altogether in cases of a pronounced disparity in size. Courtship is apparently initiated by a pheromone secreted by the freshly moulted female. When a male is drawn to within a metre or so he responds by advancing on the very tips of his walking legs, a continuous rapid movement of the maxillipeds and side to side movement of the antennae. On making contact, the two lobsters stroke each other with their antennae for up to 30 minutes after which the male mounts the female, gently rolls her over, and copulation takes place in a head-to-head position. Males usually mate with one or two females within a spate of a few days while females may also occasionally mate with more than one male. Sperm is stored in a female's seminal receptacle for 9 to 13 months, when 5,000 to 125,000 eggs are extruded, fertilised, and cemented to the non-plumose hairs of the swimmerets. The female carries the eggs for 10 to 12 months before they are hatched, the eggs changing colour from nearly black to green and ultimately brown in the interval. Thus, the total time between mating and hatching may be as much as two years. However, culturists have succeeded in reducing this period to as little as 11 months mainly through temperature control. Hatching has been recorded at temperatures as low as 9.4°C, but it usually takes place in spring or early summer, at 15 to 20°C. The newly hatched mysis larvae, which make their first moult almost immediately, assume a planktonic existence near the surface. After a few more moults they acquire well-developed claws and other external features of tiny lobsters, when they acquire the ability to swim, though still occasionally remaining planktonic. However, they soon take up a benthic, nocturnal existence. Development to this stage takes from nine to 33 days, depending on the temperature, and the survival rate from hatching to the benthic form is very low (<0.1 per cent). Post-larval lobsters ordinarily undergo four more moults in

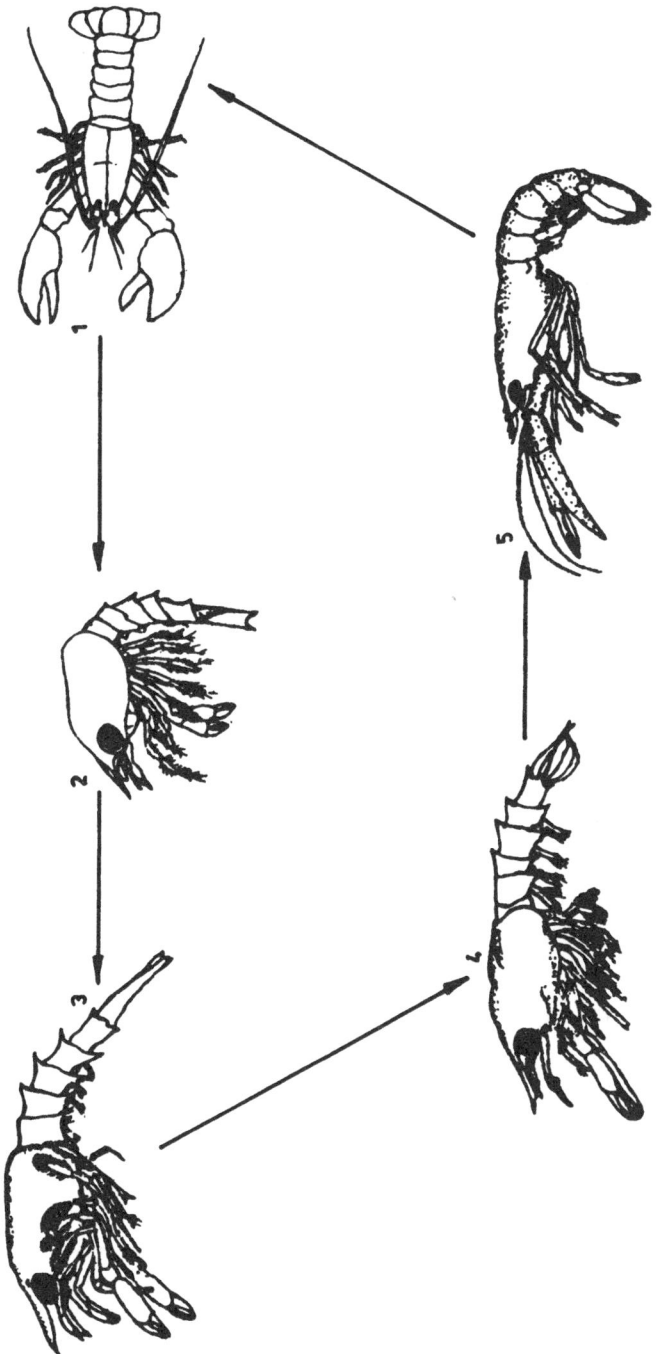

Fig. 3.18. Lifecycle of lobster (*Homarus* sp.) 1—Adult, 2–4—Stage I—III zoea larvae, 5—Stage IV megalopa, often called post-larvae. (Source: Wickin, 1982)

the first season of life, from the date of hatching to the end of that calendar year, a total span of 10 moults counting larval moults (Fig. 3.18). Thereafter moulting becomes less frequent, and at the time of attainment of sexual maturity at about six years of age, moulting occurs no more than once annually. Sexual maturity coincides more or less with the attainment of commercial size (about 20 cm in length and 0.5 kg in weight), though the precise rate of growth depends upon the temperature. Since any individual identification mark is removed with the shedding of the exoskeleton, data on the growth rate of lobsters are scarce. Lobsters kept individually under captivity have been observed to grow 9.5 to 17.4 per cent in length and up to 50 per cent in weight at a moult. There seem no significant differences in the rate of growth of the male and the female.

In Massachusetts, lobsters have been found to moult as early as April 25 and as late as December 29 at temperatures as low as 3.3°C, but most moultings occur from May to October at temperatures of 15 to 20°C. Two seasonal moulting peaks have been observed — one in the early summer when the water temperature first reaches 15 to 20°C and a second one in early fall, the latter probably a response to the substantial increase in the volume of body tissue due to heavy feeding during the summer. Lobsters have been known to live up to 10 years in captivity. Lobsters can be fed finely ground hard clams, and as they grow larger they eat fresh fish, clams and the viscera of bay scallops.

Homarus americanus has a life span of 50 to 100 years and attains weights as much as 19 kg (Bardach *et al.*, 1972).

Spiny lobster
There are seven genera and about 30 species of spiny lobsters, of which at least 15 are commercially important (Table 3.1) and have very complex though similar life histories.

The mating season depends upon the species and the age of the females and takes place in inshore waters. Older females mate earlier than younger ones and usually twice a year. The males extrude sperm in the form of a viscous fluid which attaches to the underside of the female, the outer surface hardening rapidly to form a sperm sac. The eggs are extruded afterwards and pass over the sperm sac. Simultaneously, the female breaks the sperm sac by scratching it with her legs, thus fertilising the eggs. The fertilised eggs are then transferred and attached to the swimmerets and remain there until hatching. The number of eggs produced by a female varies with its size and species and may be as low as 50,000 and as high as 4,000,000. The time interval between mating and fertilisation seems to depend on the water temperature, usually less than that for *Homarus*, and hatching is also earlier, requiring only three weeks in some species. During this period the eggs change colour from bright orange-red to a dark brown which then fades until they are almost colourless just before hatching. However, the larval development is more protracted and complex in the case of the spiny lobster. The newly hatched larva, the

Table 3.1
The commercially important species of spiny lobsters and their ranges (Bardach *et al.*, 1972)

Species	Range
Jasus lalandii	Australia, Juan Fernandez Islands (off the coast of Chile), New Zealand, South Africa, Tasmania, Tristan da Cunha
Jasus verreauxi	Southern Australia, New Zealand, Tasmania
Panulirus elphas	Great Britain to the Mediterranean
Panulirus argus	Florida, the Caribbean area, the Atlantic coast of South America, Bermuda
Panulirus gracilis	Southern Mexico to Ecuador
Panulirus inflatus	Gulf of California to Southern Mexico
Panulirus interruptus	Southern California, Lower California (Mexico)
Panulirus japonicus	Japan
Panulirus laevicauda	Bermuda to northeastern South America
Panulirus longipes	Western Australia, Ryukyu Islands, Taiwan, Philippines, New Caledonia
Panulirus marginatus	Hawaiian Islands
Panulirus ornatus	Indian Ocean, Ryukyu Islands, Taiwan, Philippines
Panulirus penicillatus	Indo-Pacific, Korea, Lower California (Mexico). Galapagos, Costa Rica
Panulirus regius	Western Mediterranean, West Africa
Panulirus versicolor	Indo-Pacific

phyllosoma, is a flat leaf-like, unusually delicate animal two to three mm long, planktonic in habit, and floats horizontally with its legs extended. Some species are negatively phototropic and exhibit extensive diurnal vertical migrations. While still remaining planktonic, phyllosoma larvae undergo numerous moults in three to six months mostly without metamorphosis, though a progressive change in form is noticeable. The next larval stage is the puerulus stage which is reached after six or more metamorphoses though the exact number seems variable being probably dependent upon its nutrition or some yet unknown factor, since under laboratory conditions *Panulirus japonicus* as shown to undergo 16 moults in 178 days without metamorphosis. The puerulus, though it resembles the adult, lacks lime, in its skeleton, is transparent and may still be planktonic. The puerulus finally moults to become a juvenile about 2.1 cm long and settles to the bottom. After two or three more moults it acquires the reddish-brown colour of adult spiny lobsters. Survival up to this point is extremely low since predation is extensive at all stages. Juvenile and adult spiny lobsters are also subject to predation but may escape frequently by hiding under rocks and other cover during the day and foraging only at night. Spiny lobsters avoid muddy bottoms and strong currents. Adult spiny lobsters normally consume a wide variety of foods including fish, worms, molluscs, and smaller crustaceans. Cannibalism may occur if their food lacks calcium carbonate, and they may even act as scavengers though usually they show a marked preference for fresh food. The rate of growth which is concomitant with the frequency of moulting varies with the food supply, water temperature and sex. Mature females moult only once or twice a year, before mating, and sometimes after hatching or disposing of their eggs, while males moult more often and any time of the year and consequently grow larger than the females. Growth between moults is about 5 to 10 per cent of body length for *Panulirus arges*. The maximum size attained with age varies in different species. *Panulirus interruptus* is said to even reach a weight of over 13 kg on rare occasions, the usual large size being around 3 kg. In California, this species is believed to take at least seven to nine years to reach the legal minimum size of 27 cm, although *P. argus* may attain this length in three years.

iii) Crabs
    Nearly all of the larger species of crabs (Brachyura) are edible, but the majority of crabs of commercial importance are confined to three families— Portunidae (swimming crabs), Xanthidae (mud crabs), and Cancridae (cancer crabs). The swimming crabs, *Scylla serrata*, have long been an incidental product of brackish-water pond culture in southeast Asia, and in the Philippines and some other countries they are sometimes even deliberately stocked in fish ponds.

Scylla serrata (Fig. 3.19)
    In most portunids, mating occurs after the female undergoes the precopulatory moult in the first year of life. Though the first of such moults may not be readily distinguishable from the usual moults, a female preparing to moult now releases a pheromone attracting the male which climbs over it clasping her with the chelipeds and the anterior pair of walking legs and carrying her about in doubler formation till the female moults in about three or four days. The male then turns the female around for copulation which

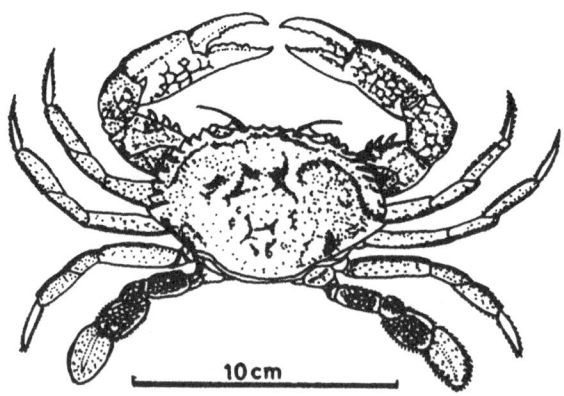

Fig. 3.19. Crab *(Scylla serrata)* (After, Chen, 1976)

may last for seven to twelve hours. Though copulation takes place immediately after the precopulatory moult of the female, since the male and female remain still together, successful copulation may occur even three days after the moulting. The female retains the sperm, and fertilisation may not take place for many weeks or even months after mating. A single copulatory act provides enough sperm for two or more successive spawnings before the female moults again. The fertilised eggs are attached to the pleopods of the female, where they hatch within a few weeks. In captive spawners, two-thirds of the total number of eggs may fail to become attached. The larvae hatch as planktonic zoeae. A few eggs may, however, hatch prematurely as 'prezoeae', but these moult and become zoeae within half an hour. After passing through several zoea stages in succession and a single megalops stage within a month or so, the larvae metamorphose to benthic juvenile crabs. The zoeae feed mainly on phytoplankton while at the megalops stage, the larvae feed on zooplanktons.

Blue crab

The most valued of the portunid crabs is the blue crab (*Callinectes sapidus*) of the Atlantic coast of the U.S.A. There are at least four portunids commonly referred to as 'blue crabs' and are also highly prized: *C. sapidus*, *C. bellicosus* of the Gulf of California, and the Japanese *Portunus trituberculatus* and *Neptunus pelagicus*.

Stone crab

The stone crab, or mud crab (*Menippe mercenaria*) is cultured because of the fine meat found in its huge claws. It is distributed from North Carolina around the southeastern United States to Mexico, and is also found in Cuba. The stone crab is an estuarine animal hiding under rocks and in crevices. Their burrows may be 30 to 60 cm long and about 15 cm wide. The young hatch at intervals throughout the spring and summer, moult successively and move to deeper channels on reaching adulthood. There are generally five zoeal stages. They require warm water with high salinity for an optimum survival rate. Larvae have been reared from eggs to crabs in the laboratory in about 27 days, using brine shrimp larvae as food.

3.3 Turtles

Turtles (Fig. 3.20) are vertebrates belonging to the class Reptilia which also includes lizards, snakes, crocodiles, tortoises etc. Since the meat and soup (prepared from plastron or the ventral shell) obtained from the green turtle, *Chelonia mydas,* are relished in many countries, an overexploitation by man has greatly reduced the natural resources of turtles, and their culture has become essential. Sea turtles are particularly suitable for aquaculture. Their omnivorous food habits allow them to eat a wide range of foods, from jelly fish to cabbage, making it easy to cultivate them and entire animals can be utilised for food or in industry. Interestingly, they migrate to the same place

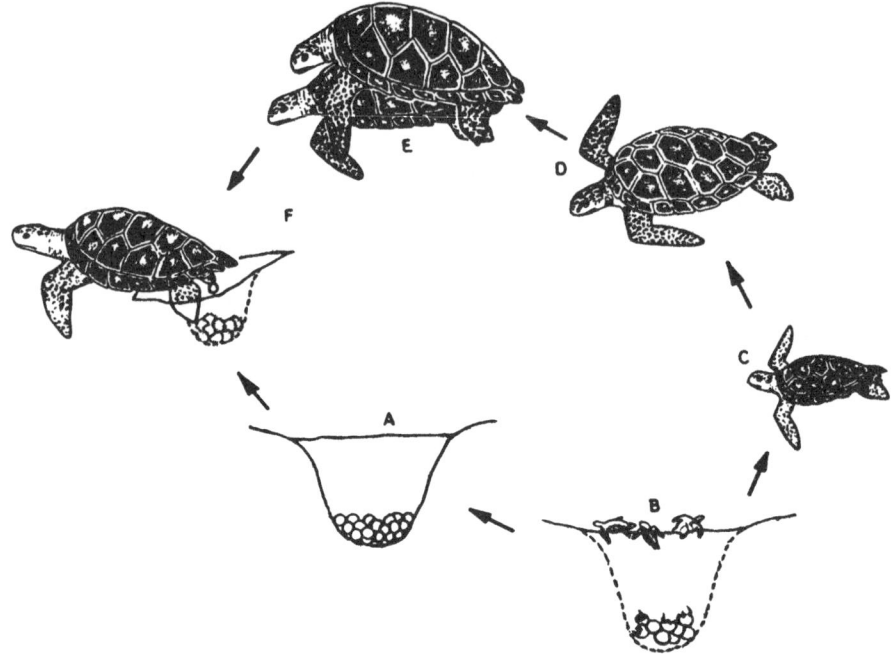

Fig. 3.20. Lifecycle of Green Turtle *(Chelonia mydas)*
    A—Incubation of eggs, B—Hatching, C—Juvenile and subadult, D—Adult,
    E—Mating, F—Egg laying (After Kafuku and Ikenoue, 1983)

for spawning as do salmon, allowing their stocks to be easily managed. The first large-scale commercial farms has been constructed on Grand Cayman Island in the British West Indies.

Green turtle *(Chelonia mydas)*
    The maximum size reached is 114 cm in carapace length and 200 kg in body weight. Its commercial size is above 60 cm in carapace length and 40 kg in body weight. The colour of the carapace is basically green or greenish-brown with prominent black markings. In the juveniles, radial brown bands are present on each shield. The dorsal surfaces of the head and flippers are predominantly black. The plastron is whitish in colour in the young specimens, but gradually turns yellow with age. The carapace is almost oval in shape with 38 shields – five central shields, four centrolateral shields, one nucal shield and 12 marginal shields on each side. Upper and lower jaws are encased in corneous sheaths. Four clawed flippers are present. Females are usually larger than the males, which in turn have longer tails and more curved claws than the females.
    Mating takes place from late February to May, off the nesting beaches. The nesting season of the green turtle is from May to early August. The

female comes to the beach at night and at first digs a roughly 40 cm deep body pit, after which she digs an egg hole of about a further 30 cm depth inside the pit with her hind legs where she lays about 10 eggs, and returns to the sea; the total activity requires around two hours. Most females nest from three to seven times per season at intervals of 10 to 15 days. Eggs are spherical, 40 to 50 mm in diameter, and 40 to 50 gm in weight. The optimum sand conditions for the incubation of eggs are a temperature of 28 to 30°C and a salinity of 365 to 730 mg Cl/kg. The incubation period is about 50 days. The hatchlings come up out of the sand three to seven days after they hatch. Many eggs do not survive to hatch since they are preyed upon, the main predator being the ghost crab (*Ocypode* sp.). The feeding migration of the turtles then takes place. In the coastal areas the main food is red algae. Green turtles also feed on drifting seaweed, crustacea, jellyfish and pelagic molluscs attached to the algae. The life span of the sea turtle has been found to be over 50 years.

3.4 Other invertebrates

i) Sponges

Sponges are lower invertebrates consisting of several specialised types of cells only, there being no formation of tissue. The different species of sponges vary greatly in colour, shape and size. When freshly removed from the sea, they are slimy and heavy and have to be pounded, rinsed and scraped to remove the living tissue. The marketable part is the exoskeleton. Commercial sponges are sold mainly for surgical purposes and also for cleaning. Sponges can tolerate temperatures of 10 to 35°C. They reproduce by regeneration, budding and also by sexual means. In sexual reproduction, larvae are produced, which drift about for a short period before attaching themselves to any clean hard object available, which is essential for their survival.

Farming of sponges is relatively simple. Large sponges are cut up into smaller parts, which are then attached to discs. Growth up to the marketable size is quicker in the case of these small cuttings than in the ones reared from larvae. For this purpose, concrete bricks or discs are placed in the water suitable for sponge growth. However sponge farming has a waning popularity due to the relatively slow growth of natural sponges and high labour costs as well as the gaining popularity of synthetic sponges. Sponge farming is common in the Mediterranean Sea, and in the Florida Keys, near Miami and the Bahamas.

ii) Bait worms

An increase in sport fishing round the world necessitates a corresponding increase in the need for bait. Bloodworms (*Glycera dibranchiata*), and sand worms (*Neanthes virens*), earthworm-like dwellers of the intertidal zone from the Canadian maritimes and New England, have become very popular as bait worms.

Bloodworms are so named because of the red coloured body fluid which contains a respiratory pigment similar to haemoglobin. These worms are burrowers, thriving in areas where the sediment is stable eı ɔugh to support their burrows, and emerge at night to feed. Females release about 1.2 to 3.0 million eggs during the summer. These hatch into free-floating larvae which after a brief drifting period sink to the bottom and burrow. They become 20 to 35 cm long by their third year of life. Reproduction and survivability of the young is optimum at water temperatures of 8 to 9°C during the spawning year.

Some research has been done at the University of West Florida on lugworms (*Arenicola cristata* and *A. brasiliensis*) to test their suitability for culture as bait worms. It has found that these worms can be induced to spawn throughout the year. The larval worms do not need food, and the post-larvae, juveniles, and adults all feed on decaying marine plants and micro-organisms such as fungi, bacteria and protozoans associated with decaying plant material, making them comparatively easy to cultivate.

3.5 Seaweeds

Species of large marine algae that are popularly and collectively known as seaweeds are of high nutritional value, though their use as human food is mainly restricted to the Orient. However these have been widely used as animal feed, as fertiliser as food additives, and also for a number of industrial purposes.

The most commercially important algae which are also consumed by man are the members of the class Rhodophyceae, the red algae. The most important of these are members of the genus *Porphyra,* known popularly as 'laver' in European and American countries and as 'nori' in Japan, where it ıs the single most commercially valuable marine product. Five species are commonly cultivated in Japan—*P. yezoensis, P. tenera, P. angusta, P. kuniedai, P. pseudolinealis.* Some other red algae are also cultured in Japan, the most important edible alga among them being funori (Gloiopeltis). The red algae cultured in Taiwan are four species of *Gracilaria,* namely, *G. gigas, G. confervoides, G. lichenoides* and *G. compressa.* The only seaweed commonly eaten in the west is the red alga *Rhodymenia palmata,* popularly known as 'dulse'. It is completely digestible by man unlike other seaweed and is eaten in the dried, or raw form or even used in cooking in the maritime provinces of Canada, Ireland, Scotland and the mediterranean countries.

The green algae commonly used as food by man are *Enteromorpha* and *Monostroma* (aonori). Both are sometimes grown together with *Porphyra.* In Japan, *Monostroma* commands the highest price for any seaweed.

Among the brown algae (class Phaeophyceae), larger species collectively known as kelp, are also of value to man in ways similar to that of red algae and a few of them are therefore cultured. In Japan, the most popular ones are wakame (*Undaria pinnatifida*) and Konbu (*Laminaria* spp.), of which the former is more widely cultured. *Caulerpa* sp. is grown in brackish-water ponds in the Philippines. The colloids derived from the tropical brown algae *Eucheuma* sp. and *Hypnea* sp. that are also cultured in the Philippines, are used as food additives.

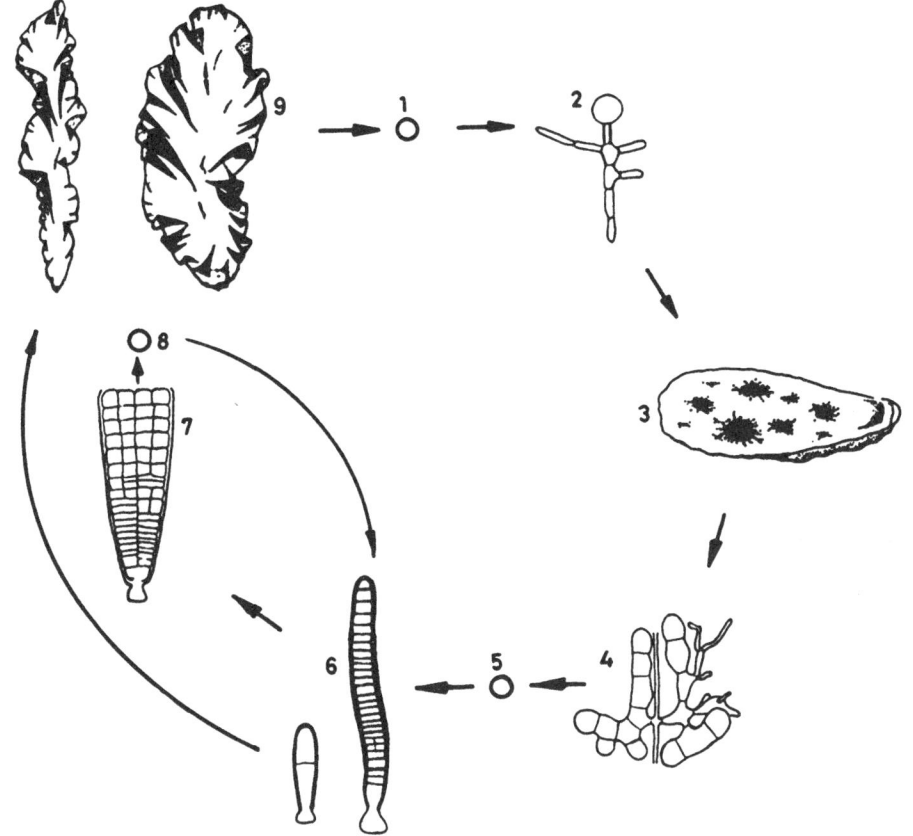

Fig. 3.21. Lifecyle of sea weed (Nori)
1—Carpospore, 2—Conchocelis, 3—Conchosporagium, 4—Parasporangium, 5—Conchospore, 6—Germination of carpospore, 7—Discharge of monospore, 8—Monospore, 9—Leafy thali (After Kafuku and Ikenoue, 1983)

Nori *(Porphyra yezoensis)*

Nori or purple laver which is dried in rectangular paper-like sheets of 19 by 21 cm is the most commonly eaten alga in Japan. Lightly broiled and seasoned with soy sauce, pieces of nori are a delicacy to be taken with rice. Also rice balls wrapped in sheets of nori form an important part of the traditional Japanese picnic lunch.

Fronds of nori are basically elliptical to broad lanceolate in shape with undulated margins, their size varying according to the locality and the prevailing environmental conditions. A fully grown frond of *P. yezoensis* is 15 to 36 cm long with a thickness of 40 to 53 μm, while that of *P. tenera* is usually 17 to 35 cm in length (sometimes even up to 1 m) and 25 to 35 μm thick. The colour is dark purple.

P. yezoensis is distributed in the areas which remain under the dominant influence of cold ocean currents round the year, and grows on rocks in tidal belts facing the open sea. *P. tenera* however grows in tidal belts around estuaries within bays and inland seas that remain under the influence of both cold and warm water currents and are bathed in water of lower salinity.

The algae germinate from spores released from *conchocelis* filaments (sporsphyte) from September to November and appear as small germlings of 1 mm length from mid to late October, when the water temperature drops to about 22°C. By mid to late November, the germlings grow rapidly to fronds (gametophyte) 15 to 20 cm long, or more. They flourish during winter in the cold water (3 to 8°C). In April, the fronds start to wither and disappear totally by May, when the water temperature rises to about 14°C. Germlings of 150 μm to 1 mm in length also form a large number of neutral spores during the period between late September and early November, which then germinate and grow into fronds. Organs for sexual reproduction are formed on the fronds when they reach a length of 3 to 5 cm during the month of November. These organs release fertilised carpospores which continue until the fronds disappear in May. The carpospores attach to shells on the sea bed to germinate and become *Conchocelis* filaments. These then bore into the shell and grow in the pearl layer of the shell to form colonies each about 1 cm in diameter during August and September, when they begin releasing spores called conchospores (Fig. 3.21).

REFERENCES

Bardach, J.E., Ryther, J.H. and McLarney, W.O. (1972): *Aquaculture. The farming and husbandry of freshwater and marine organisms,* John Wiley and Sons.

Barnes, R.D. (1980): *Invertebrate Zoology.* Saunders College.

Chen, T.P. (1976): *Aquaculture Practices in Taiwan.* Fishing News Books Ltd., Surrey, England.

Field, I.A. (1923): Biology and economic role of the Sea Mussel *Mytilus edulis.* Bulletin of U.S. Bureau of Fisheries, XXXVIII, 1921–1922. p. 127–259.

Fincham, A.A. (1983): *Basic Marine Biology.* British Museum (Natural History), Cambridge University Press.

Iversen, E.S. (1976): *Farming the edge of the sea.* Fishing News Books Ltd., Surrey, England.

Kafuku, T. and Ikenoue, H. (1983): *Modern Methods of Aquaculture in Japan.* Kodansha Ltd., Tokyo and Elsevier Scientific Publishing Company.

Milne, P.H. (1972): *Fish and Shellfish Farming in Coastal Waters.* Fishing News Books Ltd., Surrey, England.

Silas, E.G., Mohamed, K.H., Muthu, M.S., Pillai, N.N., Laxminarayana, A., Pandian, S.K., Thirunavukkarasu, A.R. and Ali, S.A. (1985): Hatchery production of penaeid prawn seed, *Penaeus indicus.* Central Marine Fisheries Research Institute Special Publication No. 23, India.

Wickin, J.F. (1982): Opportunities for farming crustaceans in Western temperate region. In: *Recent advances in aquaculture,* (eds.) J.F. Muir and R.J. Roberts, Croom Helm, London and Cambridge, West View Press, Boulder, Colorado.

Yamamoto, G. (1964): Studies on the propagation of the scallop, *Patinopecten yessoensis* (Jay), In Mutusu Bay (Japanese), Nihon Suisan Shigen Hogo Kyokai.

CHAPTER 4

Hydrology of Coastal and Deltaic Areas

4.1 Introduction

Coastal and deltaic areas generally have low elevations with respect to the mean sea level. The land slope is usually small and the type of soil is sandy, clayey, silty, or a mixture of all with either no cover or a vegetation cover in the form of mangroves or deep forest. The areas are subject to severe hydrometeorological variations in wind speed, temperature, humidity, rainfall, runoff, evaporation, as well as tidal and wave action due to hurricanes, typhoons etc. These create a number of problems particularly with respect to drainage and protection against high flood levels. Another complication arises because of the special hydraulic conditions of the lower reaches of the river discharging into the sea. The tides propagate into the estuaries and the intrusion of saline water contaminates water in the lower river reach. Low lying areas are exposed to the danger of flooding from two sides (i) the river floods, and (ii) the high back-flow levels of the sea due to storm surges.

In spite of these difficulties, the low lying areas are characterised by a favourable position, particularly from the point of view of aquacultural operation for prawn farming and other marine fishes. In these regions coastal pond culture is favoured in southeast Asian countries including India (Fig. 4.1). It is estimated that in southeast Asia there are 400,000 hectares of developed fish ponds generating an annual income of (US) $ 400 million, with scope for development of several hundred million hectares more. To develop these regions, it is necessary to go into the various hydrologic problems occurring in coastal areas particularly with regard to the designing of coastal fish ponds. These problems are primarily related to estimating surface water and ground water flow in these regions. Various aspects of the estimation of the rainfall-runoff relationship as well as the ground water potential in general have been addressed in various standard textbooks on hydrology Chow (1964), Subramanya (1984), Varshney (1974) and will not be addressed here. Only the portions relevant for engineering design of fish ponds will be discussed.

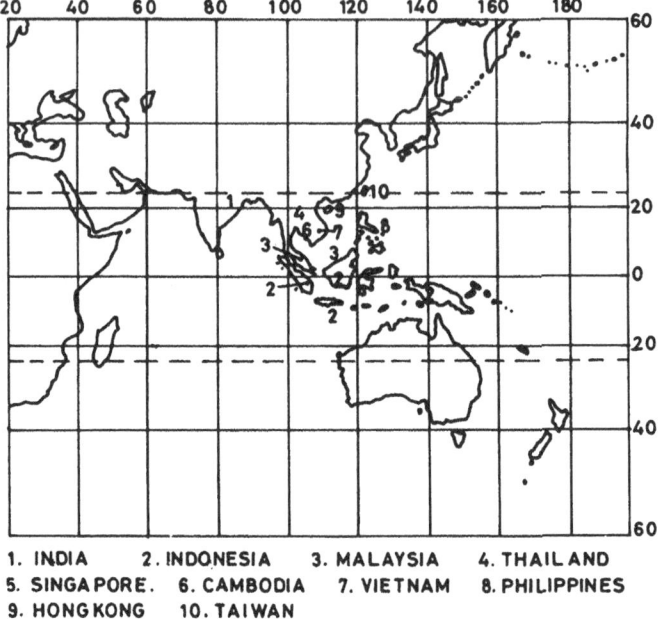

1. INDIA   2. INDONESIA   3. MALAYSIA   4. THAILAND
5. SINGAPORE.   6. CAMBODIA   7. VIETNAM   8. PHILIPPINES
9. HONGKONG   10. TAIWAN

Fig. 4.1. Countries where coastal pond culture is suitable

4.2 Topography of the coastal areas

The coastal and deltaic areas consist of both exposed and sheltered shorelines in the form of swamps, bays, estuaries and lagoons. Some portions of the shoreline may be eroding, some areas may have beaches while the remaining areas may have no beaches. There is a large area of broken coastline in India in the Sundarbans delta area which has one of the finest flat coastal swamplands ideally suited to coastal aquaculture. The Indian coastline also has one of the biggest sheltered bays in Chilka, Orissa with great potential for aquacultural operations. There are other areas of coastal lagoons, estuaries, backbays, and swamps in the coastal states of Tamil Nadu, Kerala, Goa, Maharashtra and Gujarat, including the Rann of Kutch at Saurashtra. All these areas are potentially rich sites for the development of coastal aquaculture in India besides other suitable sites in other south Asian states, as shown in Fig. 4.1. Figure 4.2 shows a typical unprotected coastal beach profile. Figure 4.3 shows the natural formations of various sheltered zones in the coastal areas. Figure 4.4 shows a typical longitudinal section of a coastal swampland as probable fish sites (Denila, 1980). Flat coastal swamplands whose elevations are within the range of ideal pond bottom elevation are generally preferred for brackish water pond culture.

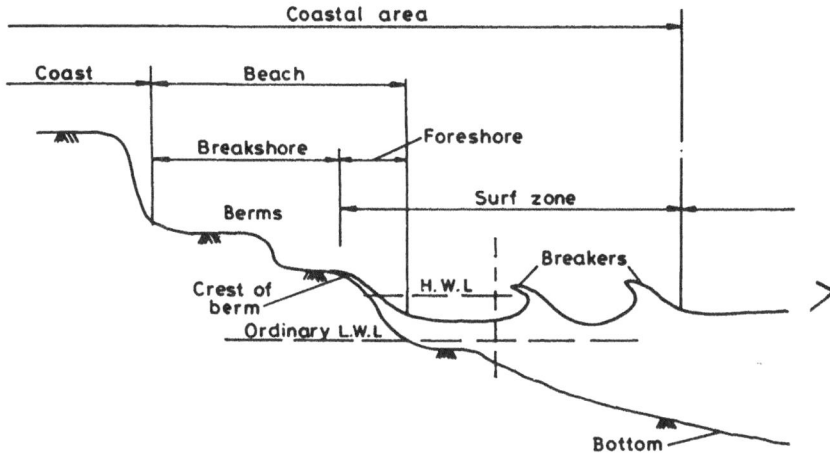

Fig. 4.2. A typical coastal beach profile
L.W.L. = Low Water Level; H.W.L. = High Water Level

4.3 Type and vegetation density

The coastal areas of the regions mentioned earlier, are in the tropical and subtropical zones. The vegetation variation depends on the elevation of the land and type of soil. The vegetation is however, mostly mangrove trees. Vegetation has several beneficial usages apart from fish culture. Trees, such as mangrove trees, are used as timber for construction, firewood, furniture, etc. In addition to the role played by them in the maintenance of various forms of biological life, mangrove forests also serve as silt collectors, promoting soil accretion, absorbers of wave impacts and buffers to storm surge levels (Menasveta, 1982). However, areas with dense mangrove tree rooting systems are difficult for fish farm construction because it takes longer to clear the area of stumps.

4.4 Climate and watershed conditions

Important climatic factors in the design of fish ponds are wind and amount of rainfall. The direction of the prevailing wind is responsible for wave action. In extreme conditions such as hurricanes and typhoons, there are very high water level fluctuations due to storm surges. Similarly, because of heavy rainfall there is a flooding affect from the upper catchment area of the river. The various rainfall factors that have to be analysed are maximum intensity, duration, frequency, and annual distribution within the watershed. The land area surrounding or lying above the fish farm site is referred to as the watershed. The runoff due to rainfall over the watershed drains into the river system which could cause flooding during the monsoon months. The volume of flood water or runoff is affected by the rainfall, type of vegetation cover,

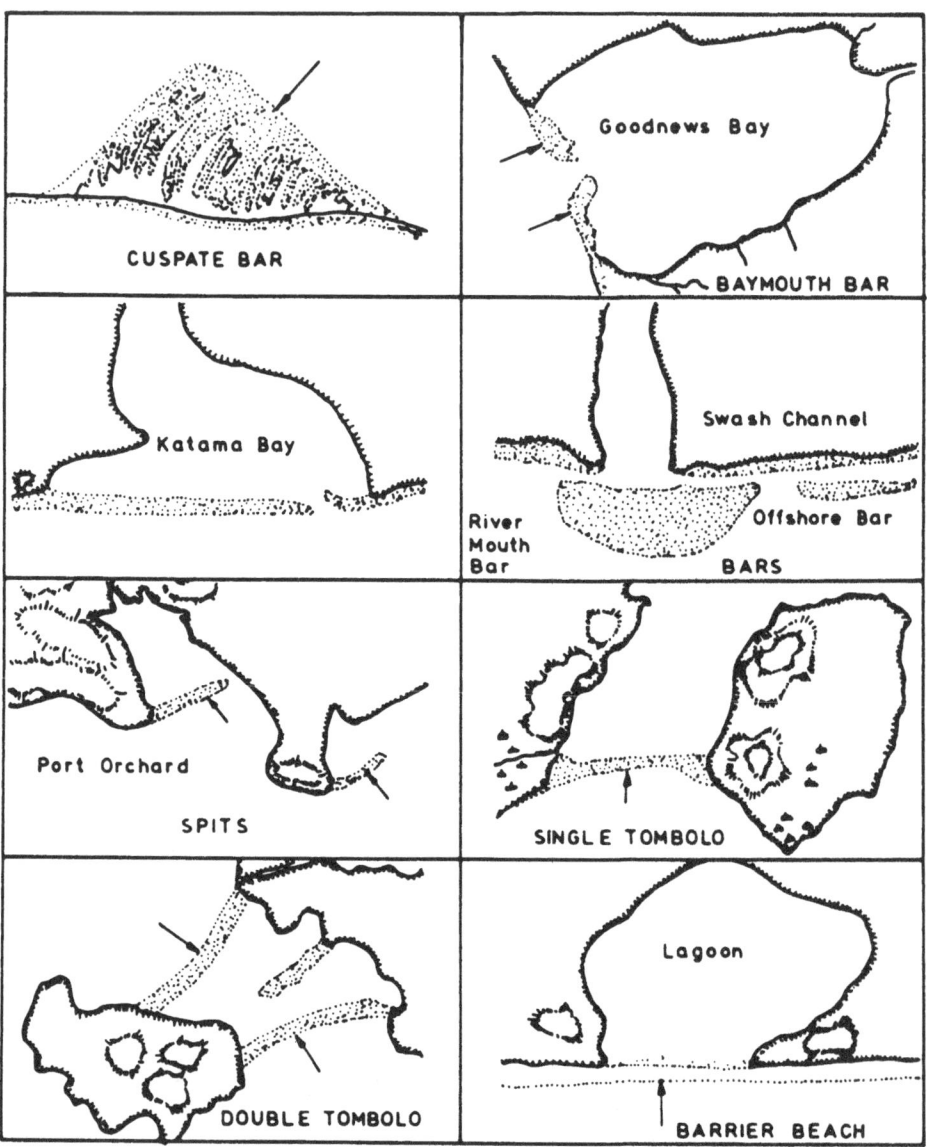

Fig. 4.3. Natural formations of various sheltered zones in coastal areas

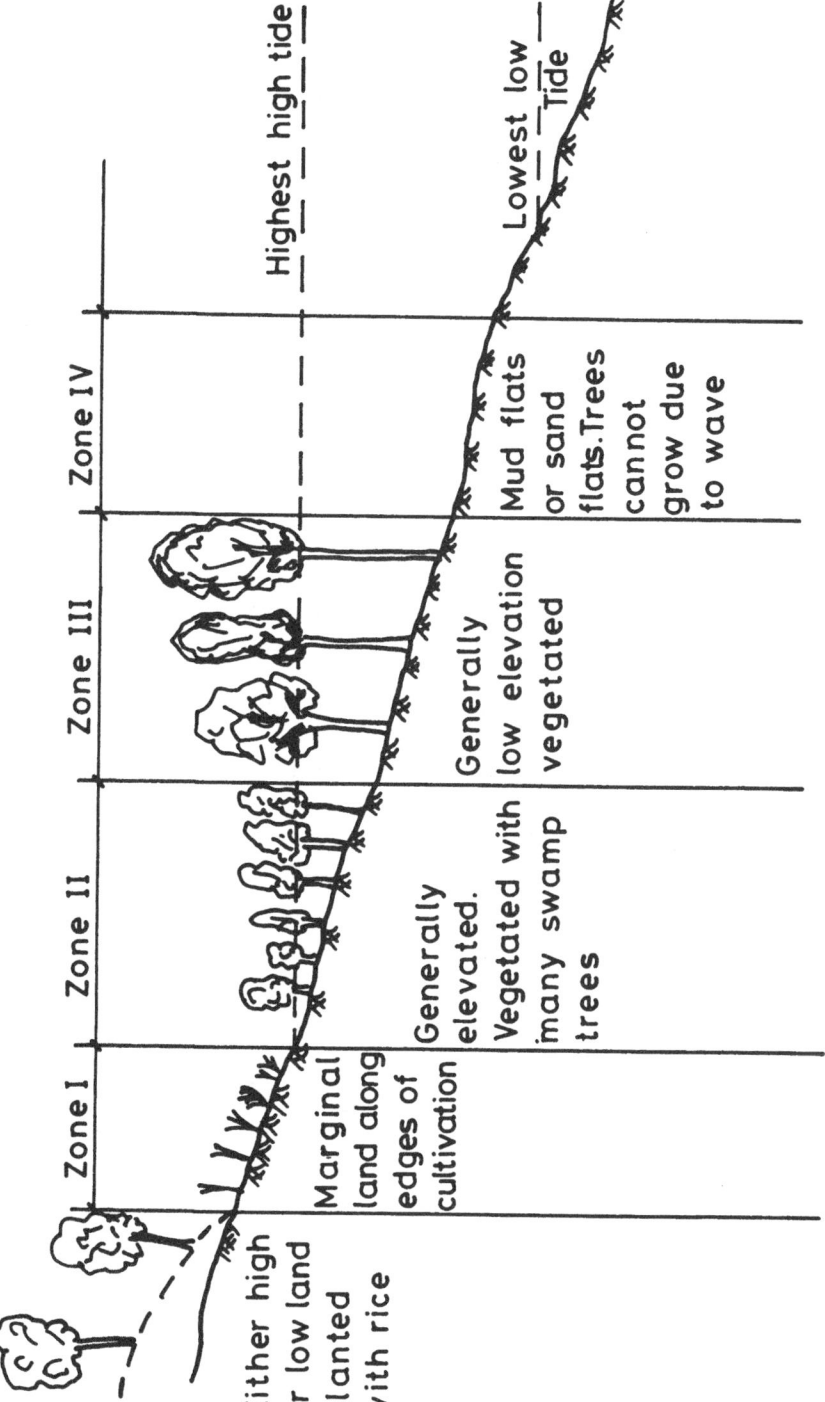

Fig. 4.4. A typical picture of a coastal swampland

soil, topography, and watershed size and shape. The estimation of flood volume can be made either from suitable empirical formulae or from the unit hydrograph concept. These aspects have been addressed in most standard textbooks on hydrology. Generally, a much higher rate of runoff can be expected in a watershed that has high rainfall intensity and duration, over a large area covered with clayey soil with a very low coefficient of permeability, and steep slopes. The nature and recurrence of peak flood near the site should be ascertained by performing an appropriate analysis of flood records, or a synthetic hydrograph. Flood records in the remote areas are generally not available for flood frequency analysis. In their absence, rough information on the flood may be obtained by making enquiries from the local residents. Usually the life of a fish farm is taken to be between 12 and 15 years. The design of the fish farm protective embankments may be based on a flood recurrence interval of at least 12 to 15 years. Details of flood estimation, flood frequency analysis, etc. may be referred to in the book by Ghosh (1986).

4.5 Nature of flood propagation

Coastal areas are exposed to floods due to the propagation of the tidal flux into the estuaries and the floods from the upstream river basin. Depending on the propagation of astronomical tides, the river can be divided into the following zones (Fig. 4.5):

Zone I: In this zone, the vertical tides occur with subsequent reversal of the current direction and intrusion of saline water.

Zone II: In this zone, the river water is fresh, although the tidal phenomena are similar to those in Zone I.

Zone III: In this zone, the water levels are still affected by the tides where the current direction remains the same but with varying velocity according to the tide.

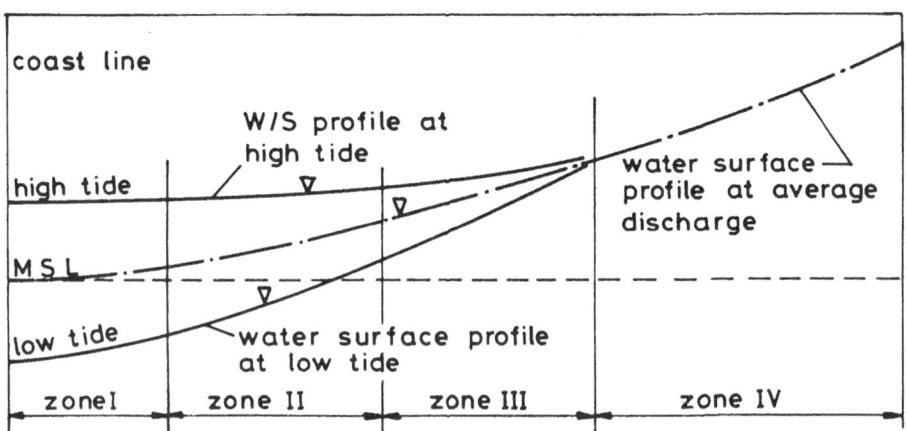

Fig. 4.5. Classification of river based on tidal propagation

Zone IV: In this zone, the water levels and the flow are only affected by the upstream discharge.

In accordance with the propagation of abnormally high sea levels caused by storm surges, a river can be divided into the following zones as shown in Fig. 4.6:

Zone I: The effect of sea levels predominates in this zone.

Zone II: A combined effect of sea and river flood occurs in this zone.

Zone III: The effect of river flood predominates in this zone.

The magnitude of superelevation of the sea level due to storm surges varies widely from one area to the other. Superelevation of 3 m or more may occur along the deltaic area of the Ganges and the deltas in Japan and Taiwan. They practically do not exist along the coasts of the Irrawaddy (Burma), Chao Phya (Thailand), Mekong (Vietnam), and Pampanga (Luzon—Philippines).

As already mentioned, river floods are generated in the upper catchment and in the deltaic area. Thus, the climatologic and physiographic conditions of the upstream river basin and the ratio between the size of the upper catchment area and the deltaic area are of primary importance. This determines to what extent the effect of floods can be damped by the construction of flood control reservoirs in the upper portion of the river basin. The characteristics of the river basin govern the type of upstream flood with respect to the rise and fall of the flood peaks, that is, gentle and flashy floods.

Figure 4.7 shows the annual hydrographs for typical years of the R. Cho Shui of Taiwan and the R. Chao Phya of Thailand. The flood hydrographs show typical differences with respect to duration of floods and rate of rise and fall of the flood peaks. The differences are mainly due to the difference in areas of the river basins. For small and short river basins the fields are flashy. For a large basin the effect of rainfall on the outflow from the basin is gentle

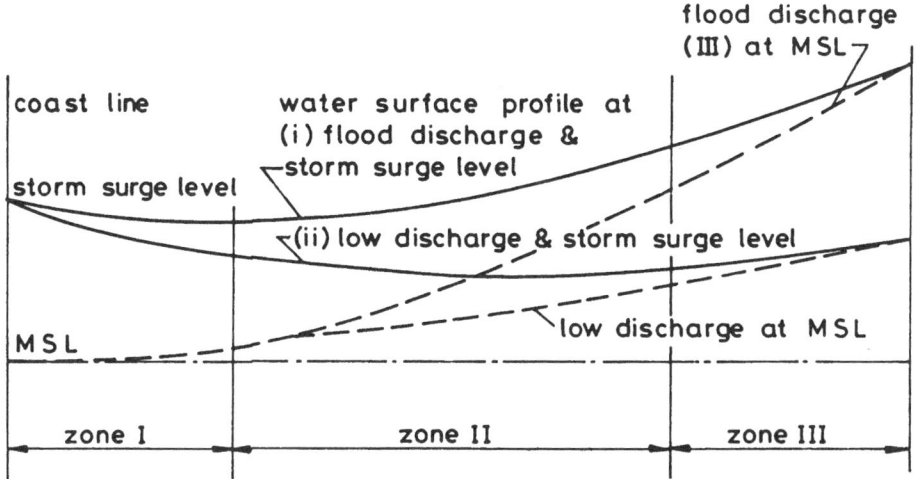

Fig. 4.6. Classification of river based on storm surge propagation

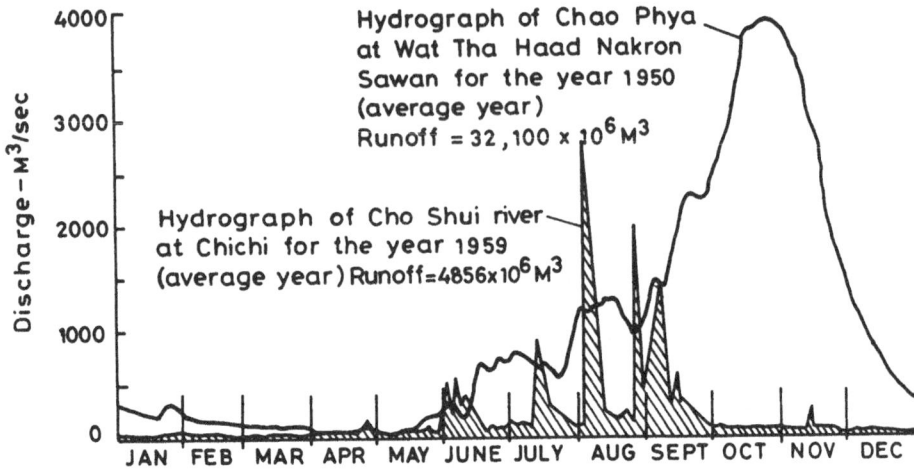

Fig. 4.7. Average hydrographs of Cho Shui and Cha Phya Rivers

because it is distributed over a large area. Embankments in the deltas of rivers in Japan, the Red River in northern Vietnam and the Pumpanga in Luzon (Philippines) are needed for protection from floods whereas in the Chao Phya and the Irrawaddy a crop of rice can be grown even though no adequate flood protection is provided.

4.6 Salt water intrusion in rivers and canals

Strict control on river water quality to maintain the proper biological environment is necessary for fish culture. One such critical area is the proper maintenance of salinity level. It is essential that the normal salinity during high and low tide at different seasons of the year at the fish farm site be known. In this regard the information of particular importance is (i) the flow rate and velocity in the rivers or canals during each season in relation to salt water intrusion and formation of salt water-fresh water wedges, (ii) the length of the wedge from the river mouth, the depth of the salt water and fresh water layers within the wedge, salinity values along the depth with wedge regions, and the frequency and duration of flooding. These are important to determine the necessity of mixing or destroying the wedge to alter the salinity, to minimise siltation along the river, or in the selection of the site for the construction of sluice gates etc.

Due to the higher density of sea water, it intrudes into the river mouth. When a river discharges into a sea with little or no tidal range like the Mediterranean Sea, or when the inflow of fresh water is large with respect to the tidal discharge, the fresh and salt water tend to remain separate. The still form of the salt water wedge is called an arrested saline wedge (Fig. 4.8). Let ρ be density of fresh water, $\rho + \triangle\rho$ the density of sea water, and ρ_m the

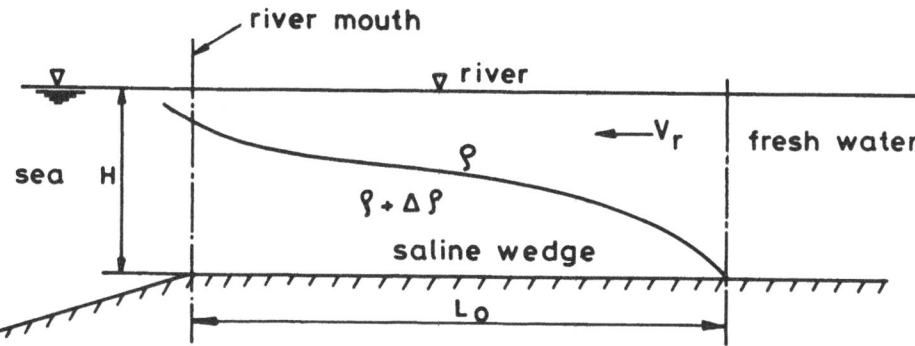

Fig. 4.8. An arrested saline wedge

average density of the two liquids. The wedge velocity then is

$$V_\Delta = \left(\frac{\Delta\rho}{\rho_m} gH \right)^{0.5} \tag{4.1}$$

where g = acceleration due to gravity

and H = depth of river water

This velocity is known as densimetric velocity and its physical significance lies in the fact that the propagation velocities of saline front moving against still water are proportional to V_Δ and nearly independent of the size of the river. The relationship between the velocity of the saline front V and the distance travelled L can be expressed in the form

$$\frac{V}{V_\Delta} = f_{nc} \left(L/H, \frac{V_r}{V}, \frac{V_\Delta H}{v}, \frac{H}{B} \right) \tag{4.2}$$

where V_r = velocity of river opposite of the advancing wedge

$\quad\ B$ = surface width of river

$\quad\ v$ = kinematic viscosity

$\quad\ L$ = travel distance of saline wedge from river mouth

Neglecting the frontal velocity of an advancing wedge, the final disposition of the arrested saline wedge can be expressed as

$$L_0/H = f_{nc} \left(\frac{V_r}{V}, \frac{V_\Delta H}{v}, H/B \right) \tag{4.3}$$

where L_o = total travel length of saline wedge

The functional form of equation (4.3) based on laboratory study is given by Ippen (1966) as

$$L_0/H = 6.0 \left(\frac{V_\Delta H}{v} \right)^{0.25} \left(\frac{2V_r}{V_\Delta} \right)^{-2.5} \tag{4.4}$$

in the avoirdupois system of units.

   Appreciable tidal currents cause vertical mixing of salt and fresh water. It is only when tidal forces predominate over the fresh water inflow that the salt and fresh water are fairly well mixed, otherwise they may only be partly mixed. In the well mixed estuary, the ratio of fresh water discharge to tidal prism is normally of the order of 0.1 or less. Due to this phenomena, the sharp boundary of the saline wedge will disappear and the effect of the penetration will become evident due to the gradient of salinity (Fig. 4.9) which will move with the tides. In Fig. 4.9 S and S_o represent local and ocean salinity respectively and (t/T) the time ratio of tide; T tidal period and t the tidal time.

An approximate estimate of the salinity level at any point X, can be determined from the following relationship proposed by Van Der Burgh, (Varshney, 1974)

$$\frac{\Delta \bar{c}}{\Delta x} = \frac{2\bar{V}\bar{c}e^{-0.00014\bar{c}}}{d^3} \tag{4.5}$$

where, $\frac{\Delta \bar{c}}{\Delta x}$ = salinity gradient; $\bar{c}$ = salinity expressed in parts per million (p/m) chlorine ion as an average over the depth of the canal or estuary at low water slack; $\bar{V}$ = average velocity of the ebb flow in m/s (not the upland flow velocity) and d = maximum depth in metres in a cross-section of the river with respect to mean tide level.

4.7 Salt water intrusion in coastal aquifers

It is necessary to have a fresh water source for fish culture. Fresh water is important for mixing with sea water to maintain the salinity level, especially in the dry season when evaporation is rapid. Fresh water is also necessary for

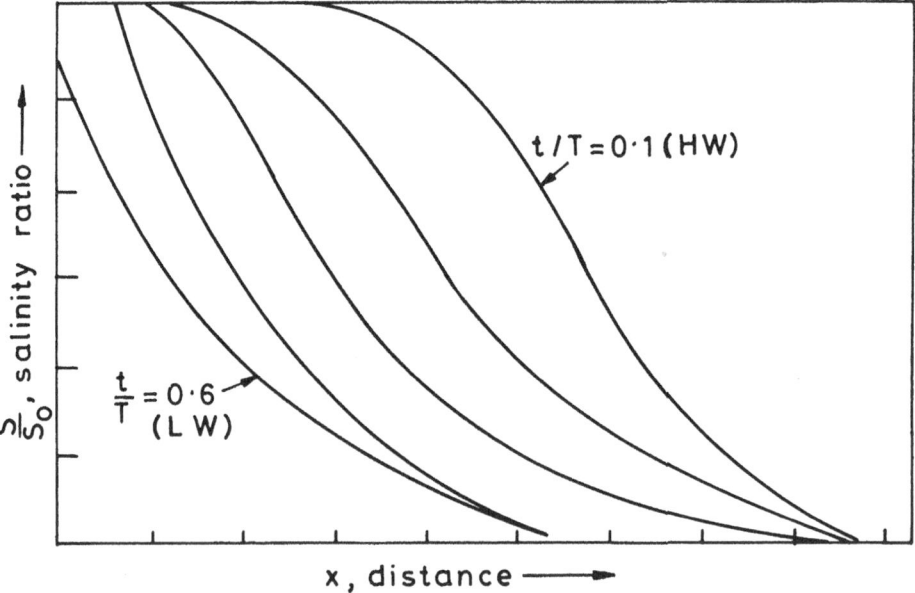

Fig. 4.9. Variation of salinity with tide and distance

the daily use of workers on the fish farm. The main sources of the fresh water in coastal and deltaic areas are (i) underground coastal aquifers, (ii) rain water and fresh drainage water collected in tanks, and (iii) fresh water reservoirs formed by closure of tidal inlets. Sources under (i) and (ii) are usually explored for meeting the requirements of fish farms. The third alternative is too costly and requires a lot of engineering investigation for a big irrigation and industrial development project in a coastal area. Consequently, we will concentrate on the first two sources. Salt intrusion implies the increase in salinity of ground water at a given location and depth. Intrusion may result from a variety of causes, primarily from connate water and sea water. Connate waters are those remaining in sedimentary rocks from the time of deposition. Increased salinity can result from the solution of rock minerals or inversions by changing sea levels. These create severe ground water problems in coastal areas (Ippen, 1966). In such situations excessive pumping may cause lateral or upward movement of saline water into the wells. Figures 4.10 and 4.11 illustrate the ground water conditions near a tidal stream carrying salt water and the arrows indicate the direction of ground water flow before and during pumping of ground water. Figure 4.11 shows that the water table is depressed around each pumping well to form a depression cone and salty water encroaches into the aquifer.

4.7.1 LOCATION OF INTERFACE AND CONFINED AQUIFER

Consider an unconfined coastal aquifer containing fresh and salt water with a common interface as shown in Fig. 4.12. The fresh and salt water densities are ρ_f and ρ_s respectively. The shape and movement of the interface

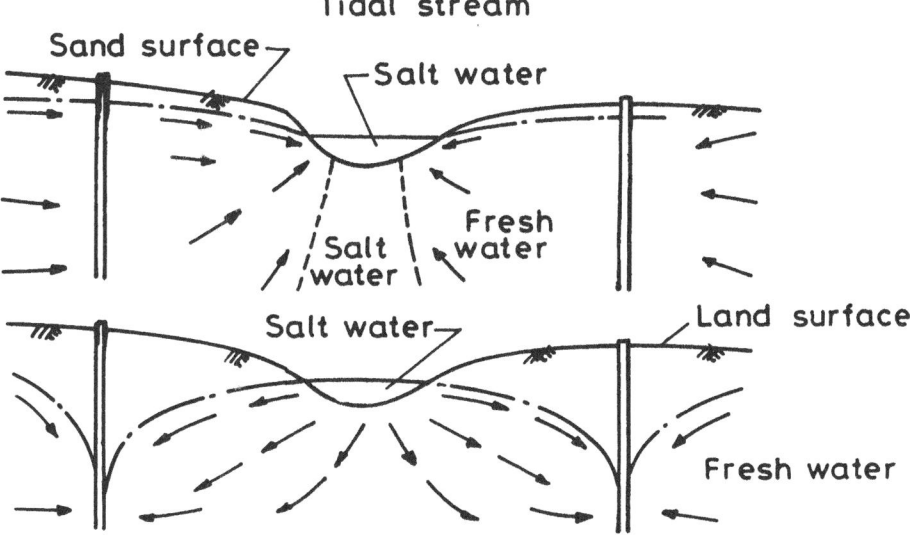

Fig. 4.10. Ground water conditions near a tidal stream under natural conditions and during pumping

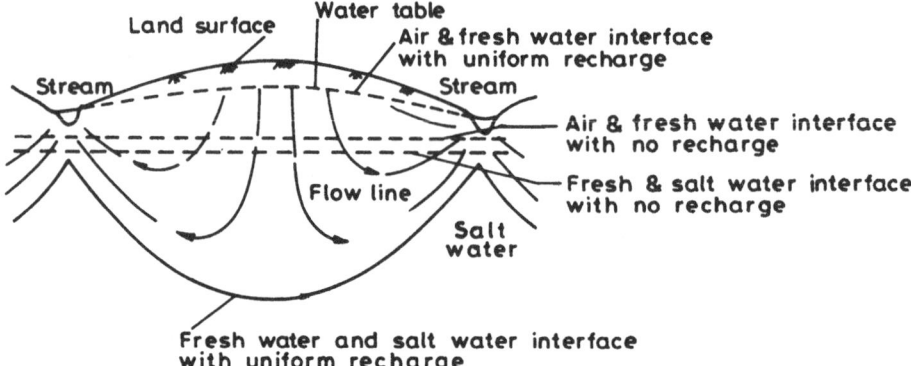

Fig. 4.11. Interface relations between fresh and salt water in a uniformly permeable aquifer

depends on the hydrodynamic balance between the two fluids. Equating hydrostatic pressure on each side of the interface at point X, gives

$$\rho_s g Z = \rho_f g h_f + \rho_f g Z$$

or

$$Z = \frac{\rho_f}{\rho_s - \rho_f} h_f \qquad (4.6)$$

This equation is known as the Ghyben-Herzberg (Chow, 1964) relation. Considering $\rho_s = 1.025$ g/cm^3 and $\rho_f = 1.00$ g/cm^3

$$Z = 40 \ h_f \qquad (4.7)$$

Although this relationship implies no flow, invariably ground water is moving in coastal areas without which a horizontal interface would develop with fresh

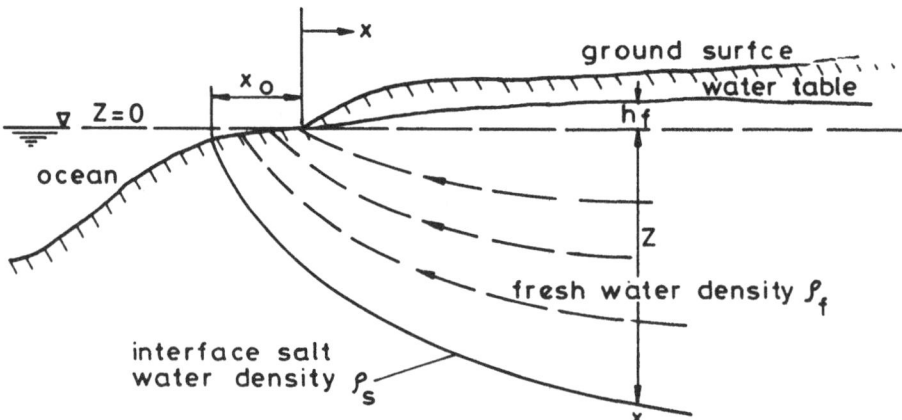

Fig. 4.12. Fresh and salt water-interface in an unconfined aquifer

water moving above salt water. Near the coast where the flow lines are sloping upwards, this relationship gives too small a depth to salt water. Further inland, the error is negligible. In many intrusion areas, the computed depth to salt water differs markedly from observations. This is due to the assumptions that the head in the salt water is at mean sea level and that the salt and fresh water are static. Considering this and taking h_f and h_s as heads in regions occupied by fluids of density ρ_f and ρ_s, then $h_f = Z + p/\rho_f$ and $h_s = Z + p/\rho_s$ where $Z =$ elevation and $p =$ pressure. At a point on the interface

$$Z = \frac{\rho_s}{\rho_s - \rho_f} h_s - \frac{\rho_f}{\rho_s - \rho_f} h_f \qquad (4.8)$$

which defines the depth to salt water by the heads and densities across the interface as shown in Fig. 4.13.

The fresh and salt water interface can be approximated by a flow net analysis. It can also be computed from the following relationship (Chow, 1964)

$$Z^2 = \frac{2\rho_f q}{K(\rho_s - \rho_f)} \left[x + \frac{q\rho_f}{2K(\rho_s - \rho_f)} \right] \qquad (4.9)$$

where $q =$ fresh water flow per unit width
$K =$ permeability
$x =$ distance from the shoreline

The length of submarine outflow section x_o is obtained by setting $Z = 0$ in equation (4.9) so that

$$x_0 = \frac{-\rho_f q}{2K(\rho_s - \rho_f)}, \text{ and the head along the line } Z = 0.$$

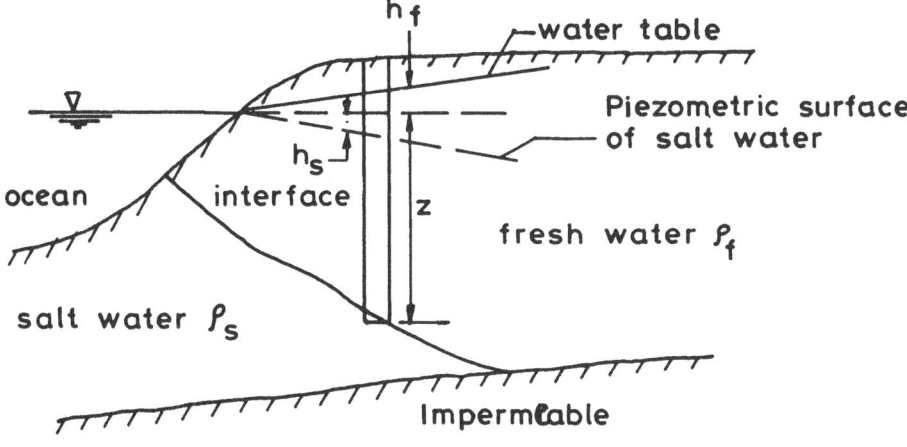

Fig. 4.13. Fresh and salt water-interface in an confined aquifer

is given by

$$h_f = \left[\frac{2qx(\rho_s - \rho_f)}{\rho_f K} \right]^{0.5}$$ (4.10)

where
h_f = height of water table at a distance x from the shoreline

Field measurements of interface indicated that the interface is not a line of zero thickness but is a zone resulting from various external influences such as natural recharge and discharge of the fresh water pumping and tidal action. These influences cause the interface to shift continuously to new equilibrium positions.

4.8 Development of fresh water and ground water flow

As has already been mentioned, the fresh water can be obtained from building tanks on horizontal or slightly sloping ground by encircling it with an embankment. The surface runoff from the adjacent areas during rainfall can be collected in the tanks either by gravity flow or by pumping. One such classical example is the tanks built in the Mekong delta of the Khmer civilisation. Such tanks are also built in the deltaic areas of the Netherlands for water supply. Construction of tanks depends on availability of suitable surface runoff and land in the vicinity of fish farms. As such, pumping from the ground aquifer, remains to be the most commonly used source of fresh water.

4.8.1 GROUND WATER FLOW

Most ground water is in motion as a result of natural and man-made changes. An understanding of hydraulic principles governing flows through aquifers is essential to quantitative estimation of ground water supply. In 1856, Darcy, (Chow, 1964) reported on experiments on the flow of water through sands. He observed that flows were proportional to the head difference and inversely proportional to the thickness of sand traversed. Consider the situation as depicted in Fig. 4.14 with a flow rate Q through a cylin-

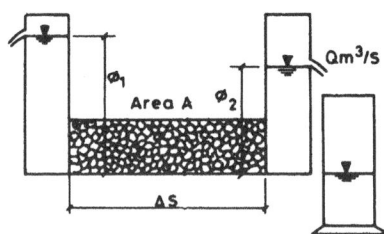

Fig. 4.14. Set up for verification of Darcy's law

der of cross-sectional area A and thickness or length ΔS. if ϕ_1 and ϕ_2 are the piezometric head at entry and exit, then

$$Q \propto A\frac{(\phi_1 - \phi_2)}{\Delta S} \quad \text{or} \quad Q = KA\frac{(\phi_1 - \phi_2)}{\Delta S} \tag{4.11}$$

where K = coefficient of permeability

$$\text{or } Q/A = q = -\frac{K(\phi_1 - \phi_2)}{\Delta S}$$

$$= -K\frac{\Delta\phi}{\Delta S} \tag{4.12}$$

$$\text{Let } \Delta S \to 0, \ q = -K\frac{\Delta\phi}{dS}$$

where q = specific discharge

This relationship is known as Darcy's Law.

The velocity of flow of water can be calculated from the relation

$$V_w = \frac{Q}{nA} = \frac{q}{n} \tag{4.13}$$

where

$$n = \text{porosity} = \frac{\text{volume of voids}}{\text{total volume}}$$

Darcy's law is valid for laminar flow which is the case for most natural ground water motion.

Usually the value of K for sand is 10^{-4} to 10^{-6} m/s and for clay K is 10^{-9} to 10^{-10} m/s

For one-dimensional flow $q = K\dfrac{d\phi}{ds}$

For three-dimensional isotropic flow the generalised Darcy's law is

$$q_x = -K\frac{\partial\phi}{\partial x}, \quad q_y = -K\frac{\partial\phi}{\partial y}, \quad q_z = -K\frac{\partial\phi}{\partial z}. \tag{4.14}$$

4.8.1.1 Ground water flow through dikes

Two general types of ground water flow problems are encountered in coastal end deltaic areas. They are seepage through dikes and location of salt and fresh water interface in a coastal aquifer which may either be confined or unconfined. The analytical treatment of these two types of problems is dealt with in this section.

Consider flow between two layers of small permeabilities as shown in Fig. 4.15

$$q_x = - K\frac{\partial \phi}{\partial x}, \quad q_y = - K\frac{\partial \phi}{\partial y}.$$

By Taylor's series,
$$q'_x = q_x + \frac{\partial q_x}{\partial x}\Delta x + \cdots$$

$$q'_y = q_y + \frac{\partial q_y}{\partial y}\Delta y + \cdots$$

The net flux due to horizontal flows is

$$\rho\frac{\partial q_x}{\partial x}\Delta x\, \Delta yH + \rho\frac{\partial q_y}{\partial y}\Delta x\, \Delta yH$$

Percolation through layer 1 is $\quad = \quad K_1\left(\frac{\phi_1 - \phi}{d_1}\right)\Delta x\, \Delta y$

and percolation through layer 2 is $\quad = \quad K_2\left(\frac{\phi_2 - \phi}{d_2}\right)\Delta x\, \Delta y$

where, ϕ_1, ϕ_2, K_1, K_2, d_1, d_2, are the piezometric heads, permeabilities, and thicknesses of layers, respectively.

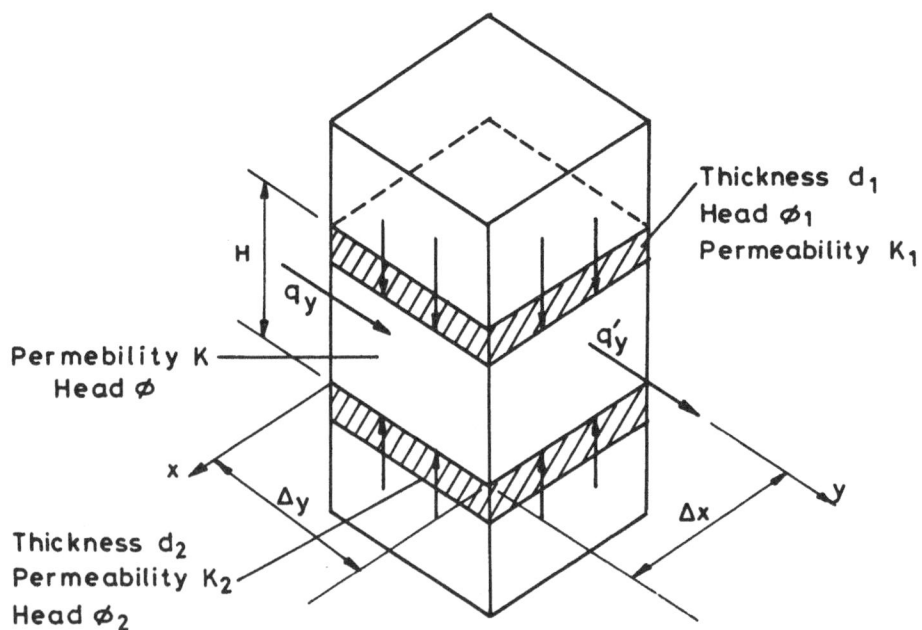

Fig. 4.15. Seepage flow between two layers

It can be shown that

$$\rho \frac{\partial q_x}{\partial x} \Delta x\,\Delta y H + \rho \frac{\partial q_y}{\partial y} \Delta x\,\Delta y H = \rho \left\{ \frac{(\phi_1 - \phi)}{C_1} + \frac{(\phi_2 - \phi)}{C_2} \right\} \qquad (4.15)$$

where $C_1 = d_1/K_1$ and $C_2 = d_2/K_2$

or,

$$- KH \left\{ \frac{\partial^2 \phi}{\partial x^2} + \frac{\partial^2 \phi}{\partial y^2} \right\} + \frac{\phi - \phi_1}{C_1} + \frac{\phi - \phi_2}{C_2} = 0. \qquad (4.16)$$

If there is no flow in the y-direction, then the equation becomes

$$- KH \frac{\partial^2 \phi}{\partial x^2} + \frac{\phi - \phi_1}{C_1} = 0 \qquad (4.17)$$

Figure 4.16 will be used to illustrate how this differential equation will now be solved to calculate the seepage flow under the dike. Assume there is no flow in the y-direction. The differential equation to be solved is

$$- KH \frac{d^2 \phi}{dx^2} + \left(\frac{\phi - \phi_1}{C_1} \right) = 0 \qquad (4.18)$$

or

$$\frac{d^2 \phi}{dx^2} = \frac{1}{KHC_1} (\phi - \phi_1), \qquad (4.19)$$

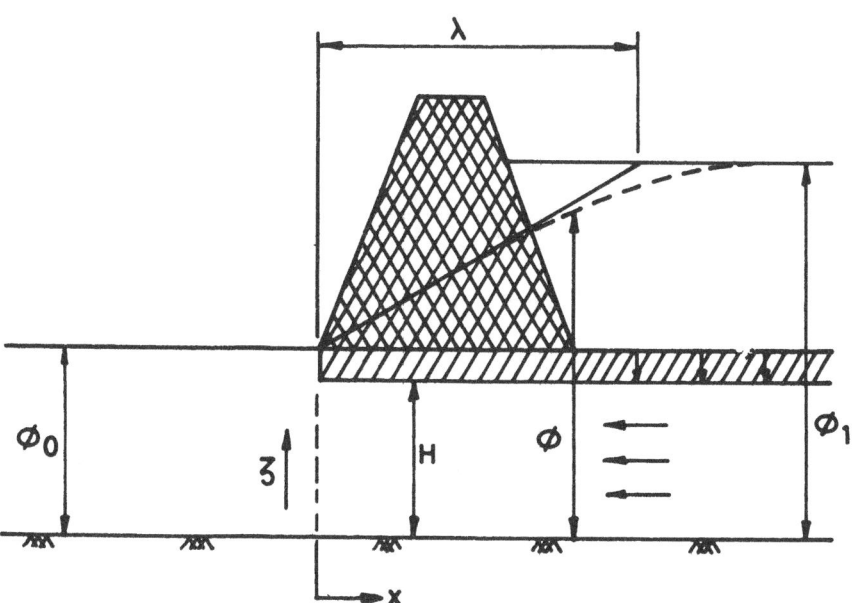

Fig. 4.16. Seepage flow under a dike

Let $\phi - \phi_1 = S$ then

$$\frac{d^2S}{dx^2} = -\frac{d^2\phi}{dx^2} = -\frac{1}{KHC_1}(-S), \quad \frac{d^2S}{dx^2} = \frac{1}{\lambda^2}S \qquad (4.20)$$

where

$$\lambda = \sqrt{KHC_1}$$

Try

$$S = e^{\mu x} \frac{dS}{dx} = \mu^\mu e^{\mu x}, \quad \frac{d^2S}{dx^2} = \mu^2 e^{\mu x}$$

or

$$\mu^2 e^{\mu x} = \frac{1}{\lambda^2} e^{\mu x}.$$

If

$$\mu^2 = \frac{1}{\lambda^2}; \quad \text{or} \quad \mu = \frac{1}{\lambda}$$

Hence $S = C_1 e^{x/\lambda} + C_2 e^{x/\lambda}$, that is, $S = C_1 e^{x/\lambda}$ or $S = C_2 e^{-x/\lambda}$ (4.21)

The boundary conditions are $x = 0$, $\phi = \phi_0$, $S = \phi_1 - \phi_0$

$$x = \infty, \phi_x = \phi_1, S = 0$$

When $x = \infty, e^{-x/\lambda} = 0, C_1 = 0, S = C_2 e^{-x/\lambda}$

When $x = 0, S = C_2 = \phi_1 - \phi_0, S = (\phi_1 - \phi_0)e^{-x/\lambda}$

or $\phi = \phi_1 - S = \phi_1 - (\phi_1 - \phi_0)e^{-x/\lambda}$

$$\therefore \frac{d\phi}{dx} = \frac{(\phi_1 - \phi_0)}{\lambda} e^{-x/\lambda} \qquad (4.22)$$

When $x = 0, \dfrac{d\phi}{dx} = \dfrac{\phi_1 - \phi_0}{\lambda}$

So $q_x = -K\dfrac{d\phi}{dx} = -\dfrac{K}{\lambda}(\phi_1 - \phi_0)e^{-x/\lambda}$

When $x = 0, q_0 = -\dfrac{K}{\lambda}(\phi_1 - \phi_0)$

So $q_x = q_0 e^{-x/\lambda}$ \qquad (4.23)

4.8.1.2 Ground water flow in confined aquifer

For a confined aquifer having horizontal fresh water flow in equilibrium with salt water the position of the interface can be derived from Darcy's Law In. Fig. 4.17.

$$\phi = \Delta(A + H) \qquad (4.24)$$

$$\frac{d\phi}{dx} = \Delta\frac{dH}{dx} \qquad (4.25)$$

$$q_0 = KH\frac{d\phi}{dx}$$

From Darcy's Law, $$q_0 = KH\Delta\frac{d\phi}{dx}$$

$$HdH = \frac{q_0}{\Delta K}dx,$$

or

$$H^2 = \frac{2q_0}{\Delta K}x + c$$

$$C = 0, \quad \text{at} \quad x = 0, \quad H = 0 \qquad (4.26)$$

$$\therefore L = \frac{H_0^2 \Delta K}{2q_0}$$

Because $$\phi = \Delta(A + H) = \Delta A + \Delta H = \Delta A + \sqrt{\frac{2q_0}{K}}\Delta x \qquad (4.27)$$

When $$x = L, \phi = \Delta A + \Delta H_0$$

Gradient of fresh water aquifer is $i = \dfrac{q_0}{kH_0}$ $\qquad (4.28)$

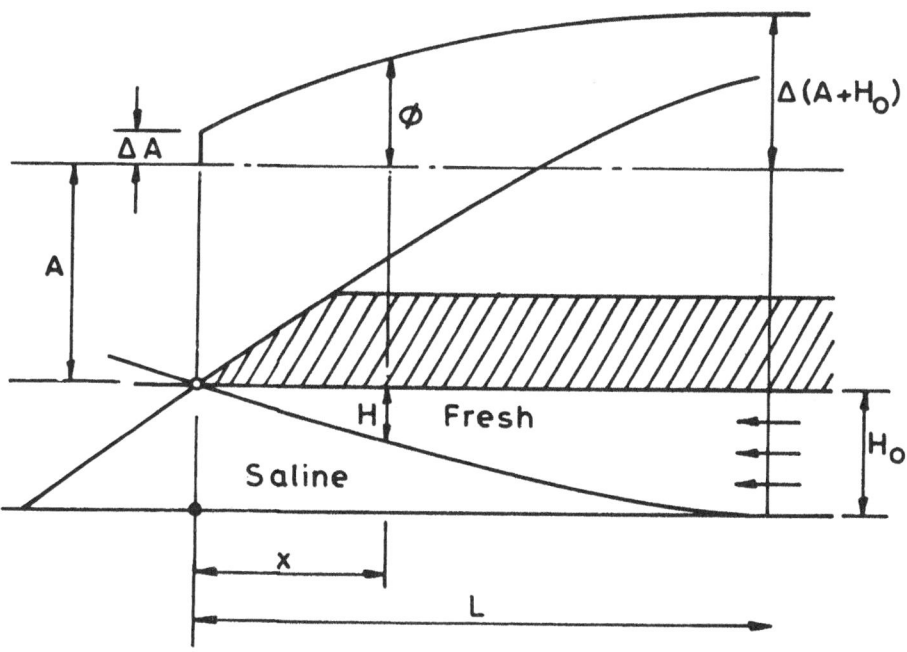

Fig. 4.17. Ground water flow in a confined aquifer

Average gradient of salt fresh water interface is

$$\frac{\Delta H_0}{L} = \frac{\Delta H_0}{\dfrac{H_0^2 \Delta K}{2q_0}} = \frac{2q_0}{kH_0} \tag{4.29}$$

or

$$q_0 = \frac{1}{2} \cdot \frac{H_0^2 K}{L} \left(\frac{\rho_s - \rho_f}{\rho_f} \right) \tag{4.30}$$

4.8.1.3 Ground water extraction

Care must be taken to prevent or minimise the chances of saline water contamination during pumping from the ground aquifer. The rate of pumping should not interfere with the stability of salt water interface. Well fields should be dispersed to reduce excessive lowering of ground water levels at any location particularly those near the coastline. In case the intrusion has the form of a wedge lying at the bottom of the aquifer, a number of shallow wells rather than a few deep ones should be used to avoid coning up of salt water upwards into the bottom of wells.

In coastal dune areas, natural and artificial recharge can be integrated with ground water extractions to stabilise salt water. In the case of permeable oceanic islands, ground water survives solely on rainfall and floats like a lens on the underlying saline water (Fig. 4.11). In such a case, the interface for an idealised circular island can be defined approximately as

$$Z^2 = \frac{\rho_f^2 W}{2\rho_s(\rho_s - \rho_f)K}(R^2 - r^2) \tag{4.31}$$

where R = island radius, W = recharge from the rainfall; and Z = depth of the interface below sea level at radius r.

To avoid upward movement of saline water minimum drawdown should be maintained during ground water extractions. In such cases infiltration galleries and horizontal collecting tunnels have proved useful, to keep drawdown to a minimum.

4.9 Control measures for salt water intrusion

Improving the salinity conditions in coastal regions requires a series of measures. The measures, however, should not interfere with drainage and navigation. To check the propagation of the salt water wedge beyond a certain point in a river, a given amount of fresh water discharge will be necessary. The formula which can be used to estimate approximate fresh water requirement is given as follows

$$Q = 15 \times (H i \cdot \frac{\Delta \rho}{\rho})^{0.5}$$

$$\tag{4.32}$$

where
A = cross-sectional area of the river (m²)
H = depth in metres
i = slope of the interface between the fresh and salt water
$\triangle \rho$ = density difference (kg/m³)

The thickness of the wedge should be within the limits of 0.12 to 0.45 H, the average values of $\triangle \rho / \rho$ and i being 0.01 and 0.0002, respectively. By application of the formula, one can find the required discharge necessary to prevent the wedge advancing up to a certain distance from the river mouth.

The other approach is based on what is called the non-tidal draft in an estuary. The non-tidal draft is defind as

$$\frac{\text{River flow}}{F \times \text{cross-sectional area}}$$

where F = average proportion of river water in a sample.

$$F = \frac{S - S_o}{S}$$

where S = salinity of open sea,
S_o = salinity of estuary water sample.

This formula gives the net distance per day that water must move seaward in order to carry the water from the river out to the sea. A knowledge of F permits the calculation of the volume of fresh water in a given segment of the estuary. The flushing time is the time required for the river flow to supply this volume. For the whole estuary the total fresh water requirement is the summation of that in each individual segment.

There are also preventive methods which can be used, such as, through special designs of lock and sluice gates, direct pumping of the salt water in the sump to the sea, salt water drains, etc.

Once the intrusion has occurred in a coastal aquifer it is not easy to control and may take years to remove the contamination. The suggested methods to overcome intrusion problems are:
  i) Reduction of pumping or modification of pumping rate thereby eliminating the overdraw which causes intrusion.
 ii) Artificial recharge. This method may be uneconomical in terms of capital and annual costs.
iii) Pumping by a line of wells parallel to the coast.
 iv) Recharging of fresh water into a line of wells parallel to the coast.
  v) Construction of an impermeable barrier to prevent inflow of sea water. This has many technical difficulties and economical limitations.

REFERENCES

Chow, V.T. (Editor-in-Chief). (1964): *Handbook of Applied Hydrology,* McGraw-Hill Book Company.

Denila, L. (1980): Site selection and pond construction, Mimography .paper SEAFDEC, Philippines.

Ghosh, S.N. (1986): *Flood Control and Drainage Engineering*, Oxford IBH Publishing Co. Pvt. Ltd.

Ippen, A.T. (1966): *Estuary and Coastline Hydrodynamics*, McGraw Hill Book Co. Ltd.

Menasveta, P. (1982): Environmental considerations for the development of coastal fish farms in the Indo-Pacific region. *Seminar Coastal Fishpond Engineering*, Indonesia.

Subramanya, K. (1984): *Engineering Hydrology*, Tata-McGraw Hill Publishing Co. Limited.

Varshney, R.S. (1974): *Engineering Hydrology*, Nemchand and Brothers, Roorkee.

USDA (1955): *WATER. The Year Book of Agriculture.* Oxford and IBH Publishing Co. Pvt. Ltd.

CHAPTER 5

Tides and Tidal Computations

5.1 Introduction

An aquacultural engineer has to plan and design fish farms in coastal, estuarial, and deltaic areas which are subjected to tidal actions. He may also design facilities such as jetties and fishing harbours for the movement of cargo, people and fish. For proper planning, design, and execution, a complete study of the natural phenomena such as tides and its effect at or near a fish farm as well as the effect of construction of other engineering structures on tides has to be made. The present chapter is devoted to the understanding of tidal phenomena, sources of tidal information, method of tidal computations, and various characteristics of tidal propagation in estuaries.

5.2 Tidal phenomenon

5.2.1 DEFINITION

The word 'tide' is derived from the Anglo-Saxon 'tyd' which actually means 'seasons' in a general sense (Russel and MacMillan, 1952).

However, *tide* today means the periodical rise and fall of the water surface of large bodies of water twice in the period of a day as a result of gravitational attraction of celestial bodies in particular the moon and the sun. When the body of water swells, the water level rises and moves some distance into the land, and when there is a contraction it recedes from the shore. These are respectively called the flood tide and ebb tide.

5.2.2 HISTORY OF TIDES

From time immemorial tidal phenomena have been of vital importance to those who depend on fish culture for livelihood. A clear idea of what has been thought about tides in the past centuries can be had from scientists' observations as ennumerated as follows.

Early in 450 B.C. Herodotus recorded the tidal phenomena in the Red sea, which served as the eastern terminus to the ships of Tarshish.

Hililco, the Carthaginian admiral who visited British shores about 500 B.C. noted the strong ebb and flow off the British coast. Aristotle, in 350 B.C. gave an account of the fact that the ebbs and floods followed a definite pattern in relation to the moon. Pytheas, in 325 B.C. noticed the alteration in tidal range between springs and neaps and also measured accurately the height of the tide. Plury, in 23 to 79 A.D. noted for a number of places the rough fixed relationships between the time of full moon and the occurrence of high water, and that the difference in the amplitude range between high and low water is least when the moon shows half its disc and the greatest when it shows its full disc at full moon or only a slender crescent at new moon.

The following observations made by different scientists also give a clear idea that tidal studies remained active during the 18th century (Deacon, 1971).

Newton and Bernoulli between 1700 and 1782 developed the idea of Maclaurin and Euler, and calculated the variations in tide-raising forces due to the varying annular distance of the sun and moon, their distance from the earth and declination from the place of the equator. They then expressed their results in tables which could be used for predicting the tides.

In the period 1749 to 1827, Laplace developed the concept of a dynamic theory of tides which he claimed to be complementary to the equilibrium concepts of Newton.

Later, in the year 1797 Murdoch Mackenzie surveyed the coast of Ireland and the west coast of Scotland for the Admiralty and published information about the tides, giving the interval between the moon's transit and the time of high water at new moon, the direction of tidal streams and the heights of spring and neap tides.

In the late 18th and early 19th centuries, the ideas which prevailed regarding tide movements were contained in the equilibrium theory, according to which the attractive forces of the sun and moon cause the ocean to approach the shape of a spheroid with the major axis approximately aligned so that the greatest elevations of water occur one, below the moon, and the other, on the opposite side of the earth. This was the position of equilibrium but because the earth rotates on its axis the two lumps of water appear to travel round the world as the crests of the tide wave, 180° apart, once in every 24 hours as shown in Fig. 5.1.

Between 1803 and 1865, Lubbock tried to relate records of the heights and times of high water over a long period to calculate the different variations to be detected in the tides, the semi-diurnal rise and fall, the twice monthly springs and neaps.

Thereafter, in 1880 to 1881, William Waldern and Captain Henry Denham adopted the practice of equating mean sea level with the half tide level, the mid-point of the combined range of rise and fall.

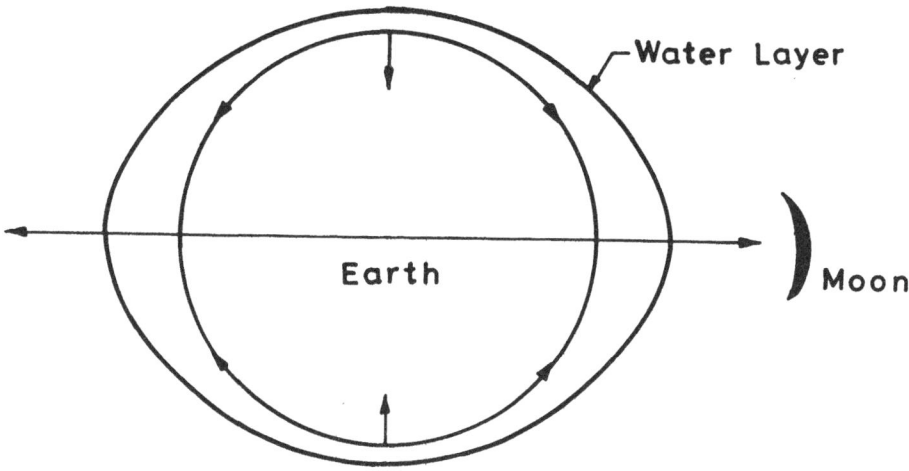

Fig. 5.1. Equilibrium theory of tides

5.2.3 TIDAL THEORY

Newton's Law of gravitation states that every body in the universe attracts every other body with a force acting along the straight line that connects the bodies and at a magnitude proportional to the product of their matter and inversely proportional to the square of the distance between them. The main assumptions made in the application of this theory are:

a) The earth is completely surrounded by an ocean of uniform depth.

b) The inertia and viscosity of the water and its attraction for parts of itself are negligible so that the ocean may be supposed capable of assuming instantaneously the figure of equilibrium.

5.2.4 TIDE PRODUCING FORCES

The moon is the principal tide generator. The earth and the moon attract each other. The resultant forces acting are shown in Fig. 5.2. These separate positions are maintained by the earth and the moon revolving monthly about their common centre of gravity. Let this be the only motion taking place with the daily rotation of the earth on its axis supposedly nullified since it exercises no influence on the tide producing force (Defant, 1961).

The consequence of the earth's revolution about the common centre of gravity is that all particles on the earth are subjected to a centrifugal force of uniform intensity and the direction of these forces are parallel to the line joining the centres of the earth and the moon at any instant. While the total centrifugal force thus generated is balanced by the moon's attraction the latter is not uniformly distributed over the earth. Since attraction is inversely proportional to the square of the distance, particles nearer the moon are subject at a greater force and those at the opposite end to a smaller force. On adding the two systems vectorally the resultant forces are obtained.

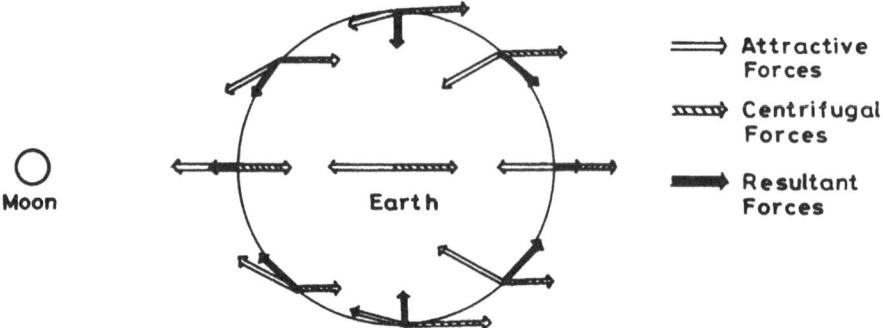

Fig. 5.2. Tide producing forces

These locally unbalanced forces are the tide producing forces of the moon (Defant, 1961).

Apart from moon the sun is the only other celestial body that produces tide-generating forces. However in view of the considerable distance of the sun from the earth the tide producing forces due to sun is much smaller compared to moon. To formulate mathematical relationship of the tide generating forces the factors to be considered are as follows:

(i) rotation of the earth (ii) the revolution of the moon around the earth in an orbit inclined to the equator of the earth (iii) the motion of the earth around the sun along the ecliptic which is also inclined to the equator.

The tidal forces generating the tides on the earth can be determined from the known movements of the sun and the moon with respect to the earth. As the movements are periodic in nature the tides may also be developed into periodic constituents. However it is not possible to deduce from these known forces the theoretical periodical constituents of the tide on the surface of the earth under actual terrestrial conditions. It is only possible to determine tides under idealised conditions as mentioned under equilibrium theory. In the case of equilibrium theory it is assumed that the free surface maintains its equilibrium form at all times. In reality the sun and the moon change their positions with respect to earth continuously, thereby causing a corresponding motion of water particles. Thus the equilibrium surface will change continuously with time.

From practical experience it has been found that real tide has constituents analogous to those predicted by the equilibrium tides and that the period of the constituents are the same as those of the equilibrium constituents. However the terrestrial component lags behind the corresponding equilibrium constituents. Additionally its amplitude is also different. The propagation of tides generated by external forces on waters penetrating into the shelf gets distorted due to shallow water of coastal areas and estuaries. The shallow water constituents resulting from the distortion can be distinguished in over tides and compound tides. The over tides have an angular speed which is an exact multiple of one of the astronomical constituents. In the case of a compound tide its angular speed equals the sum of the differences of the angular speeds of two or more astronomical constituents.

Types of tides

Due to factors mentioned above various types of tides are found along the coasts. They are predominantly of Semidiurnal types produced mainly by the M_2, S_2, N_2 and K_2 constituents, having nearly equal high and low waters in a lunar day, where

M_2 = the principal lunar semi-diurnal components

N_2 = lunar elliptic semi-diurnal constituents (it depends on the variation of the distance)

S_2 = principal solar semi-diurnal components

K_2 = lunar declination semi-diurnal constituents (depends mainly on the variation in the declination of the moon and not on the variation of its distance).

Diurnal tides are produced mainly by the K_1, O_1 and P_1 constituents. where

K_1 = luni-solar diurnal component

O_1 = principal lunar diurnal component

P_1 = solar declination diurnal constituents

They have one high water and one low water on each lunar day and they occur mainly in the oceans between the equator and tropics.

At a time during a lunar month, the moon crosses the equator having very small declination, when the tide is mainly semi-diurnal. With the declination of moon at maximum, the diurnal component may be sufficient to produce one daily high or low water. These tides are given the name as mixed tide.

In the year 1897 Vander Stok (Defant, 1961), adopted an F ratio, designated the 'Formzhal' of tides, using the four components M_2, S_2, K_1 and O_1. The tides can be clearly classified based on the range of this ratio F.

$$\text{Formzhal, } F = \frac{K_1 + O_1}{M_2 + S_2} \tag{5.1}$$

The classification based on this F range can be grouped as:

		F range
a)	Semi-diurnal type	0 −0.25
b)	mixed type	
	i) semi-diurnal dominance	0.25−1.5
	ii) diurnal dominance	1.5 −3.0
c)	diurnal type	3.0 −

a) Semi-diurnal type

It has two high waters and two low waters daily of approximately the same height. Thus, the time period is approximately 12.42 hours. The interval between the transit of the moon and high water at a locality is nearly constant. The mean range at spring tide is $2(M_2 + S_2)$.

b) *Mixed type*
  i) Semi-diurnal dominance
  There are daily two high and two low waters showing inequalities in height and time, which attain their maximum when the declination of the moon has passed its maximum. The main spring tide range is $2(M_2 + S_2)$.
  ii) Diurnal dominance
  Occasionally only one high water a day, following the maximum declination of the moon. At other times, there are two high waters in the day showing strong inequalities in height and time, especially when the moon has passed through the equator. The mean spring tide range is $2(K_1 + O_1)$.

c) *Diurnal type*
  In this type, high water occurs only once in a day. It varies with the declination, especially that of the moon. The time period of this tide approximates 24 hours and 50 minutes (24.83 hours). The mean spring tide range is $2(K_1 + O_1)$. At neap tide, when the moon has passed through the equatorial plane, there can also be two high waters. Figure 5.3 shows graphically the three different types of tides.

5.3 Definitions of terms used in tidal computations
  i) Flood tide is the incoming or rising tide.
  ii) Ebb tide is the outgoing or the falling tide.
  iii) Tidal period is the time interval between two successive high waters.
  iv) Tidal range refers to the difference between the higher high water (HHW) and the lower low water (LLW) as shown in Fig. 5.3 in a mixed tide. In diurnal and semi-diurnal tides it is the difference between the high water and the low water.
  v) Mean range is the average of the difference between all high waters and all low waters.
  vi) Diurnal inequality refers to the differences in two high waters or of the two low waters of each day.
  vii) Spring tides: in every 14.3 days or roughly 15 days there are several days when the high waters are much higher and low waters much lower than usual. These tides are called spring tides. The occurrence of spring tides is being associated with full moons or new moons, that is, when the sun, moon and the earth are aligned either in opposition or in conjunction as shown in Fig. 5.4.
  viii) Neap tides: in every 14.3 days or approximately 15 days, there are several days when the high waters are lower than usual and the low waters are

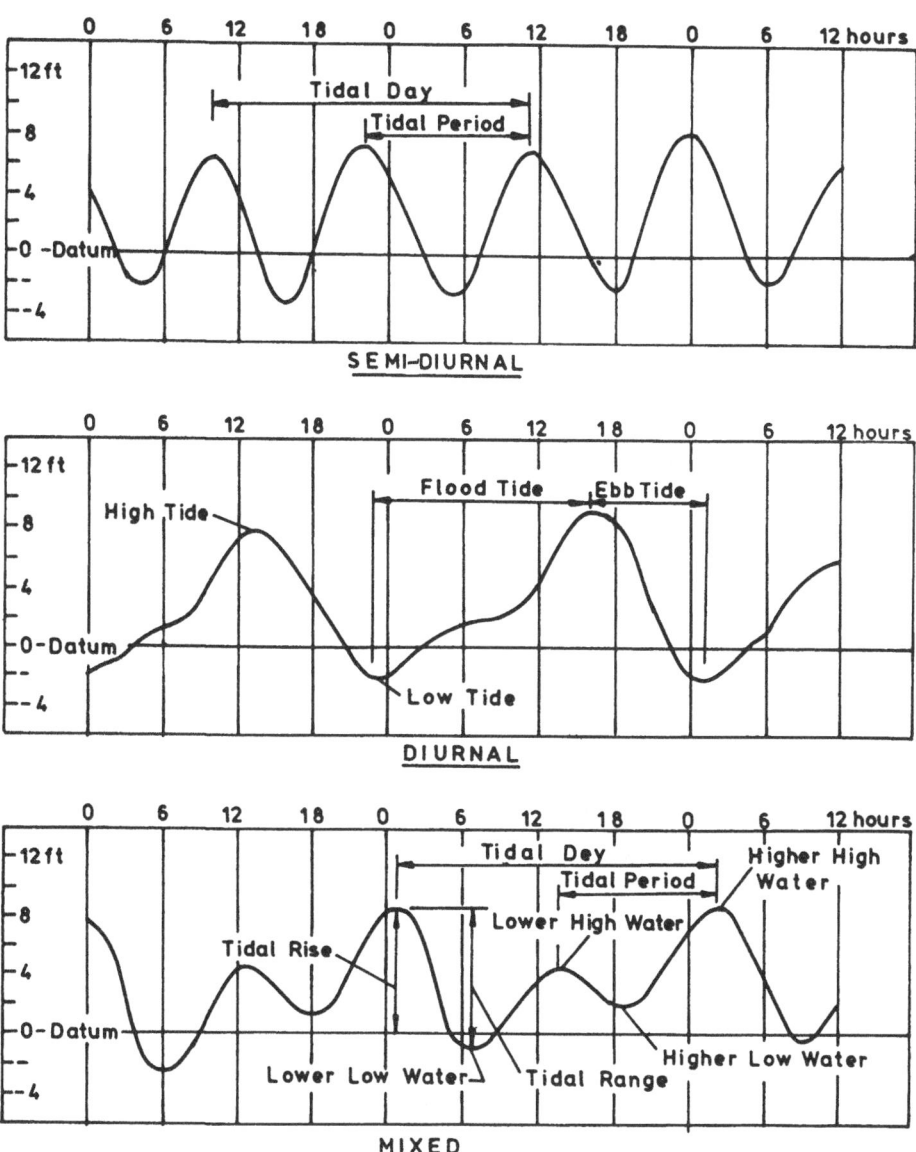

Fig: 5.3. Types of tides

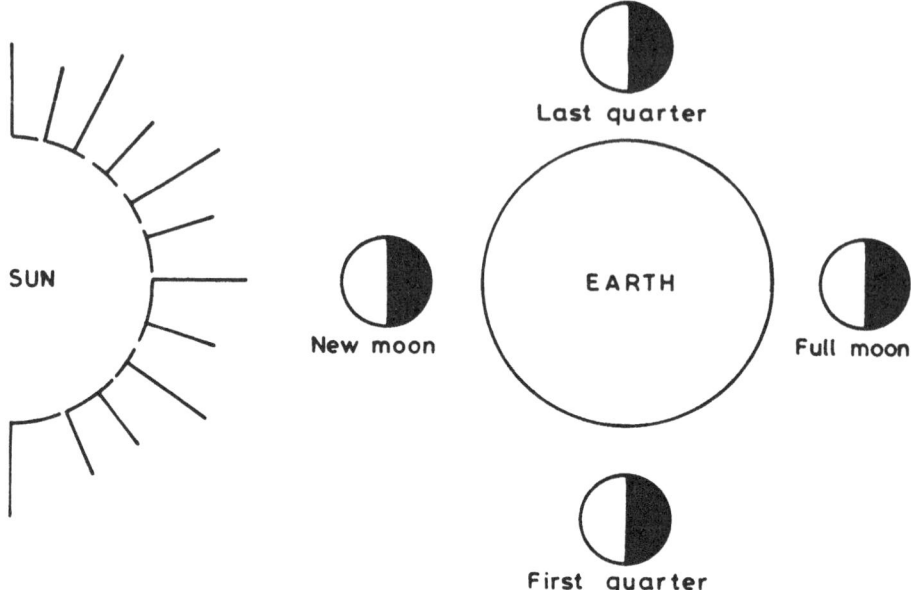

Fig. 5.4. Tide phenomenon—Solar and lunar system

higher than the usual. These tides are called neap tides. The spring and neap tides can be seen from Fig. 5.5.

ix) Equinoctial, spring tides: are the extra high spring tides that occur twice a year at the time of vernal and autumnal equinoxes. When the sun and the moon are actually or nearly vertical over the equator. In this case the influences off the sun and the moon are most closely coincident and direct.

x) Tidal datum: some reference datum is selected or chosen, with reference to tidal elevation. This datum is usually the long term average of some property of tides such as the MLLW (mean lower low water). MLLW is the average of the lower low water over a 19 year interval. MLLW is usually the zero datum being used as reference in fish farms.

xi) Tidal currents: the alternating horizontal movement of water associated with the rise and fall of the tide caused by the astronomical tide producing forces. *Ebb current* is defined as the current away from the shore or down a tidal stream. It is usually associated with the decrease in height of the tide. *Flood current* is defined as the current towards the shore or up a tidal stream. It is associated with the increase in height of the tide. In the open sea, the orbital path of the oscillatory tidal wave is undisturbed hence flood and ebb movement persists for sometime after high and low tide. At the same time it has been shown by measurements

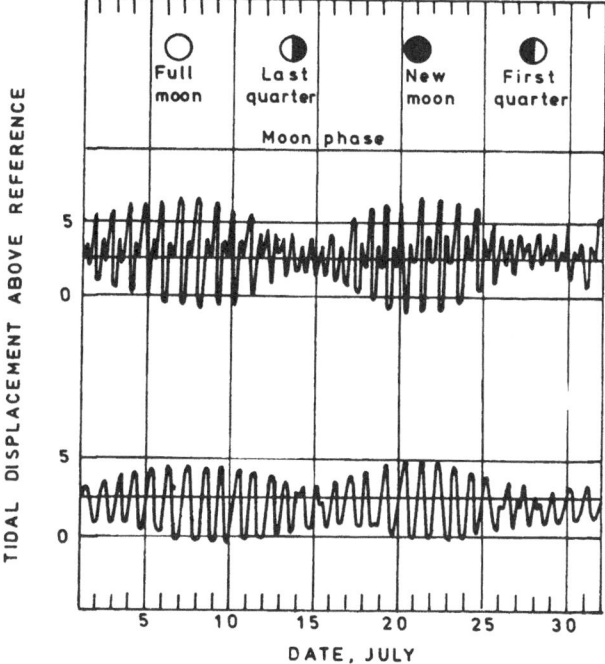

Fig. 5.5. Predicted tides at two different latitudes, correlation of Spring and Neap tides with moon Phase

on ships that tidal currents in the open sea normally rotate. The velocities of the currents are low and the velocity of the maximum flow current is approximately equal to the velocity of maximum ebb current.

xii) Tidal prism: is the total amount of water that flows into a harbour or estuary or out again with the movement of the tide excluding any fresh water flow.

xiii) Tide station: is a place at which tide observations are taken. It is called a primary tide station when continuous observations are taken over a number of years to obtain basic tidal data for the locality. A secondary tide station is one which is operated over a short period of time to obtain data for a specified purpose.

xiv) Tidal flats: are marshy or muddy land areas which are covered and uncovered by the rise and fall of tides.

xv) Tidal inlet: is a natural inlet maintained by tidal flow. Or any inlet in which the tide ebbs and flows.

xvi) Tide gauges: the vertical movements of tides are measured by means of a tide gauge. When a prolonged record of tides at a given locality is required, the gauges at that station become permanent gauges.

Gauges may be classified as non-self registering and self registering. The former requires the attendance of an observer to read the elevations of water surface. The latter is automatic and produces automatic records. The non-registering gauges are

a) The staff gauge
b) The float or box gauge
c) The weight gauge

These gauges record the water level in a stilling basin by (1) using a staff fixed in the chamber, (2) an indicator connected to a floating body in the chamber, and (3) the lowering of a weight whose contact with water is electrically indicated.

Mean and spring tidal ranges for some of the major ports of the world are listed in Table 5.1.

Table 5.1

Ports	Mean range	Spring range
	m	*m*
(1)	(2)	(3)
Anchorage, Alaska	8.15	9
Antwerp, Belgium	4.8	5.4
Auckland, New Zealand	2.44	2.81
Baltimore, U.S.A	0.36	0.39
Bilboa, Spain	2.74	3.6
Bombay, India	2.66	3.6
Boston, U.S.A	2.9	3.35
Buenos Aires, Argentina	0.67	0.73
Burnteoat Head, Nova Scotia (Bay of Fundy)	12.7	14.5
Canal Zone, Atlantic Side	0.214	0.36
Canal Zone, Pacific Side	3.74	5.0
Capetown, Union of South Africa	1.16	1.58
Cherbourg, France	3.96	5.5
Dakar, Africa	1.0	1.34
Dover, England	4.43	5.7
Galveston, (Texas) U.S.A.	0.305	3.18
Genoa, Italy	0.183	0.244
Gibraltar, Spain	0.7	0.94
Hamburg; Germany	2.32	2.44
Havana, Cuba	0.305	3.64
Hong Kong	0.94	1.62
Honolulu, Hawaii	3.64	0.58
Juneau, Alaska	4.25	4.87
LaGuaira, Venezuela	—	3.05
Lisbon, Portugal	2.56	3.32
Liverpool, England	6.45	8.24
Manila, Philippines	—	1.0
Marseilles, France	0.122	0.183
Melbourne, Australia	0.517	0.58

Ports	Mean range m (2)	Spring range m (3)
Murmansk, U.S.S.R.	2.41	3.02
New York, U.S.A.	1.34	1.62
Osaka, Japan	0.76	1.0
Oslo, Norway	0.305	0.336
Quebec, Canada	4.20	4.72
Rangoon, Burma	4.10	5.20
Reykjavik, Iceland	2.80	3.80
Rio de Janeiro, Brazil	0.76	1.07
Rotterdam, Netherlands	1.52	1.65
San Diego U.S.A.	1.28	1.77
San Francisco U.S.A.	1.21	1.74
San Juan, Puerto Rico.	0.36	0.396
Seattle, U.S.A.	2.32	3.44
Shanghai, China	2.2	2.7
Singapore, Malaya	1.7	2.21
Southampton, England	3.05	4.16
Sydney, Australia	1.1	1.37
Valparaiso, Chile	0.915	1.2
Vladivostok, U.S.S.R.	0.183	0.214
Yokohama, Japan	1.07	1.43
Zanzibar, Africa	2.66	3.8

5.4 Sources of tidal information

For planning and selection of fish farm sites, tidal heights are obtained from tide and current tables published by each maritime country annually in the region. Values in the tide table are fairly accurate specially when the site is very close to the shore. However, if it is located several kilometers inland (for example, 10 km) actual tide gauging becomes essential.

Each country has established tide gauging stations located around important ports. In the Philippines for example, there are six stations, and in India there are several around the coast. Reference stations for other stations are listed under the section for Tidal Difference' as well as constants of the tide and current tables. The predicted daily time and height of high and low waters for the tidal stations can be read directly from the tide tables.

Tides for other places are predicted by applying tidal differences and ratios found in the tide and current tables. Unless very accurate data are required, corrections on tide tables may be neglected.

In addition to the preceding tidal information, data for a particular site may be obtained by tidal computations.

5.5 Tidal computations

Tidal computations serve two main purposes. The first is computation for prediction of tides and storm surges in coastal areas, estuaries and along rivers, over intervals ranging from several days to a year. This information is necessary for the design of fish farms. The second concerns computations relating to big engineering projects. The tidal computations for the first group refer mainly to the existing geographic and hydraulic conditions in maritime waters, etc. Those of the second group involve prediction of tides at a certain location during or after the construction of engineering works.

Two groups of projects require tidal data. They are respectively (1) projects which do not effect tidal motion (2) projects which influence the tidal motion in the river system. The first group relates to structures along the banks of tidal inlets such as drainage sluices, dykes, jetties etc. In these cases tidal computations must provide informations about water level variations and the currents along the structures.

The second group relates to projects concerning closure of river branches, construction of a barrage on the river or new navigational channel, etc. In these cases it is necessary to study the changes in the tides in the whole river system thoroughly.

Before execution of the project, computations are necessary for the determination of the changes in tides as well as the currents as for example increasing velocity means that the structure will be subjected to scour during construction. Furthermore the level of extreme high tide and low low-water levels are needed to find out the elevations of dike and the sill level of the drainage sluices.

However it can be said that lesser number of tidal computations are required for technical works than for the prediction of tides. It, therefore, follows that the tidal computations for the purpose of technical structures are different from those required for the purpose of tide prediction.

5.6 Methods of tidal computations

The prediction of the astronomical tides in coastal areas is made by harmonic analysis to an observed series of data. This will enable one to determine the tidal constants. On entering the shallow water the tides get distorted. Accordingly it will be necessary to consider shallow water constituents. The importance of the constituents can be determined from the physical properties of the river course such as the variation in stream width, depth and storage. The magnitudes of the constants can be obtained by harmonic analysis of the actual observations or by computation of the tidal propagation in that river. Computation of tidal propagation in the river system can be done by numerical calculation of the propagation of tidal cycle as a whole or by computing the harmonic constituents. There are no general rules as to which method is preferable as each method emphasises certain special features of tidal motion. In the harmonic method, propagation of a series of sinusoidal waves representing the almost periodical and continuous character of tidal motion is considered. In the characteristic method, tidal motion is considered to be the propagation of a succession of small disturbances from an initial state, and the periodic character of the tidal motion is placed in the background. The propagation of a small disturbance may be considered to be identical to the propagation of a discontinuity. Generally speaking, the harmonic method may be considered a macroscopic method, in which the tide as a whole is presented by a limited group of parameters such as amplitudes and phases. The method of characteristics, on the other hand, is a microscopic method in which the tide is built up step by step during the computation.

The method of characteristics does not appear to be as attractive as the finite-difference method on a computer especially in complicated situations usually encountered in practical applications. In the finite-difference method, the differential equations for propagation of long waves are replaced by difference equations and distinct points are selected, at intervals of time and distance. From curves representing the boundary and initial conditions, solutions can be obtained by proceeding in small time steps. In both finite-difference and characteristic methods the distinct points (x, t), where the water level or velocity must be computed, form a grid or net in the x, t plane. It is usually an irregular grid in the characteristic method while the finite-difference method uses a regular grid.

5.7 Characteristics of tidal propagation in an estuary

An estuary is the part of a river that is affected by tides. It is also the region near a river mouth where fresh water of the river mixes with salt water of the sea. Figure 5.6 shows the plan and cross-section of an idealised estuary with the composition of bed material and salinity variation along its length.

There is a shoreward movement of water particles in the estuary. The movement continues until the time of high water after which a seaward movement takes place. The change of tidal wave as it passes into shallow water estuary is that the period of flood tends to become shorter and that of the ebb longer. When a river flows directly into a tidal sea, the tide characteristics are modified compared with those of oceanic tides in three ways:

1) The speed of the tide as it travels upstream depends of the depth of the channel, that is, $C = \sqrt{gd}$
    where d = average depth of flow, and
    where g = acceleration due to gravity.
2) With increasing distance in the upstream direction the duration of ebb is increased while that of the flood is decreased.
3) The range of tides tends to decrease with increasing distance from the open sea.

Tidal current classification is made in accordance with estuary geomorphology. When there is no obstruction at the entrance to the estuary, the tidal wave length, L can be estimated as:
  i) $L \approx 48.1\sqrt{d}$ where L = wave length in miles, and d = mean depth in feet. The above parameter controls the scale and duration of the estuarian tidal currents.

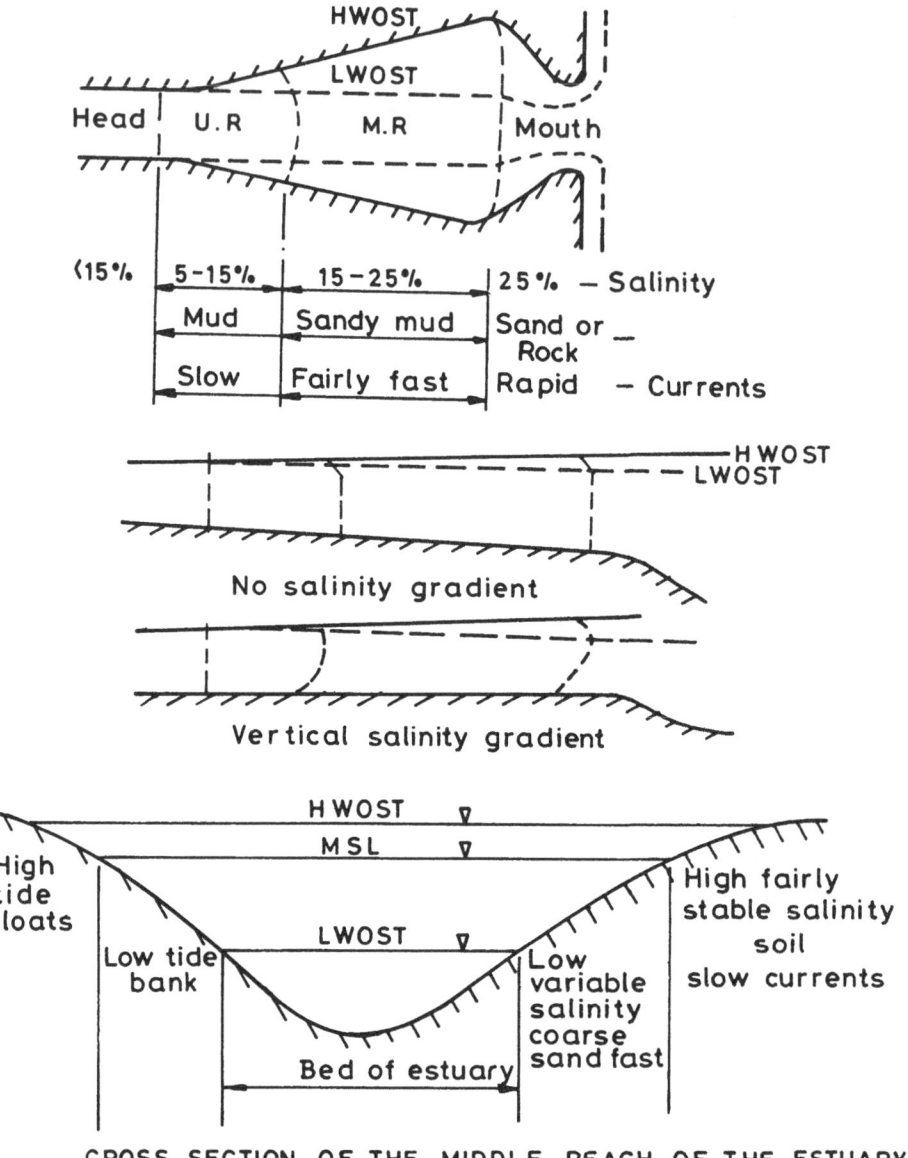

Fig. 5.6. Cross-section of an idealised estuary

Highest water of spring tide = HWOST
Lowest water of spring tide = LWOST
Mean sea level = MSL
Upper reach = UR, Middle reach = MR

ii) Where the length of the tidal reach exceeds 0.25 *L*, then the maximum currents and slack water may be expected at high tide and mid tide.

iii) When the tidal reach length does not exceed 0.25 *L* then flood current will accumulate water at its head and will themselves be diminished before high water. Here slack water occurs at high water and the maximum current occurs around mid tide.

iv) Estuaries with narrow mouths constitute a third group. In this group the tidal flow is restricted and the tidal range is reduced.

In addition to the above the differences in depths at different points affect the duration of flood and ebb. Further river flows helps to prolong the ebb and shorten the flood. In the case of estuaries with an unrestricted mouth, the runoff from the river is negligible in comparison with the total volume of the estuary. Hence the length of the ebb and flood periods increases and decreases with increasing distance in the upstream direction. The change in ebb or flood duration is accompanied by marked disparity between the maximum ebb tide and flood tide current velocity. The asymmetry in the tidal flow has a significant bearing on the accumulation of sediment. A decrease in tidal range in the upstream direction may result from a loss of energy due to friction within the channel. However, energy loss and reduction in tidal range may be modified by a decrease of channel size. In estuaries which are characterised by a large tidal range, shoals, and rapid decrease in width upsteam such as in the Hooghly estuary, a phenomenon known as tidal bore may develop. A tidal bore occurs when the tidal rise is so rapid that the water advances as a wall which may be several feet high. Generally a shoaling channel steepens the tidal curve and a bore results if it becomes vertical. The decrease in width increases the tidal range which is further increased by spring tides. It is at these times that bores are more generally seen. Though a bore is a very striking feature, the tide continues to rise to a higher level after the bore has passed.

The duration of an open flood tide is shorter than of an ebb. To accompany this difference, the tidal currents lose the rotational behaviour characteristics of the open sea and become rectilinear. With a significant shallowing of depth and a marked lengthening of the tidal excursion an increase in current velocity occurs. The tidal excursion *E* equals the distance a particle of water would travel from low to high water along the axis of the estuary. This may be calculated as

$$E = \frac{2}{\pi} \times \text{average maximum velocity} - 6.2 \text{ hours.}$$

The average maximum velocity is equal to 0.85 maximum spring tide velocity.

REFERENCES

Deacon, M. *Scientists and the Sea, 1660–1900, A Study of Marine Science*, Academic Press, London, 1971.

122   *Coastal Aquaculture Engineering*

Defant, A. *Physical Oceanography, Vol. II*, Pergamon Press, Oxford, 1961.
Dronkers, J.J. *Tidal Computations in Rivers and Coastal Waters*, North Holland Publishing Company. Amsterdam, 1964.
Dronkers, J.J. Tidal Computation for rivers, coastal areas and seas, *Proceedings, ASCE*, 95: HY 1, Jan. 1969.
Ippen, A.T. *Estuary and Coastline Hydrodynamics*, McGraw Hill Book Corpn. New York, 1966.
Russel, R.C.H., and MacMillan D.H. *Waves and Tides*, Hutchinson's Scientific and Technical Publications, London, 1952.

CHAPTER 6

Water Waves and
Their Computation

6.1 Introduction

The effects of water waves are of paramount importance in the field of coastal aquacultural engineering. Waves are primarily responsible for determining the geometry and composition of beaches. Consequently, they significantly influence the planning and design of harbours, coastal inlets and waterways, shore protection measures, coastal structures, and other coastal works.

It is necessary to understand the fundamental physical processes in surface wave generation and propagation before one can understand the complex water motion in near-shore areas of large bodies of water. In this chapter, an introduction to surface wave theories, surface and water particle motion, wave energy, wave transformation due to interaction with the bottom and with structures are made. This introduction will facilitate understanding of wave motion and limitations of selected theories.

6.2 Wave fundamentals

A simple wave is one which can be described in simple mathematical theories. Waves which are difficult to describe in form or motion because they comprise several components, are called complex waves.

Sinusoidal or Simple harmonic waves are examples of simple waves. A wave is periodic if its motion and surface profile recur in equal intervals of time. A wave which moves relative to a fluid is called a progressive wave. The direction in which a wave moves is the direction of wave propagation. If a wave merely moves up and down at a fixed position, it is called standing wave or a clapotis. A progressive wave will be of a permanent form if it propagates without experiencing any change in a free-surface configuration.

Water waves are considered oscillatory or nearly so if the water particle motion is described by orbits which are closed or nearly closed for each period. Most finite amplitude wave theories describe nearly oscillatory waves because a fluid particle in each wave moves a small distance in the direction of wave advancement. This motion is called mass transport of waves. When water particles advance with waves and do not return to their original positions, the wave is called a wave of translation. A solitary wave is a wave of translation.

It is necessary to distinguish between various types of water waves that may be generated and propagated. One way of classification is by wave period T. Wave period, T is the time required to travel a distance of one wave length. Frequency F is $1/T$. Wave classifications given by Kinsman (1965) are shown in Fig. 6.1. They show that the relative amount of energy contained in ocean waves are related to their frequencies. Gravity waves, with a period of 1 to 30 seconds are of particular interest to engineers. A narrower range of wave periods from 5 to 15 seconds is usually of more importance in coastal engineering. Waves in this range are referred to as gravity waves. Gravity is the principal restoring force for waves. Gravitational force attempts to bring the fluid back to its equilibrium position. Because a large portion of the total wave energy is associated with gravity waves, gravity waves are of the greatest importance in dealing with design of coastal and offshore structures. Gravity waves can be further divided into two states:

i) When the waves are under the influence of wind in a generating area, they are called seas.

ii) When the waves move out of the generating area and are no longer subjected to significant wind action, they are called swells.

6.3 Wave mechanics

Waves in the ocean are a complex and constantly changing sea of crests and troughs on the water surface because of the irregularity of wave shape and variability in the direction of propagation. When waves move out of the area where they are directly affected by the wind, they assume a more orderly

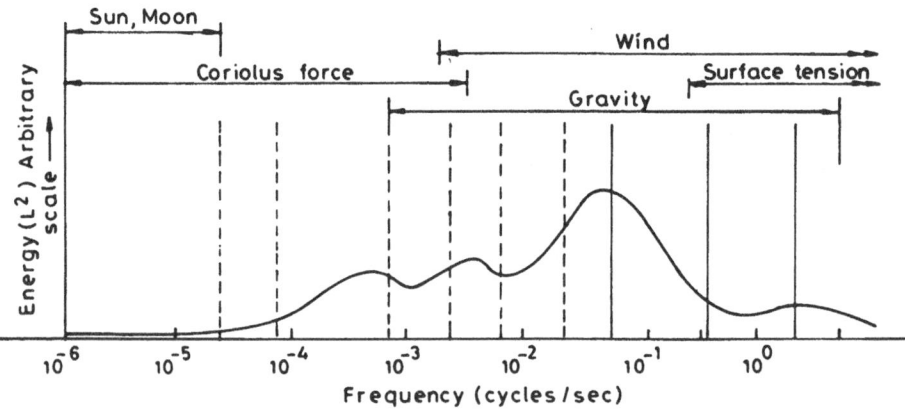

Fig. 6.1. Classification of Surface Waves by wave band

state with the appearance of definite crests and troughs with a more rhythmic rise and fall. These waves may travel thousands of miles after leaving the area in which they were generated. Wave energy is dissipated internally within the fluid by interaction with the air above, by turbulence on breaking, and at the bottom in shallow depths. Waves which reach coastal regions expend a large part of their energy in the seashore region. As a wave moves near shore, wave energy may be dissipated as heat through turbulent fluid motion induced by breaking, through bottom friction, and percolation. To coastal engineers, breaking is important, because it affects both beaches and man-made shore structures. Shore protection measures and coastal structure designs are dependent on the ability to predict wave forces and fluid motion beneath waves and the reliability of such predictions. Simple waves, where elementary functions can be used to describe wave motion, are generally used for predicting wave forces and currents. Two classical theories developed by Airy (1845) and Stokes (1880) describe simple waves. The predictions are better where water depth relative to wave length is not too small. For shallow water, the solitary wave theory often provides an acceptable approximation of simple waves.

For an elementary sinusoidal progressive wave

$$\eta = a \, \cos\left(\frac{2\pi x}{L} - \frac{2\pi t}{T}\right), \tag{6.1}$$

where η = elevation of water surface relative to still water level
L = horizontal distance between corresponding points on two successive waves,
H = vertical distance to its crest from the preceding trough
T = the time for two successive crests to pass a given point
a = amplitude of the wave
x = distance in the direction of wave propagation
With origin $x = 0$ the wave profile for $t = 3T/4, \, 7T/4, \, 11T/4$ is shown in Fig. 6.2.

In Figure 6.2 $\eta = a = H/2$ at wave crest and $\eta = -a = H/2$ at wave trough. Because wave celerity is defined as $C = L/T$ (L = distance travelled by a wave during one wave period = wave length)

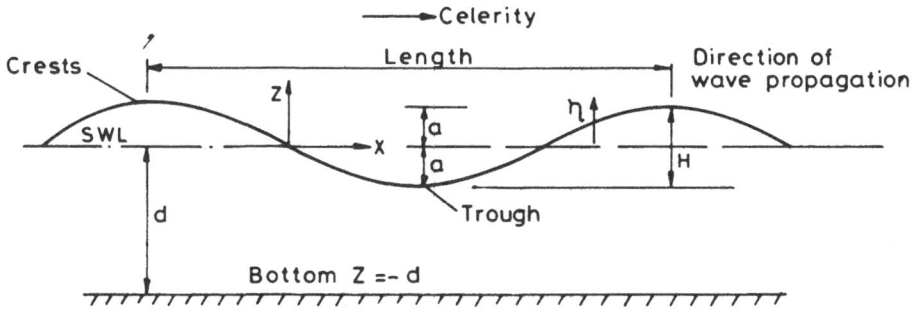

Fig. 6.2. Sinusoidal progressive wave

and
$$C = \sqrt{\frac{gL}{2\pi} \tanh\left(\frac{2\pi d}{L}\right)}, \tag{6.2}$$

where d = distance from sea bed to still water level.

It can be shown that

$$C = \frac{gT}{2} \tanh\left(\frac{2\pi d}{L}\right) \tag{6.3}$$

The equation (6.1) sometimes is expressed as $\eta = a\cos(Kx - \omega t)$, where the quantities $K = 2\pi/L$, $\omega = 2\pi/T$ are designated as wave number and wave frequency respectively.

The expression for wave length is

$$L = \frac{gT^2}{2\pi} \tanh\left(\frac{2\pi d}{L}\right) \tag{6.4}$$

6.3.1 GRAVITY WAVES CLASSIFICATION

The classification of gravity waves is done in accordance with the table (Kinsman, 1965).

Classification	d/L	$2\pi d/L$	$\tanh 2\pi d/L$
Deep	$> 1/2.$	$> \pi$	$\simeq 1$
Transition	1/25 to 1/2	1/4 to π	$\tanh 2\pi d/L$
Shallow	$< 1/25$	$< 1/4$	$\simeq 2\pi d/L$

For deep water the wave celerity is given as

$$C_0 = \sqrt{\frac{gL_0}{2\pi}} = L_0/T \quad \text{or} \quad C_0 = \frac{gT}{2\pi} \tag{6.5}$$

Although deep water actually occurs at infinite depth, $\tanh 2\pi d/L$ for most practical purposes is approximately 1 at a much smaller d/L. For $d/L = 1/2$, $\tanh 2\pi d/L \simeq 0.9964$.

Thus when $d/L > 1/2$, the wave characteristics are independent of depth. The period T remains constant and independent of depth for oscillatory waves. Hence $C_0 = 5.12T$ (ft/s) and $L_0 = 5.12T^2$ (ft). If the above equation is used to compute wave celerity when the relative depth is 0.25, the resulting error will be 9 per cent. For transitional depth the original equation should be used.

When relative depth is shallow, that is, $2\pi d/L < 1/4$ or $d/L < 1/25$ *then*

$$C = \sqrt{gd} \tag{6.6}$$

The above equation is useful in dealing with long period waves. Thus when a wave travels in shallow water, wave celerity depends on water depth. Some of the useful wave functions are:

$$C/C_0 = L/L_0 = \tan h \frac{2\pi d}{L} \qquad (6.7)$$

$$d/L_o = \frac{d}{L} \tan h \frac{2\pi d}{L} \qquad (6.8)$$

6.3.2 LOCAL FLUID VELOCITY AND ACCELERATIONS

In wave force studies it is often desired to know the local fluid velocities and accelerations for various values of z and t during the passage of a wave. The horizontal and vertical components u and w of the local velocities are given by:

$$u = \frac{H}{2} \frac{gT}{L} \frac{\cosh\,[2\pi(z+d)/L]}{\cosh\,[(2\pi d/L)]} \cos\left(\frac{2\pi x}{L} - \frac{2\pi t}{T}\right) \qquad (6.9)$$

$$w = \frac{H}{2} \frac{gT}{L} \frac{\sinh\,[\{2\pi(z+d)/L]}{\cosh\,(2\pi d/L)} \sin\left(\frac{2\pi x}{L} - \frac{2\pi t}{T}\right) \qquad (6.10)$$

They describe the velocities at any distance $(z + d)$ above the bottom. The veolcities are harmonic in both x and t for a given value of phase angle $\theta = \left(\frac{2\pi x}{L} - 2\pi t/T\right)$, the hyperbolic functions, cosh and sinh as functions of result in an approximate exponential decay of the magnitude of velocity as shown in Fig. 6.3. The local fluid particle accelerations can be obtained from the aforementioned equations by differentiating with respect to t, that is,

$$\alpha_x = \frac{+g\pi H}{L} \frac{\cosh\,[2\pi(z+d)/L]}{\cosh\,(2\pi d/L)} \sin\left(\frac{2\pi x}{L} - \frac{2\pi t}{T}\right) \qquad (6.11)$$

$$\alpha_z = \frac{-g\pi H}{L} \frac{\sinh\,[2\pi(z+d)/L]}{\cosh\,(2\pi d/L)} \cos\left(\frac{2\pi x}{L} - \frac{2\pi t}{T}\right) \qquad (6.12)$$

The positive and negative values of the horizontal and vertical fluid accelerations for various Q values $= 2\pi x/L - 2\pi t/T$ are also shown in Fig. 6.3.

6.3.3 SUBSURFACE PRESSURE

Subsurface pressure equals the sum of two contributing components, that is, dynamic and static pressures, and is given by

$$p' = \rho g \frac{\cosh\,[2\pi(z+d)/L]}{\cosh\,(2\pi d/L)} \frac{H}{2} \cos\left(\frac{2\pi x}{L} - \frac{2\pi t}{T}\right) - \rho gz + p'_a \qquad (6.13)$$

where p' = total or absolute pressure, ρ = mass density for saline water $\rho = 2.0$ lbs sec^2/ft^4 = 2.0 slugs/ft^3, and $\rho = 1.94$ slugs/ft^3, for fresh water.

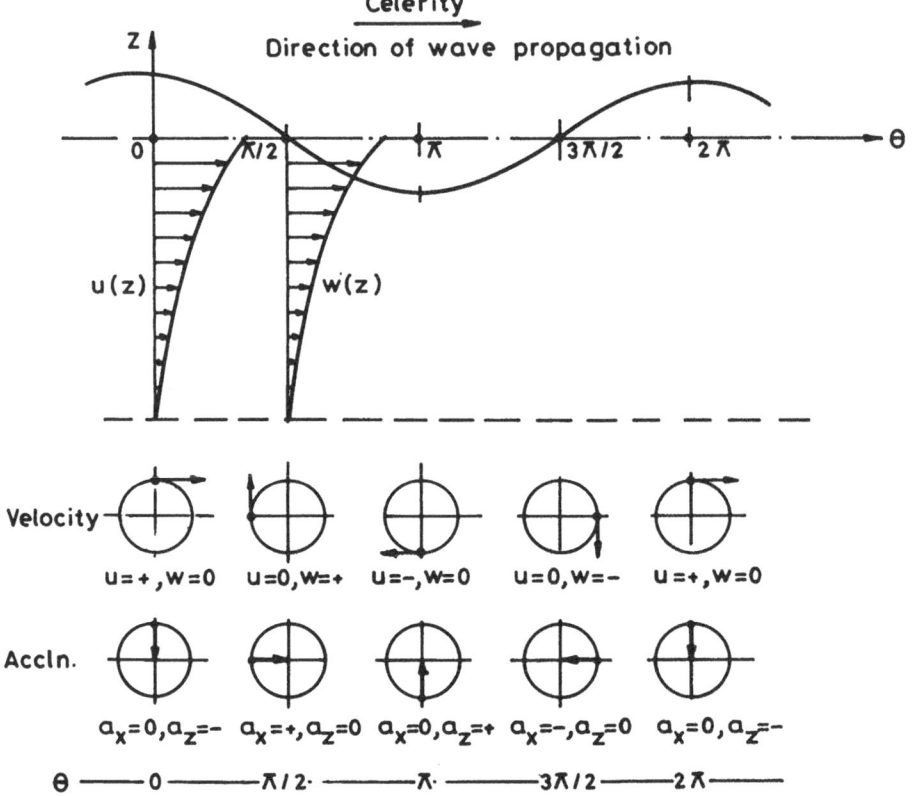

Fig. 6.3. Local fluid velocities and accelerations

The first term on the right hand side of equation (6.13) is the dynamic component due to acceleration. The second term is the static component of pressure, and the third is that due to atmospheric pressure. For convenience, the pressure is usually taken as the gauge pressure defined as

$$p = p' - p_a = \rho g \frac{\cosh\left[2\pi(z+d)L\right]H}{\cosh\left(2\pi d/L\right)2}\cos\left(\frac{2\pi x}{L} - \frac{2\pi t}{T}\right) - \rho g z \qquad (6.14)$$

Equation (6.14) can be written as

$$p = \rho g \eta \frac{\cosh\left[2\pi(z+d)/L\right]}{\cosh\left[2\pi d/L\right]} - \rho g z \qquad (6.15)$$

Since

$$\eta = H/2 \cos\left(\frac{2\pi x}{L} - \frac{2\pi t}{T}\right)$$

the ratio

$$K_z = \frac{\cosh\left[2\pi(z+d)/L\right]}{\cosh\left(2\pi d/L\right)} \qquad (6.16)$$

is the pressure response factor. Hence, equation (6.15) can be written as

$$p = \rho g(\eta K_z - z) \qquad (6.17)$$

The pressure response factor K, for the pressure at the bottom when $z = -d$, is

$$K_z = K = \frac{1}{\cosh\left(2\pi d/L\right)} \qquad (6.18)$$

This factor is tabulated as a function of (d/L_0) and (d/L) (Shore Protection Manual, 1977).

It is often necessary to determine the height of surface waves based on subsurface measurement of pressure. One can write

$$\eta = \frac{N(p + \rho g z)}{\rho g K_z} \qquad (6.19)$$

where z = depth below still water level of the pressure gauge and N = correlation factor = 1 if the linear theory of Airy is applicable. Several empirical studies have found N to be a function of wave period, water depth, wave amplitude and other factors. In general, N decreases with decreasing period. N is greater than 1.0 for long period waves and less than 1.0 for short period waves.

Note: The pressure response equation is not valid for a positive value of z

At z = 0, $K_p = 1$ and $p/\gamma = \eta$

It would thus be inconsistent to assume that pressure is hydrostatic for positive z

$$\text{At } z = -h, \quad K_p = \frac{1}{\cosh kh} \text{ and}$$

$p/\gamma = (\eta/\cosh \; kh) + h$. The dydrostatic bottom pressure is $p/\gamma = h + \eta$ For wave phases at which η is negative (that is, under the trough) the bottom pressure is greater than hydrostatic pressure, because

$$h - \frac{\eta}{\cosh kh} > h - \eta$$

conversely the bottom pressure is less than hydrostatic value

$$h + \frac{\eta}{\cosh kh} < h + \eta$$

6.3.4 VELOCITY OF A WAVE GROUP

The speed of a group of waves or a wave train is generally not identical to the speed of individual waves within the group. The group speed is termed the group velocity C_g. The individual speed is the phase velocity or wave celerity as given by equation (6.2), that is,

$$C = \left[\frac{gL}{2\pi} \tanh\left(\frac{2\pi d}{L} \right) \right]^{1/2}$$

For waves propagating the deep or transitional water with gravity as the primary restoring force, the group velocity will be less than the phase velocity.

In deep water, $(4\pi d/L)/\sinh(4\pi d/L)$ is approximately zero. The group velocity becomes

$$C_g = \frac{1}{2}\frac{L_0}{T} = 1/2 C_0 \tag{6.20}$$

where $C_0 =$ phase velocity in deep water condition. It shows that group velocity $= 1/2$ phase velocity. In shallow water

$$\sinh 4\pi d/L \simeq 4\pi d/L \quad \text{and} \quad C_g = 1/T = C \simeq (gd)^{1/2} \tag{6.21}$$

Because the wave celerity is fully determined by depth in shallow water all component waves in a wave train will travel at the same speed precluding alternate reinforcing and cancelling of components. In deep and transitional water, wave celerity depends on wave length, hence longer waves travel slightly faster and produce small phase difference resulting in wave groups. Group velocity is important because wave energy is propagated, at this velocity.

6.3.5 WAVE ENERGY AND POWER

The total energy of a wave system is the sum of kinetic and potential energy. Kinetic energy is that part of the total energy due to water particle velocities associated with wave motion. Potential energy is that part of the energy due to fluid mass above the trough, that is, wave crest. According to the Airy theory if potential energy is determined relative to mean water level, and all waves are propagated in the same direction, potential and kinetic energy components are equal, that is

$$E = E_k + E_p = \frac{PgH^2L}{16} + \frac{PgH^2L}{16} = \frac{PgH^2L}{8} \tag{6.22}$$

Total wave energy per unit suface area or specific energy or energy density is given by

$$\bar{E} = \left(\frac{PgH^2}{8}\right) \tag{6.23}$$

Wave energy flux equals the rate at which energy is transmitted in the direction of wave propagation across a vertical plane perpendicular to the direction of wave advance and extending down the depth. The average energy flux per unit wave crest width transmitted across a plane perpendicular to wave advance is

$$\bar{P} = \bar{E}\, nC = \bar{C}_g E \tag{6.24}$$

This is frequently called wave power and

$$n = \frac{1}{2}\left[1 + \frac{4\pi d/L}{\sinh 4\pi d/L}\right], \tag{6.25}$$

n, being the transmission is efficient
If a plane is taken other than perpendicular to the direction of wave advance
$\bar{P} = \bar{E}\, C_g \sin\emptyset$ (6.26)

where $\emptyset$ = angle between the plane (across which the energy is being transmitted) and the direction of wave advance. The equations for deep and shallow water are

$$\bar{p}_0 = 1/2\, E_o C_o \text{ (deep)} \tag{6.27}$$

$$\bar{p} = \bar{E}\, C_g = \bar{E}C \text{ (shallow)} \tag{6.28}$$

respectively for steady state when the waves are moving so that their crests are parallel to the bottom contours.

$$\bar{E}_o n_o C_o = \bar{E}\, nC \text{ or since, } n_o = 1/2,\ 1/2\, \bar{E}_o C_o = \bar{E}\, nC \tag{6.29}$$

When the wave crests are not parallel to the bottom contours some parts will be travelling at different speeds, the wave will be diffracted and equation (6.29) does not apply.

6.3.6 SOLITARY WAVE THEORY
The waves considered were oscillatory or nearly oscillatory waves. The water particles move backwards and forwards with the passage of each wave

which has a distinct wave crest and wave trough. A solitary wave is neither oscillatory nor does it exhibit a trough, it lies entirely above the still water level. It is a wave of translation relative to the water mass.

Consider a solitary wave of height H in a depth of water d travelling in the *Positive x* direction as shown in Fig. 6.4. The surface profile η given by Ippen (1966) is

$$\eta = H\left[\operatorname{sech} 3/4 \frac{H}{d^3}(x - Ct) \right]^2 \qquad (6.30)$$

where the origin of x is at wave crest.

The volume of water within the wave above the still water level per crest width is

$$V = [\tfrac{16}{3}d^3 H]^{0.5} \qquad (6.31)$$

An equal volume of water per unit crest length is transported forward past a vertical plane that is normal to the direction of wave advance. The celerity of solitary wave is

$$C = \sqrt{g(H + d)} \qquad (6.32)$$

The water particle velocities u,w in the horizontal and vertical direction for a solitary wave are given by Munk (1949).

$$u = CN \frac{1 + \cos(My/d)\cosh(Mx/d)}{[\cos(My/d) + \cosh(Mx/d)]^2} \qquad (6.33)$$

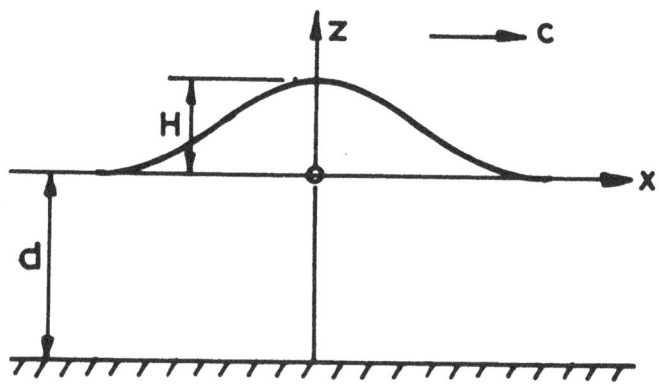

Solitary wave

Fig. 6.4. A solitary wave

$$w = CN \frac{\sin(My/d)\sinh(Mx/d)}{[\cos(My/d) + \cosh(Mx/d)]^2} \quad (6.34)$$

where $y =$ elevation from the bottom

M and N are functions of (H/d) as shown in Fig. 6.5 and y is measured from the bottom. The horizontal component is important as it is often used to predict wave forces on marine structures in shallow water. The U_{max} occurs when x and t are both zero, hence

$$U_{max} = \frac{CN}{1 + \cos(My/d)} \quad (6.35)$$

The total energy is evenly divided between kinetic and potential energy. The total energy per unit crest width is

$$E = \frac{8}{3\sqrt{3}} \rho g H^{3/2} d^{3/2} \quad (6.36)$$

The pressure beneath a solitary wave depends on the local fluid velocity. However, it may be approximated by

$$p = \rho g \, (Y_s - Y) \quad (6.37)$$

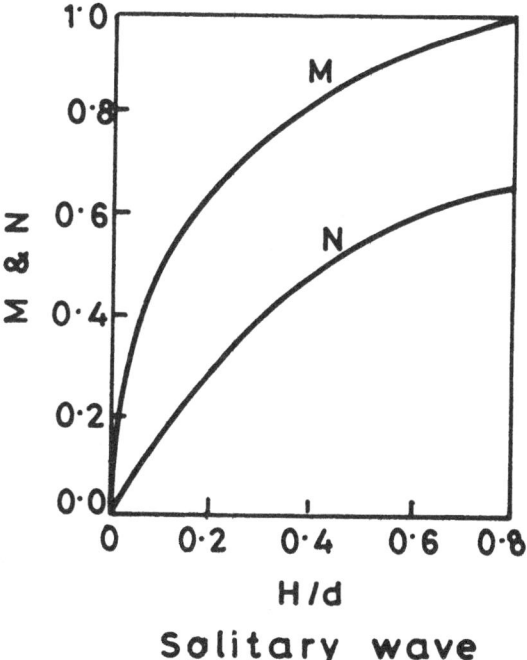

Solitary wave

Fig. 6.5. Functions M and N in solitary wave

where

Y_S = elevation of the water surface measured from bottom, Y = location from bottom where pressure is p.

As the wave moves into shoaling water it eventually becomes unstable and breaks. McCowan (1891) assumed that a solitary wave breaks when the water particle velocity at the wave crest equals wave celerity, that is

$$(H/d)_{max} = 0.78 \qquad (6.38)$$

6.3.7 WAVE REFLECTION

Water waves may be either partially or totally reflected from both natural and man-made barriers such as a wall,cliff or steep beach. Reflection of waves implies a reflection of wave energy. Consequently, multiple reflections within a harbour complex can result in a build up of wave energy which appears as wave agitation and surging in the harbour. These surface fluctuations may cause excessive motion of moored ships and other floating facilities and result in considerable strain on mooring lines. Figure 6.6 shows the variation of water particle motions in a standing wave system, which is the cause of stresses in mooring lines. For this purpose, the breakwaters should be so aligned that the waves reflected from them are not directed towards piers or other harbour installations. The interior slope should be designed to break the waves.

6.3.8 WAVE DIFFRACTION

Diffraction of water waves is a phenomenon in which energy is transferred laterally along a wave front. It is noticeable where an otherwise regular train of waves is intercepted by a barrier such as a breakwater. In the case of harbours protected by breakwaters, the opening in the breakwater serves as a channel for shipping. The opening admits a certain amount of waves into the calm waters of the harbour. These waves expand in the form of wave fronts having their centre at the gap. The energy of this wave front decreases as the waves travel away from the gap. This means that the further away a structure is from the gap, the smaller is the wave acting on it. The wave height within a harbour can be approximately determined by using Stevenson's formula (Bindra, 1978)

$$h = H\left[\sqrt{B/L} - 0.027 \sqrt{D} \ (1 + \sqrt{B/L}) \right] \qquad (6.39)$$

where h = height of reduced waves at any point in the harbour, in m;

   B = breadth of entrance, in m;
   L = breadth of harbour at the required point, in m, which equals length of area enclosed with radius D and centre at middle of entrance; and
   D = distance from entrance to the required point, in m.

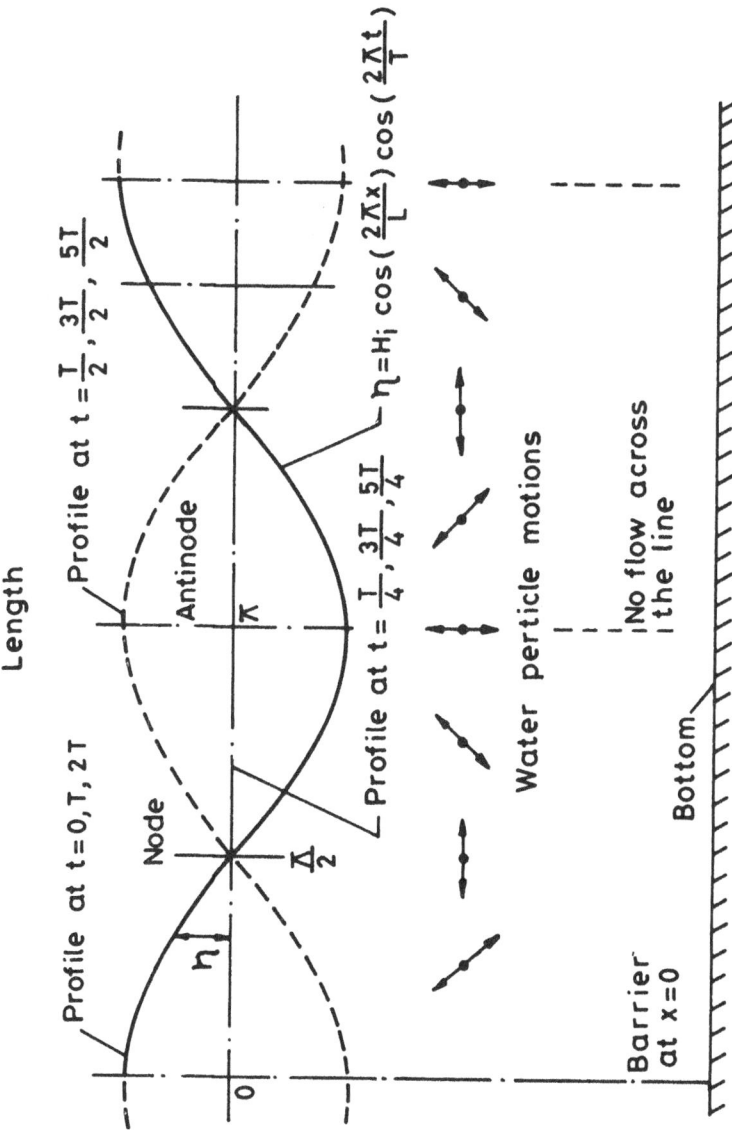

Fig. 6.6. Standing wave system reflection from a vertical barrier.

6.3.9 WAVE REFRACTION

In the open sea, the crest of an oscillatory wave is always normal to the direction of propagation. Near the coast however, contact with shoaling ground off the headlands occurs before that of the bays. The velocity of the wave off the headland decreases relative to that part of the wave off the bays. Consequently, the wave begins to follow the shape of the coast. Where the coastline is not too exaggerated by the time the shoreline is reached the wave crest will adjust itself to become parallel with all major features of the shoreline, that is, the phenomenon of wave refraction. By definition, this is a change in direction due to a change in wave length. The law relating to this change is the same as that for refraction of light

$$\frac{\sin r}{\sin i} = \frac{L_2}{L_1} \qquad (6.40)$$

where i = angle of incidence, r = angle of refraction,
L_1, L_2 = incident and refracted wave lengths, respectively.

In general, coastal configurations are curved. The morphology of the sea bed ultimately influences the direction of the wave which can only be deduced step-wise by means of a wave refraction diagram. Evidently concentration of wave energy occurs at the headlands and the reverse occurs in the bays, that is, convergence and divergence occur respectively as shown in Fig. 6.7. Waves gain height at a convergence and lose it at a divergence, as a result of this phenomenon.

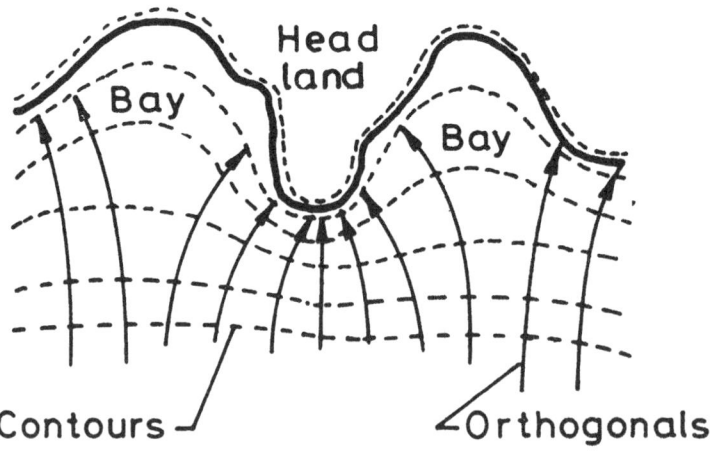

Fig. 6.7 Refraction along an irregular shore line

REFERENCES

Airy, G.B. (1845): *On Tides and Waves*, Encylopaedia Metropolitana.

Bindra, S.P. (1978): *Docks and Harbour Engineering*, Dhanpat Rai & Sons.

Ippen, A.T. (1966): *Estuary and Coastline Hydrodynamics*, McGraw-Hill Book Company, Inc.

Kinsman, B. (1965): *Wind Waves, Their Generation and Propagation on the Ocean Surface*, Prentice- Hall.

Kinsman, B. (1977): *Shore Protection Manual,* Volume I, U.S. Arms Coastal Engineering Research Center, Corps of Engrs.

McCowan, J. (1891): *On the Solitary Waves*, London, Edinburgh and Dublin Phil. Mag. and J. Sc.: Vol. 32(5): 45.

Munk, W.H. (1949): *The Solitary Waves and Its Application to Surf Problems*, Ann. N.Y. Acad. Sci., 51: 376–424.

Stokes, G.C. (1880): *On the Theory of Oscillatory Waves Mathematical and Physical Papers*, Vol. I, Cambridge University Press, Cambridge.

CHAPTER 7

Mechanics of Sediment Transport and Coastal Fluvial Processes

7.1 Introduction

The mechanics of sediment transport is one of the most complicated and difficult subjects in the study of river and coastal fluvial processes. Most of the transport functions are empirical or semi-empirical in nature. Analytical analyses are mainly limited to simplified one-dimensional flow conditions. Consequently, most of the transport functions were developed for sediment transport in rivers. Classical transport equations were developed mainly from the consideration of forces. However, the energy approach to the study of sediment transport has gained increasing acceptance in recent years. This approach can be applied to the study of river and coastal sediment transport processes.

This chapter explains the basic mechanics of incipient motion as well as bed, suspended and total load transport. Comparisons are made to provide an assessment of the assumptions used and the accuracy of different transport equations. The energy concept is used for the study of coastal fluvial and sediment transport processes.

7.2 Incipient motion

When the flow condition exceeds certain limitations, sediment particles will start to move. The forces acting on a spherical sediment particle at the bottom of an open channel are shown in Fig. 7.1. For most natural rivers, the channel slopes are small enough so that the component of gravitational force in the direction of flow can be neglected in comparison with other forces acting on a spherical sediment particle. The forces to be considered are the drag force, F_D; lift force, F_L; submerged weight, W_s; and resistance force, F_R. A sediment particle is at a state of incipient motion when one of the following conditions is satisfied:

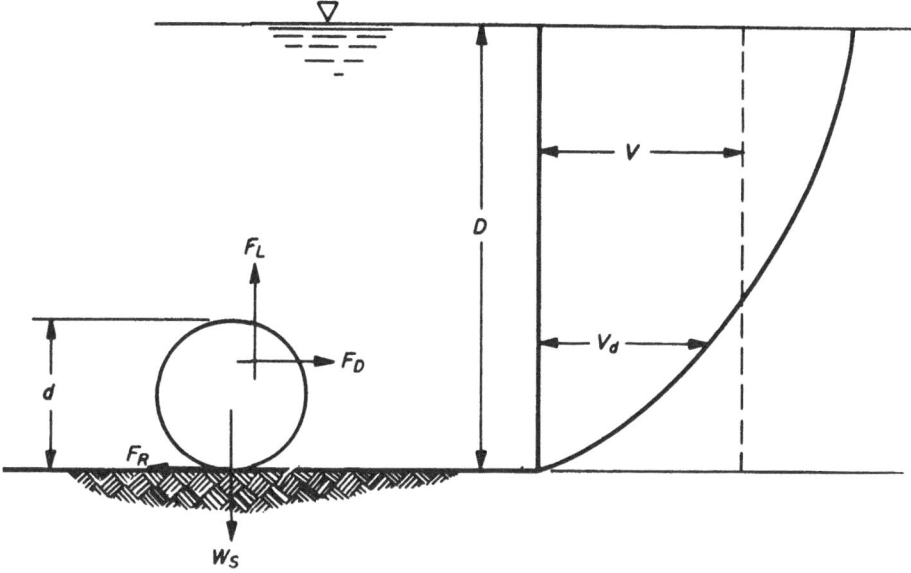

Fig. 7.1. Diagram of forces acting on a sediment particle in open channel flow

$$F_L = W_s \tag{7.1}$$
$$F_D = F_R \tag{7.2}$$
$$M_o = M_R \tag{7.3}$$

where M_o = overturning moment due to F_D and F_R; and
M_R = resisting moment due to F_L and W_s.

Most incipient motion criteria are derived either from shear stress or from velocity approach.

7.2.1 SHEAR STRESS APPROACH

One of the most widely used incipient motion criterion is Shields' (1936) diagram based on shear stress. Shields considered the factors important to the determination of incipient motion as shear stress τ, the difference in density between sediment and fluid $(\rho_s - \rho_f)$, the diameter of the particle d, the kinematic viscosity v, and gravitational acceleration g. These five quantities can be grouped into two dimensionless quantities.

$$\frac{d\sqrt{\tau_c/\rho_f}}{v} = \frac{dU_*}{v} \tag{7.4}$$

and

$$\frac{\tau_c}{d(\rho_s - \rho_f)g} = \frac{\tau_c}{d\gamma[(\rho_s/\rho_f) - 1]} \tag{7.5}$$

where ν = kinematic viscosity
    ρ_s and ρ_f = density of sediment and fluid, respectively
    $\gamma = \rho g$ = the specific weight of fluid
        g = gravitational acceleration
        U_* = the shear velocity
        τ_c = critical shear stress at initial motion

The relationship between these two parameters was then determined from experiments. Figure 7.2 shows the experimental results obtained at incipient motion At points above the line, the particle will move. At points below the line, the flow is unable to move the particle. One of the disadvantages of using Shields' diagram is that both shear stress τ and shear velocity U_* in Shields' diagram are considered as dependent and independent variables, yet they are interchangeable by $U_* = \sqrt{\tau/\rho_f}$. Consequently, the critical shear stress cannot be determined directly from Shields' diagram; it must be determined through trial and error. To resolve this problem, the American Society of Civil Engineers Task Committee on Preparation of Sediment Manual (Vanoni, 1975) uses a third parameter

$$\frac{d}{\nu}\sqrt{0\cdot1\left(\frac{\gamma_s}{\gamma}-1\right)gd}$$

as shown in Fig. 7.2, where γ_s is the specific weight of sediment. The use of this parameter would enable us to determine its intersection with the Shields' diagram and its corresponding values of shear stress.

7.2.2 VELOCITY APPROACH

Fortier and Scobey (1926) made an extensive field survey on maximum permissible value of mean velocities in canals. The permissible velocities for canals of different materials are summarised in Table 7.1. Although there is no theoretical study to support or verify the values shown in Table 7.1, these results are based on inputs from experienced engineers and should be useful for preliminary designs.

Hjulstrom (1935) made a detailed analysis of data obtained from the movement of uniform materials. Because the channel bottom velocity, which is directly responsible for sediment movement, is difficult to measure, his study is based on average flow velocity. Figure 7.3 gives the relationship between sediment size and average flow velocity for erosion, transportation, and sedimentation.

Yang (1973) considered all the forces acting on a sediment particle as shown in Fig. 7.1. His incipient motion criteria are based on the flow conditions which satisfy equations (7.1) and (7.2). The basic equation thus obtained is given on page 142.

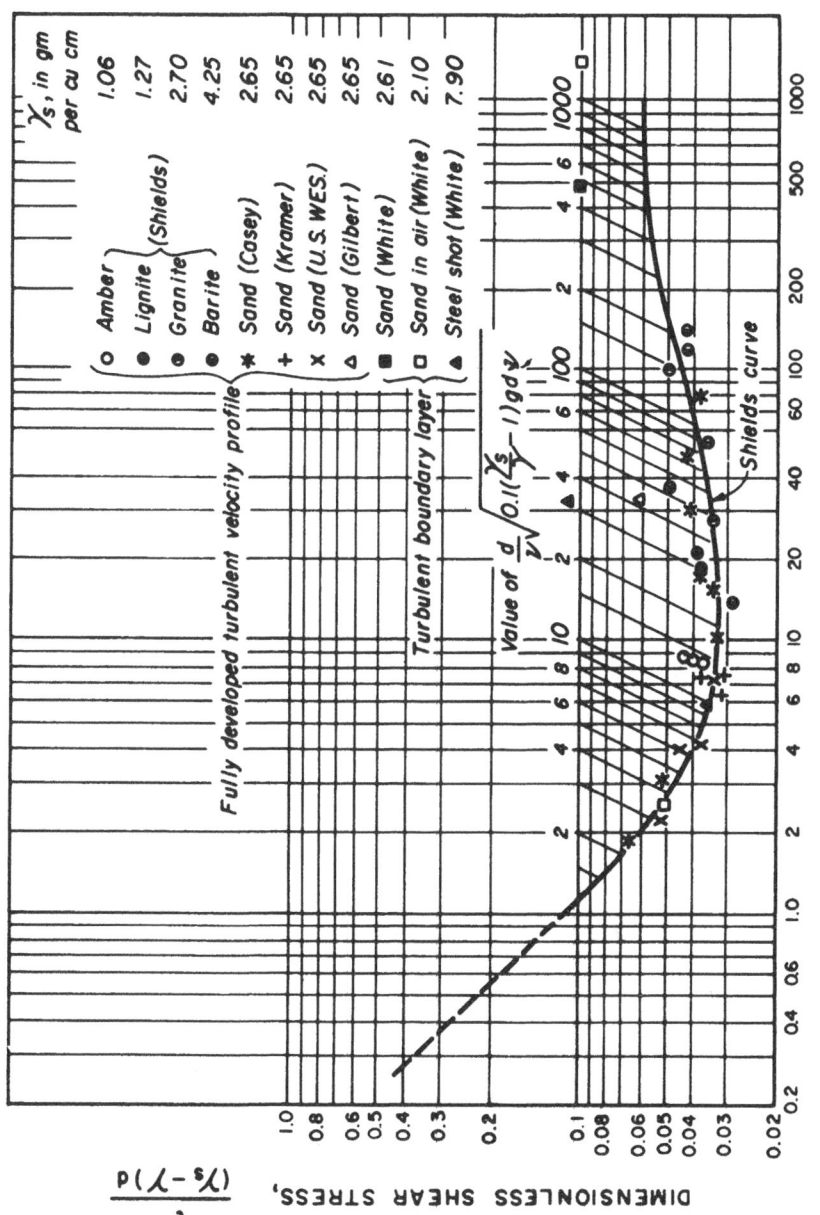

Fig. 7.2. Shields' diagram for incipient motion (Vanoni, 1975)

Table 7.1
Permissible canal velocities (Fortier and Scobey, 1926)

Original material excavated for canal (1)	Velocity, ft/s, after aging, of canals carrying*		
	Clear water, no detritus (2)	Water-transporting colloidal silts (3)	Water-transporting noncolloidal silts, sands, gravels, or rock fragments (4)
Fine sand (noncolloidal)	1.50	2.50	1.50
Sandy loam (noncolloidal)	1.75	2.50	2.00
Silt loam (noncolloidal)	2.00	3.00	2.00
Alluvial silts when noncolloidal	2.00	3.50	2.00
Ordinary firm loam	2.50	3.50	2.25
Volcanic ash	2.50	3.50	2.00
Fine gravel	2.50	5.00	3.75
Stiff clay (very colloidal)	3.75	5.00	3.00
Graded, loam to cobbles, when noncolloidal	3.75	5.00	5.00
Alluvial silts when collodial	3.75	5.00	3.00
Graded, silt to cobbles, when colloidal	4.00	5.50	5.00
Coarse gravel (noncolloidal)	4.00	5.50	5.00
Cobbles and shingles	5.00	5.50	6.50
Shales and hard pans	6.00	6.00	5.00

* Depth of 3 feet or less.

$$\frac{V_{cr}}{\omega} = \left[\frac{5 \cdot 75 \left(\log \frac{D}{d} - 1 \right)}{B} + 1 \right] \sqrt{\frac{\psi_1 \psi_2 \psi_3}{\psi_2 + \psi_3}} \qquad (7.6)$$

where B = roughness coefficient
D = water depth
d = sediment particle diameter
V_{cr} = the average critical velocity at incipient motion
ω = the average fall velocity of sediment particles

Ψ_1, Ψ_2, Ψ_3 = coefficients.

The values of Ψ_1, Ψ_2, and Ψ_3 have to be determined from experiments. The value of the roughness function, B, depends on whether the boundary is

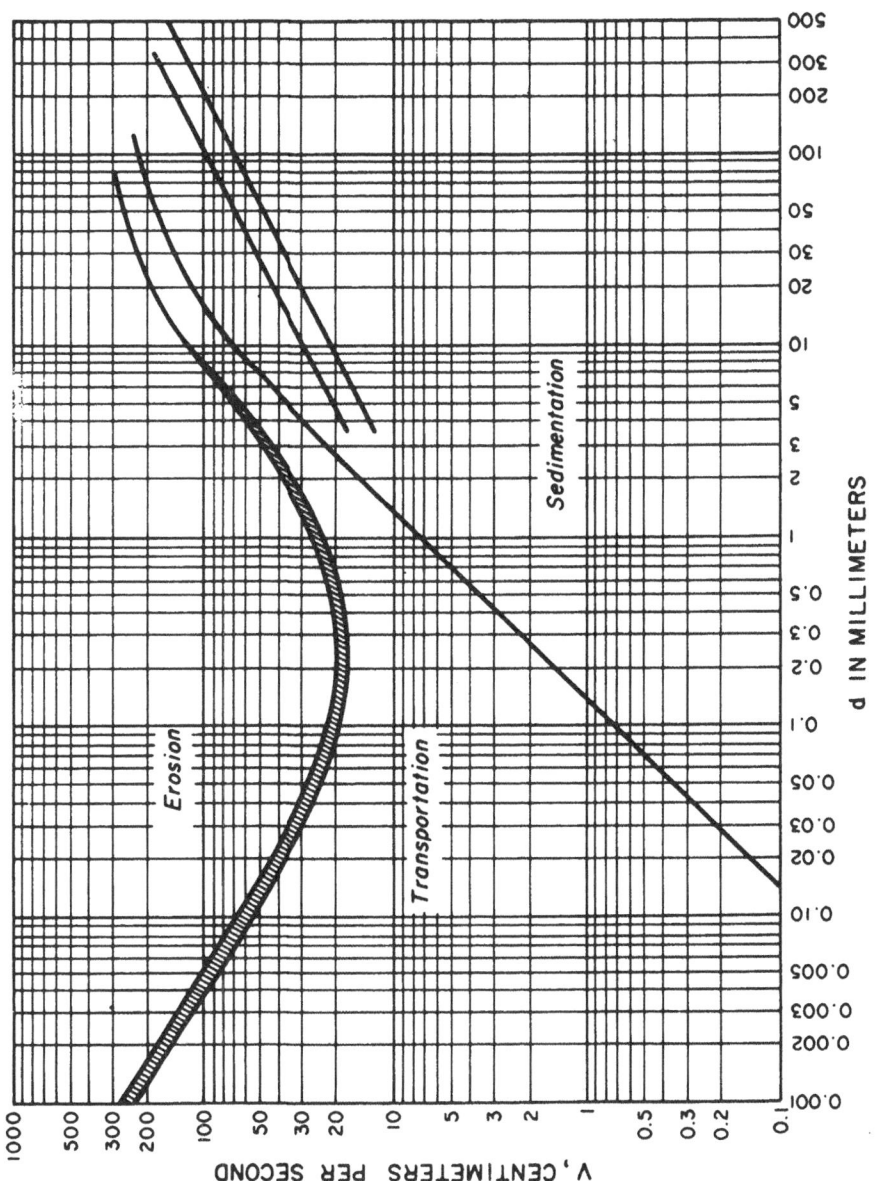

Fig. 7.3. Erosion-deposition criteria for uniform particles (Hjulstrom, 1935)

in a hydraulically smooth, transition, or completely rough regime.

In the hydraulically smooth regime, B is a function of only the shear velocity Reynolds number, U_*d/v (Schlichting, 1962), that is

$$B = 5.5 + 5.75 \log \frac{U_*d}{v}, \quad 0 < \frac{U_*d}{v} < 5 \tag{7.7}$$

where $U_* =$ shear velocity
$v =$ kinematic viscosity of water

Then equation (7.6) becomes

$$\frac{V_{cr}}{\omega} = \left[\frac{\log \dfrac{D}{d} - 1}{\log \dfrac{U_*d}{v} + 0.956} + 1 \right] \sqrt{\frac{\psi_1 \psi_2 \psi_3}{\psi_2 + \psi_3}} \tag{7.8}$$

which is a hyperbola on a semilog plot between V_{cr}/ω and U_*d/v.

In the completely rough regime, the protrusions reach outside the laminar sublayer. The laminar friction contribution can be neglected, and B is a function of only the relative roughness d/D, that is,

$$B = 8.5, \frac{U_*d}{v} > 70 \tag{7.9}$$

then equation (7.6) becomes

$$\frac{V_{cr}}{\omega} = \left[\frac{\left(\log \dfrac{D}{d} - 1 \right)}{1.48} + 1 \right] \sqrt{\frac{\psi_1 \psi_2 \psi_3}{\psi_2 + \psi_3}} \tag{7.10}$$

Equation (7.10) indicates that in the completely rough regime, the plot of V_{cr}/ω against U_*d/v is a straight horizontal line. In the transition regime with the shear velocity Reynolds number between 5 and 70, protrusions extend partly outside the laminar sublayer. Both the laminar friction and turbulent friction contributions should be considered. In this case, B deviates gradually from equation (7.7) with increasing U_*d/v. It is reasonable for us to expect that, basically, equation (7.8) is still valid but with the relative roughness, d/D, playing an increasingly important role as U_*d/v increases.

Laboratory data collected by different investigators were used by Yang (1973) for the determination of coefficients in equations (7.8) and (7.10). The incipient motion criteria thus obtained are

$$\frac{V_{cr}}{\omega} = \frac{2.5}{\log \dfrac{U_*d}{v} - 0.06} + 0.66, \quad 1.2 < \frac{U_*d}{v} < 70 \tag{7.11}$$

and

$$\frac{V_{cr}}{\omega} = 2.05, \quad 70 \leqslant \frac{U_* d}{\nu} \tag{7.12}$$

Comparisons between equations (7.11), (7.12), and laboratory data are shown in Fig. 7.4. It should be pointed out that although equation (7.10) indicates that V_{cr}/ω should be a function of relative roughness d/D, available data are insufficient to determine its effect on incipient motion. Consequently, this effect was ignored and a constant value was used by Yang in equation (7.12).

Laboratory tests were made recently by Talapatra and Ghosh (1983) to compare the incipient motion criteria proposed by Shields (1936) and Yang (1973). They found that their data agree very well with Yang's incipient motion criteria.

7.3 Bedload transport

When the flow conditions satisfy or exceed the criteria for incipient motion, sediment particles along an alluvial bed will start to move. If the motion of settlement particles is rolling, sliding, or sometimes jumping along the bed, it is called bedload transport. Generally, the basic approach used in the study of bedload movement is to assume that the bedload can be determined by shear stress, energy slope, discharge, or velocity of the flow. Due to the stochastic nature of bedload movement, probabilistic approach was also used. Numerous equations were proposed in the past. Only a few will be introduced herein to illustrate different approaches used in the study of bedload transport.

7.3.1 SHEAR STRESS APPROACH

DuBoys (1879) assumed that sediment particles move in layers along the bed as shown in Fig. 7.5. These layers move because of tractive force τ acting along the bed. The thickness of each layer is ϵ. Under equilibrium condition, the tractive force should be balanced by the total resistance force between these layers, that is,

$$\tau = \gamma DS = C_f m \epsilon (\gamma_s - \gamma) \tag{7.13}$$

where C_f = friction coefficient
              m = total number of layers
              ϵ = layer thickness
              D = water depth
              S = channel slope
     γ_s and γ = specific weight of sediment and water, respectively.

If the velocity varies linearly between the first and the mth layer, the total bedload discharge per unit channel width is

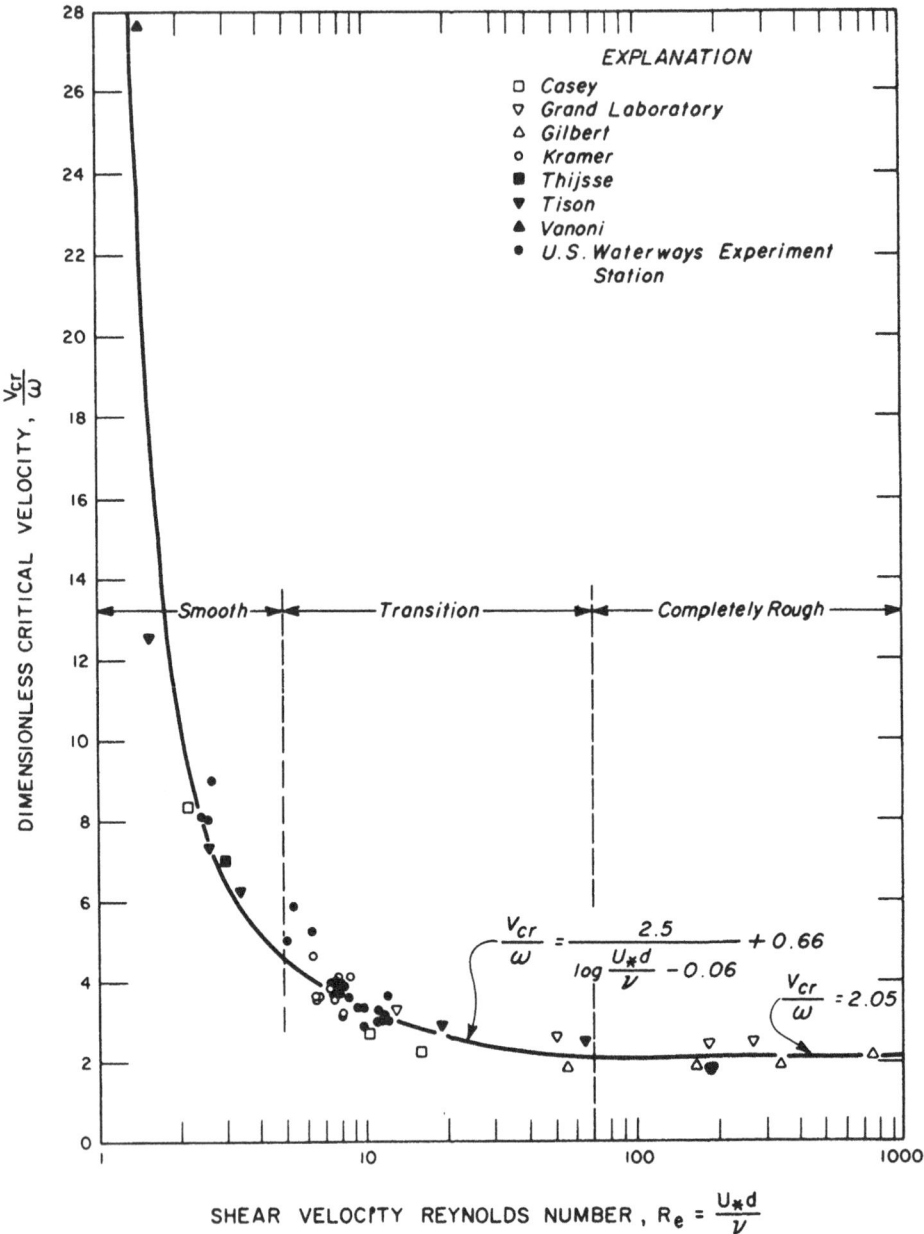

Fig. 7.4. Relationship between dimensionless critical average velocity and Reynolds number (Yang, 1973)

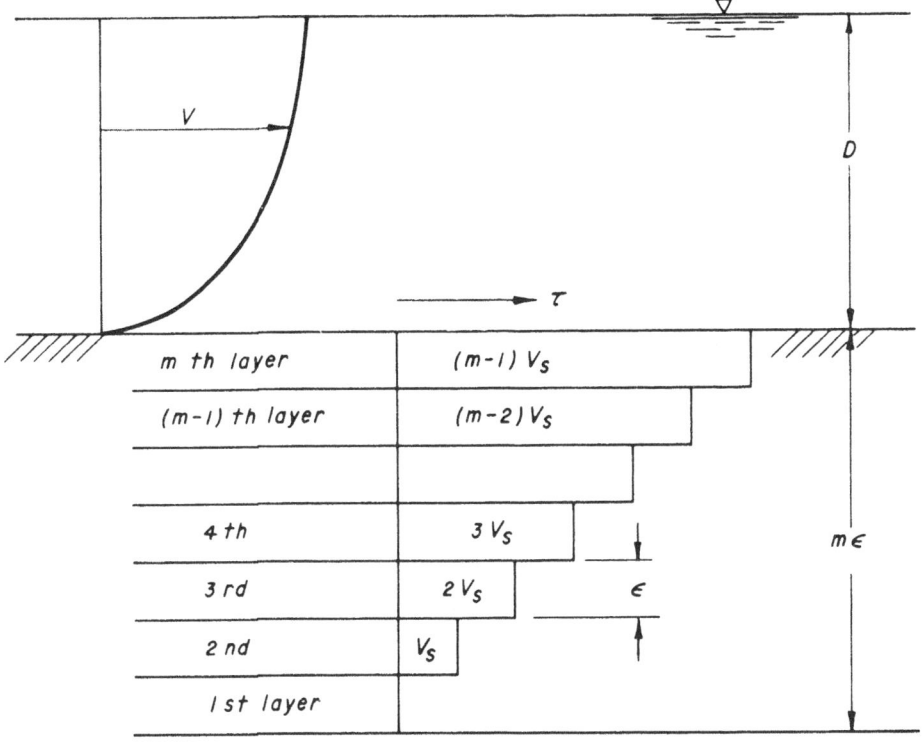

Fig. 7.5. A sketch of DuBoys bedload model

$$q_b = \varepsilon V_s \frac{m(m-1)}{2} \tag{7.14}$$

where V_s = velocity of the second layer as shown in Fig. 7.5.

At incipient motion, $m = 1$, then equation (7.13) becomes

$$\tau_c = C_f \varepsilon (\gamma_s - \gamma) \tag{7.15}$$

and

$$m = \tau/\tau_c \tag{7.16}$$

where τ_c = critical tractive force along the bed.
From equations (7.14) and (7.15)

$$q_b = \left(\frac{\varepsilon V_s}{2\tau_c^2} \right) \tau (\tau - \tau_c) \tag{7.17}$$

$$= K\tau(\tau - \tau_c)$$

The coefficient K in equation (7.17) is related to the characteristics of sediment particles.

Straub (1935) found that the K value in equation (7.17) is related to particle size d, that is,

$$K = \frac{0.173}{d^{3/4}} = \text{ft}^6/\text{lb}^2 - \text{s} \tag{7.18}$$

or

$$K = \frac{0.391}{d^{3/4}} = \text{m}^6/\text{kg}^2 - \text{s} \tag{7.19}$$

Equation (7.18) is for the English system, while equation (7.19) is for the metric system. The d value has to be in mm for both systems. Thus, the DuBoys equation becomes:

$$q_b = \frac{0.173}{d^{3/4}} \tau(\tau - \tau_c) = \text{ft}^3/\text{s}/\text{ft} \tag{7.20}$$

or

$$q_b = \frac{0.391}{d^{3/4}} \tau(\tau - \tau_c) = \text{m}^3/\text{s}/\text{m} \tag{7.21}$$

The relationship between τ_c, K, and d is shown in Fig. 7.6.

7.3.2 ENERGY SLOPE APPROACH

Meyer-Peter *et al.* (1934) conducted extensive laboratory studies on sediment transport. The importance of an energy slope to sediment transport was recognised by them. The Meyer-Peter formula for bedload using the metric system is

$$\frac{0.4q_b^{2/3}}{d} = \frac{q^{2/3}S}{d} - 17 \tag{7.22}$$

where q_b = bedload in kg/sec/m
q = water discharge in kg/sec/m
S = slope
d = particle size in meters

The constants 17 and 0.4 are valid only for sand with a specific gravity of 2.65. The Meyer-Peter formula can be applied only to coarse material with particle size greater than 3 mm. For mixture of non-uniform material, d should be replaced by d_{35}, that is, 35 per cent of the mixture is finer than d_{35}.

7.3.3 DISCHARGE APPROACH

Schoklitch introduced two bedload equations based on discharge approach. His first equation (Schoklitsch, 1934) in metric units is

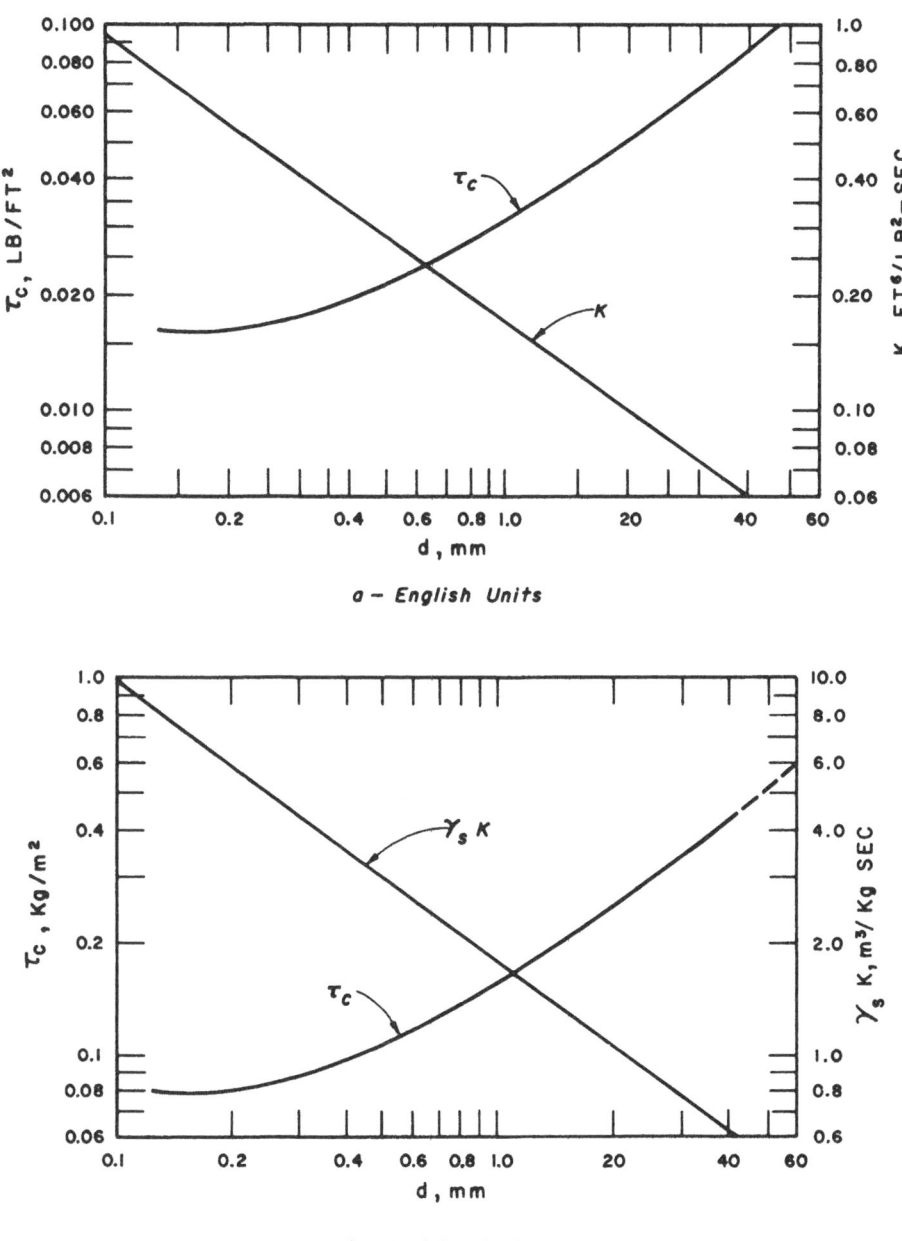

Fig. 7.6. Sediment parameters and critical tractive force for DuBoys' bedload equation (Straub, 1935)

$$q_b = 7000 \frac{S^{3/2}}{d^{1/2}}(q - q_c) \tag{7.23}$$

where q_b = bedload in kg/s/m
  d = particle size in millimetres
  s = slope
  q and q_c = water discharge and critical discharge at incipient motion, respectively, in m³/s/m

The critical water discharge in equation (7.23) for sediments with specific gravity of 2.65 is given by

$$q_c = \frac{0.00001944d}{S^{4/3}} \tag{7.24}$$

Schoklitsch in the year 1943 proposed his second equation in metric units as

$$q_b = 2500 S^{3/2}(q - q_c) \tag{7.25}$$

For sediments with specific gravity of 2.65, the critical discharge in equation (7.25) is

$$q_c = \frac{0.6d^{3/2}}{S^{7/6}} \tag{7.26}$$

where d = the particle size in metres.

7.3.4 VELOCITY APPROACH

The basic DuBoys formula, that is, equation (7.17) can be rewritten as

$$q_b = K\,(\gamma SD)\,(\gamma SD - \gamma SD_c)$$
$$= K(\gamma S)^2 (D)\,(D - D_c^2) \tag{7.27}$$

where D and D_c = normal and critical water depth at incipient motion, respectively.

Donat (1929) assumed that Chezy's equation can be used and the Chezy's C value remains the same for D and D_c. Equation (7.27) can then be changed to

$$q_b = \frac{K}{C^4}\,\gamma^2\,V^2\,(V^2 - V_c^2) \tag{7.28}$$

Where C = Chezy's roughness coefficient
  γ = specific weight of water
  V and V_c = average and critical velocity at incipient motion, respectively

The K value can be obtained from Fig. 7.6.

7.3.5 PROBABILISTIC APPROACH

Einstein (1942, 1950) had two basic ideas which broke with the concept used in the past: (1) the critical criterion for incipient motion was avoided, because it is difficult to define, (2) the bedload transport is related to the turbulent flow fluctuations rather than to the average values of forces the flow exerts on sediment particles. Consequently, the beginning and ceasing of sediment motion is expressed with the probability concept. Based on experiments, Einstein found that

  i) A steady and intensive exchange of particles exists between bed material and bedload.
 ii) The movement of bedload is in a series of steps. The average step length is about 100 times the particle diameter.
iii) The rate of deposition per unit bed area depends on the transport rate past a given section, as well as the probability that the hydrodynamic forces which allow the particle to deposit. The rate of erosion depends on the number and properties of particles in a unit area and the probability that the instantaneous hydrodynamic lift force on the particle is large enough to move it. For a stable bed condition, the rate of deposition must be equal to the rate of erosion.

From the preceding concepts, Einstein (1950) developed his bedload transport function from a probabilistic approach. The final relationship is shown in Fig. 7.7. In Fig. 7.7, the value of Φ_* is related to bedload transport rate by

$$\phi_* = \frac{i_{Bw}q_{bw}}{i_{bw}\rho_s g} \sqrt{\frac{\rho}{(\rho_s - \rho)gd^3}} \tag{7.29}$$

where i_{Bw} = percent of bedload by weight in size d
$\quad\quad i_{bw}$ = number of sediment particles available on the bed
$\quad\quad \rho,\rho_s$ = density of water and sediment, respectively
$\quad\quad g$ = gravitational acceleration
$\quad\quad d$ = sediment particle diameter
$\quad\quad q_{bw}$ = bedload discharge by weight per unit channel width with sediment of size d

The value of ψ_* is a function of bed material and flow conditions. Complicated and detailed computational procedures are given in Einstein's (1950) original paper.

7.4 Suspended load transport

Suspended load refers to sediment that is supported by the upward components of turbulent currents and stays in suspension for an appreciable length of time. Most fine sediment particles are transported as suspended load. Suspended load transport rate can be defined as on page 153.

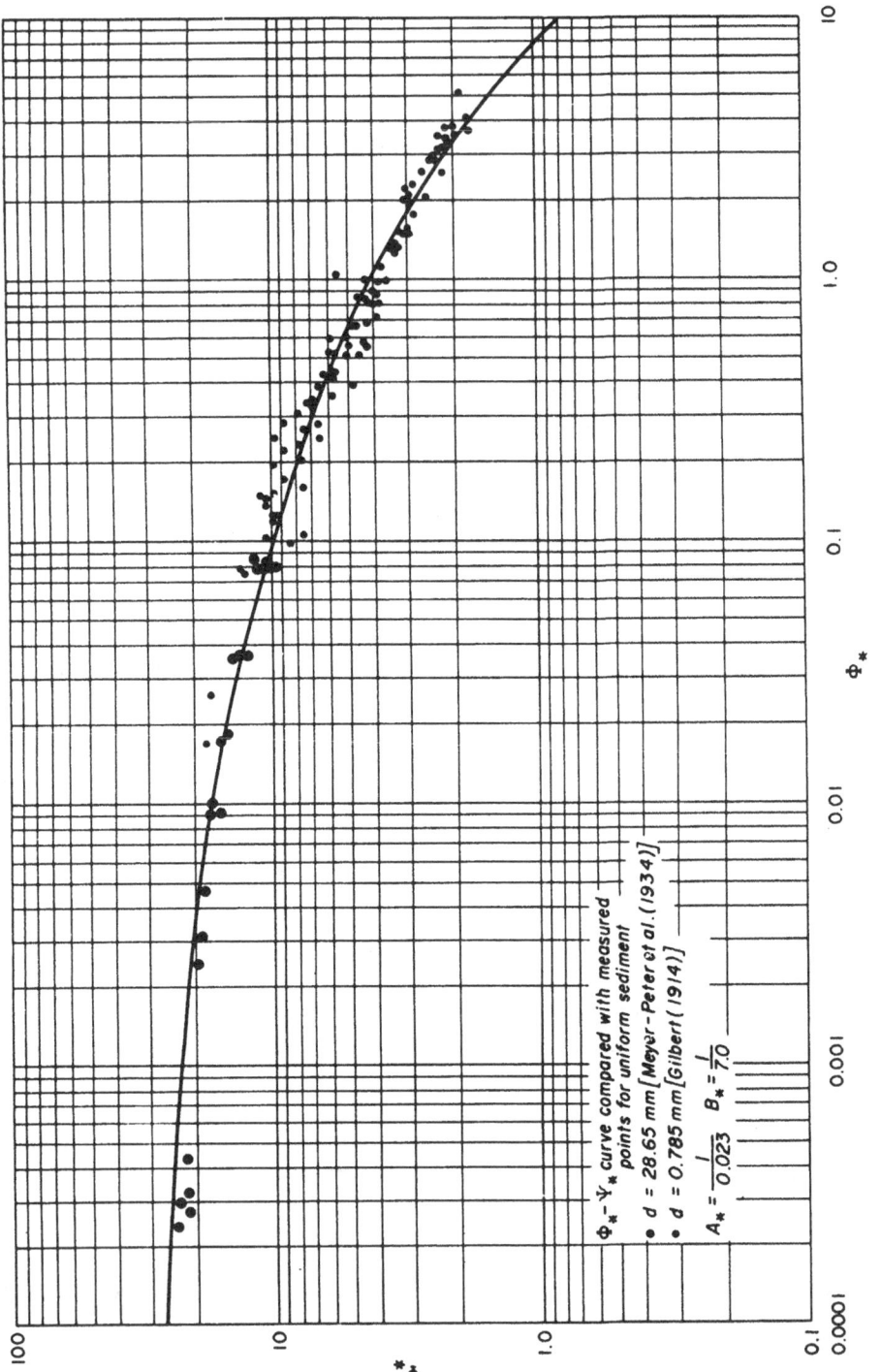

Fig. 7.7. Relationship between ψ_* and Φ_* for Einstein's bedload function (Einstein, 1950)

$$q_{sv} = \int_a^D \bar{u} \bar{c} \, dy \qquad (7.30)$$

or

$$q_{sw} = \gamma_s \int_a^D \bar{u} \bar{c} \, dy \qquad (7.31)$$

where q_{sv} and q_{sq} = suspended load transport rate in terms of volume and weight, respectively

$\bar{u}$ and $\bar{c}$ = time-averaged velocity and sediment concentration at a distance y above the bed, respectively

a = thickness of bedload transport

D = water depth

γ_s = specific weight of sediment

Before equations (7.30) or (7.31) can be integrated, $\bar{u}$ and $\bar{c}$ must be expressed mathematically as functions of y. The basic theory used for the determination of suspended load concentration and equations which can be obtained from this theory are presented herein.

7.4.1 EXCHANGE THEORY

Under steady equilibrium condition, the downward movement of sediment due to fall velocity must be balanced by the net upward movement of sediment due to turbulent fluctuation. The net exchange of sediment must be zero, that is,

$$\omega C + \varepsilon_s \frac{dC}{dy} = 0 \qquad (7.32)$$

where ε_s = sediment diffusion coefficient, which is a function of y

ω = fall velocity of sediment particles

C = sediment concentration

Integration of equation (7.32) yields

$$C = C_a \exp\left(-\omega \int_a^y \frac{dy}{\varepsilon_s} \right) \qquad (7.33)$$

where C and C_a = sediment concentration at a distance y and a above the bed, respectively.

Assuming that the Prandtl-von Karman velocity distribution is valid and diffusion coefficients for water and sediment have the same value, it can be shown that

$$\frac{C}{C_a} = \left(\frac{D-y}{y} \frac{a}{D-a} \right)^z \qquad (7.34)$$

Where $Z = \omega / k U_*$

U_* = shear velocity

k = Karman-Prandtl universal constant = 0.4 for clear water

Equation (7.34) is known as the Rouse (1937) equation. Comparison between experimental results obtained by Vanoni (1946) and equation (7.34) is shown in Fig. 7.8. Equation (7.34) provides the basic equation based on exchange theory for the development of suspended sediment transport equations.

7.4.2 SUSPENDED LOAD EQUATIONS

Lane and Kalinske (1941) assumed that diffusion coefficients for water and sediments in suspension are the same. The average value of ϵ_s over the flow depth is

$$\bar{\epsilon}_s = \frac{\int_0^D \epsilon_s dy}{D} = \frac{kU_*}{D^2} \int_0^D (yD - y^2) dy \tag{7.35}$$

For $k = 0.4$

$$\bar{\epsilon}_s = \frac{U_* D}{15} \tag{7.36}$$

Introducing equation (7.36) into equation (7.33) yields

$$C = C_a \exp\left[-\frac{15\omega}{U_*}\left(\frac{y-a}{D}\right) \right] \tag{7.37}$$

where C and C_a = suspended sediment concentration at y and a above the
bed, respectively
ω = fall velocity corresponding to d_{50}

Equation (7.37) can be integrated through the depth of flow to obtain the average suspended concentration, provided that the concentration at $y = a$ is known.

Based on his bedload transport function and Rouse equation, Einstein (1950) developed his suspended load equation. The suspended load per unit channel width for each sediment size fraction is

$$i_{sw}q_{sw} = i_{Bw}q_{bw}\left[2.303 \log\left(\frac{30.2D}{\Delta}\right)I_1 + I_2 \right]$$

$$= i_{Bw}q_{bw}[P_E I_1 + I_2] \tag{7.38}$$

where $P_E = 2.303 \log \dfrac{30.2D}{\Delta}$

Δ = a correction factor
i_{sw} = percentage of material in a given size range

Graphical solutions of I_1 and I_2 for different values of Z are given by Einstein (1950) as shown in Figs. 7.9 and 7.10. Total suspended load is the

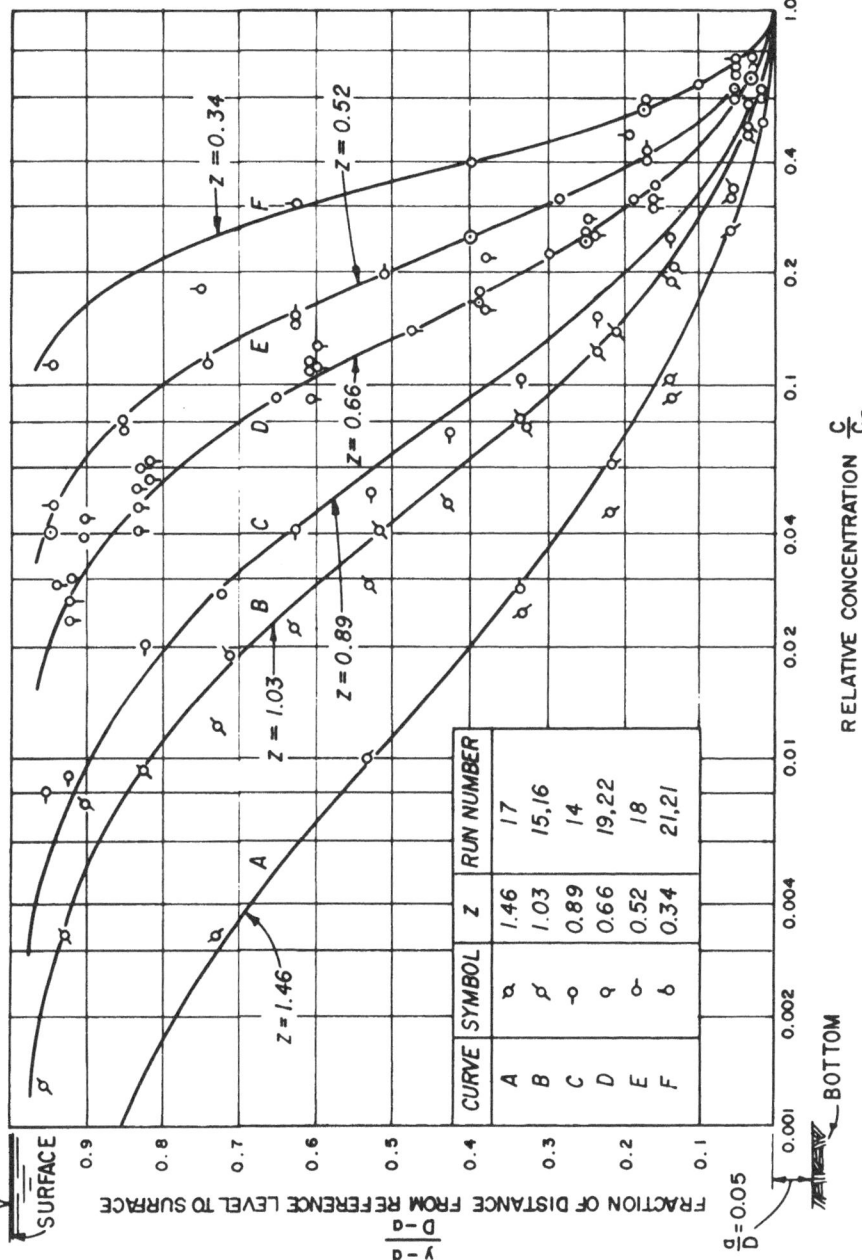

Fig. 7.8. Comparison between experimental results and Rouse's equation (Vanoni, 1946)

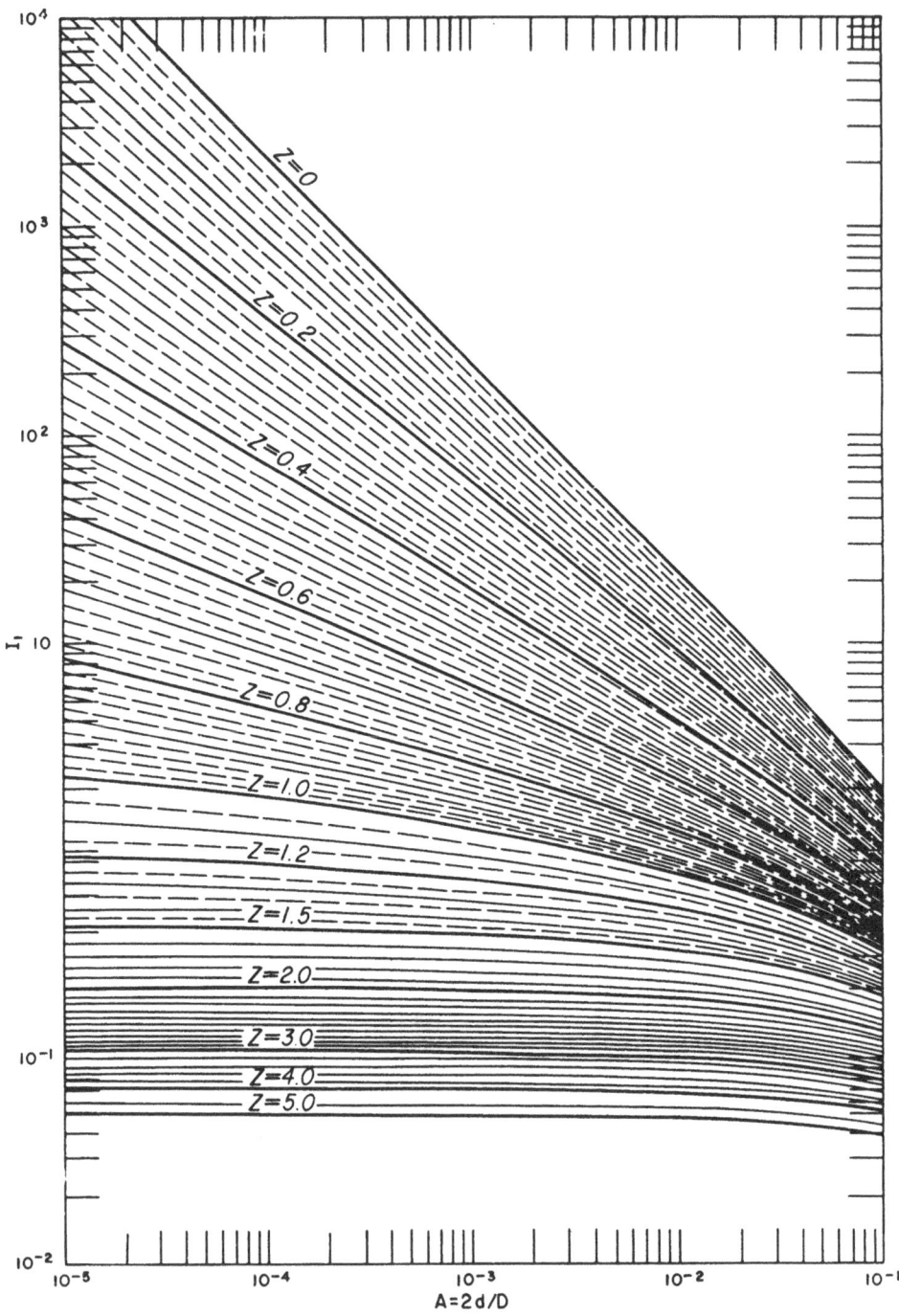

Fig. 7.9. Function I_1 in terms of A for values of Z (Einstein, 1950)

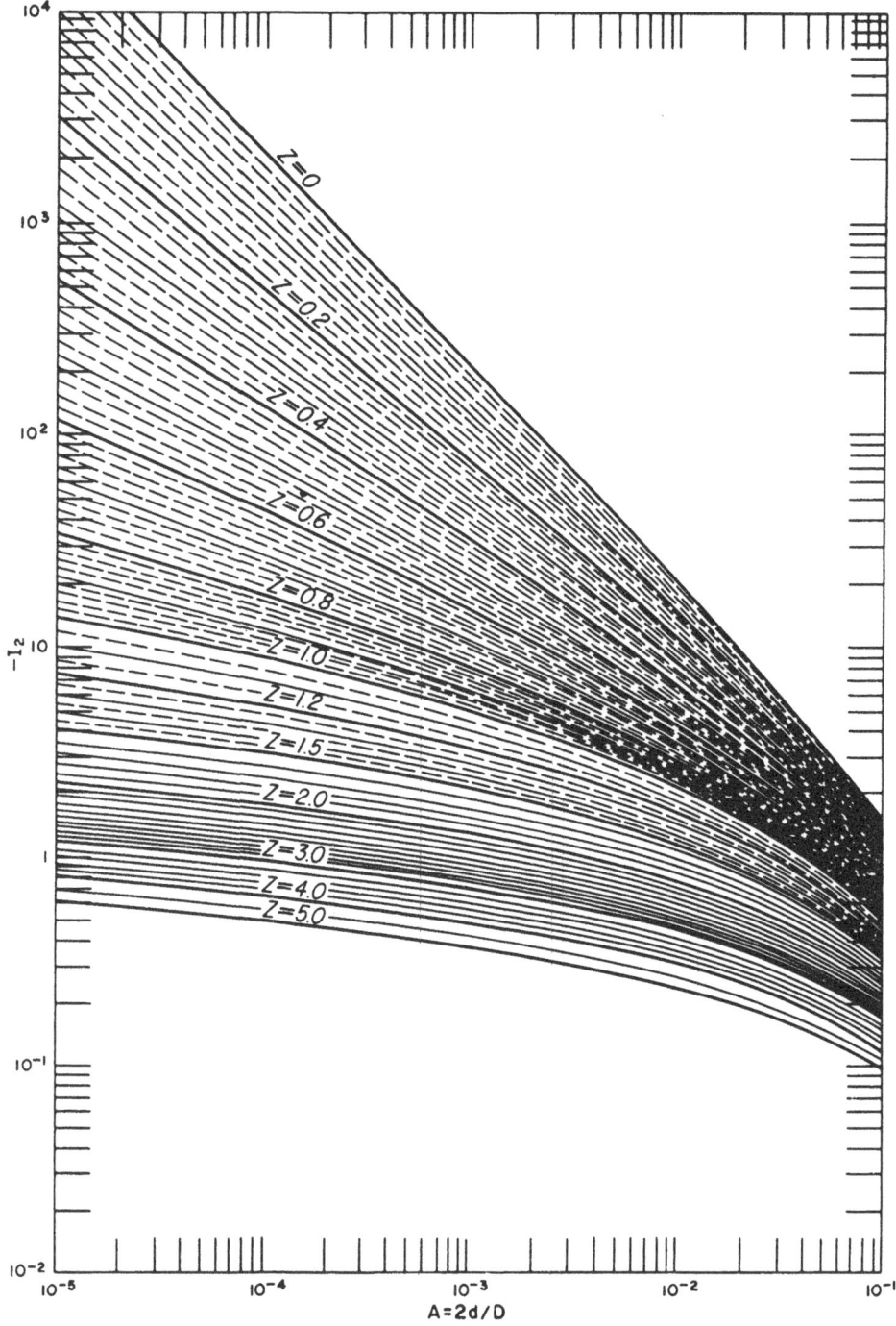

Fig. 7.10. Function I_2 in terms of A for values of Z (Einstein, 1950)

sum of those computed by equation (7.38) for each size fraction over the size range of the suspended material.

7.5 Total load transport

Total load is the sum of bedload and suspended load. Suspended load consists of those materials which can be found in the bed material and those which are finer than the bed material. The transportation of those fine materials in suspension is called wash load. The amount of wash load depends mainly on the supply and has very little to do with the hydraulics of the flow. Consequently, wash load has to be excluded from total load computation, and most total load equations actually compute total bed material load.

7.5.1 GENERAL APPROACH

There are two general approaches in the determination of total load. The first approach is to compute bedload and suspended load separately, and then add them together to obtain total load. The second approach is to develop total load function directly without dividing them into bedload and suspended load. A sediment particle may be transported as bedload at one time and as suspended load at another time or location. With the exception of coarse materials, which are mainly transported as bedload, total bed material load equations should be used for the determination of sediment transport capacity in most cases.

Einstein's (1950) approach is a classical example of computing total bed material load from bedload and suspended load, that is

$$\begin{aligned} i_t \, q_t &= i_{Bw} \, q_{bw} + i_{sw} \, q_{sw} \\ &= i_{Bw} \, q_{bw} \, (1 + P_E \, I_1 + I_2) \end{aligned} \qquad (7.39)$$

where $i_t \, q_t$ = total bed material load per unit channel width for each sediment size fraction

i_t = percentage of bed material in a given size fraction

q_t = total bed material load per unit channel width

Total bed material load is the sum of those computed by equation (7.39) for each size fraction over the size range of the bed material. The values of $i_{bw} \, q_{bw}$ and $i_{sw} \, q_{sw}$ can be computed from equations (7.29) and (7.38), respectively.

7.5.2 POWER APPROACH

The concept that rate of sediment transport should be related to the rate of energy dissipation of flow in transporting sediment particles has gained increasing acceptance in recent years. This concept has been applied to sediment transport in rivers as well as in the coastal areas.

7.5.2.1 *Bagnold's approach*

The power concept was first introduced by Bagnold (1966). Bagnold's relationship for bedload transport is

$$\left(\frac{\gamma_s - \gamma}{\gamma}\right) q'_{bw} \tan \alpha = \tau V e_b \qquad (7.40)$$

where γ_s and γ = specific weight of sediment and water, respectively
$\quad\quad q'_{bw}$ = bedload transport rate
$\quad\quad \tan \alpha$ = ratio of tangential to normal shear force
$\quad\quad \tau$ = shear force acting along the bed
$\quad\quad V$ = average flow velocity
$\quad\quad e_b$ = efficiency coefficient

In equation (7.40), τV is the stream power or the power per unit area acting along the bed. The values of e_b and $\tan \alpha$ are given by Bagnold in Fig. 7.11.

The rate of work used in transporting suspended load is

$$\left(\frac{\gamma_s - \gamma}{\gamma}\right) q_{sw} \frac{\omega}{\bar{u}_s} \qquad (7.41)$$

where q_{sw} = suspended load discharge in dry weight per unit time and width
$\quad\quad \bar{u}_s$ = mean transport velocity of suspended load
$\quad\quad \omega$ = fall velocity of suspended sediment

The rate of energy left available for transporting suspended load is

$$\tau V (1 - e_b) \qquad (7.42)$$

Based on general physics, that rate of work being done should be related to power available times the efficiency of the system,

$$\left(\frac{\gamma_s - \gamma}{\gamma}\right) q_{sw} \frac{\omega}{\bar{u}_s} = \tau V (1 - e_b) e_s \qquad (7.43)$$

where e_s = suspended load transport efficiency coefficient.
Equation (7.43) can be rearranged to

$$\left(\frac{\gamma_s - \gamma}{\gamma}\right) q_{sw} = (1 - e_b) e_s \frac{\bar{u}_s}{\omega} \qquad (7.44)$$

Assuming $\bar{u}_s = V$, Bagnold found $(1 - e_b) e_s = 0.01$ from experimental data. Thus, the suspended load can be computed by

$$\left(\frac{\gamma_s - \gamma}{\gamma}\right) q_{sw} = 0.01 \tau V^2 \qquad (7.45)$$

The total load is the sum of bedload and suspended load, that is, from equations (7.40) and (7.45)

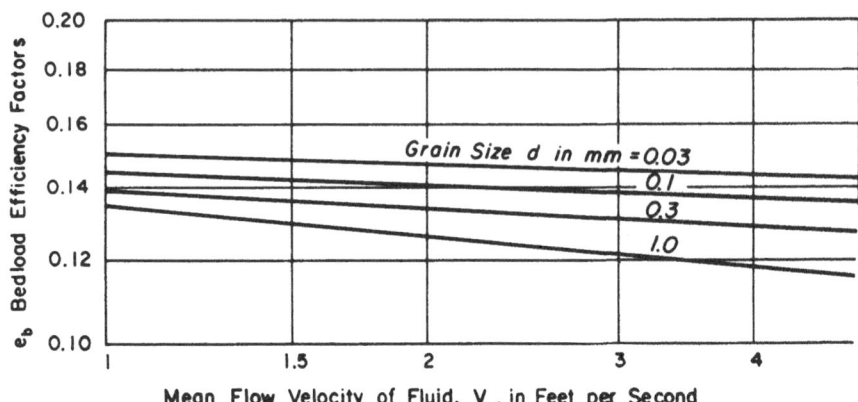

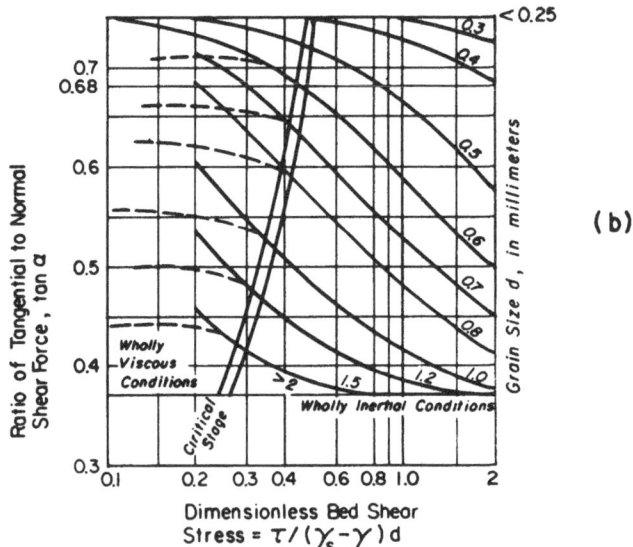

(b)

Fig. 7.11. Variation of e_b and tan α in Bagnold's bedload transport function (Bagnold, 1966)

$$q_t = q_{bw} + q_{sw} = \left(\frac{\gamma}{\gamma_s - \gamma}\right)\tau V\left(\frac{e_b}{\tan \alpha} + 0.01\frac{V}{\omega}\right) \tag{7.46}$$

7.5.2.2 Engelund and Hansen's approach

Engelund and Hansen (1972) applied Bagnold's stream power concept and the similarity principle to obtain a sediment transport function

$$f'\phi = 0.1\theta^{5/2} \tag{7.47}$$

with

$$f' = \frac{2gSD}{V^2} \tag{7.48}$$

$$\phi = \frac{q_t}{\gamma_s \sqrt{\left(\frac{\gamma_s - \gamma}{\gamma}\right)gd^3}} \tag{7.49}$$

$$\theta = \frac{\tau}{(\gamma_s - \gamma)d} \tag{7.50}$$

where g = the gravitational acceleration
S = the energy. slope
V = the average flow velocity
q_t = the total sediment discharge by weight per unit width
γ_s and γ = specific weight of sediment and water, respectively
d = the median particle diameter
τ = the shear stress along the bed

7.5.2.3 Ackers and White's approach

Based on Bagnold's stream power concept, Ackers and White (1973) applied dimensional analysis technique to express the mobility and transport rate of sediment in terms of some dimensionless parameters. They postulated that only a part of the shear stress on the channel bed is effective in causing the movement of coarse sediment. While in the case of fine sediment, suspended load movement predominates and the total shear stress is effective in causing the movement of sediment. Their mobility number for sediment is

$$F_{gr} = \frac{U_*^n}{\sqrt{gd\left(\frac{\gamma_s}{\gamma} - 1\right)}}\left[\frac{V}{\sqrt{32\log\left(\frac{\alpha D}{d}\right)}}\right]^{1-n} \tag{7.51}$$

where U_* = the shear velocity
n = the transition exponent depending on sediment size
α = the coefficient in rough turbulent equation ($=10$)
d = sediment particle size
D = the water depth

They also expressed the sediment size by a dimensionless grain diameter

$$d_{gr} = d \left[\frac{g\left(\frac{\gamma_s}{\gamma} - 1\right)}{v^2} \right]^{1/3} \tag{7.52}$$

Then a general dimensionless sediment transport function can be expressed by

$$G_{gr} = f(F_{gr}, d_{gr}) \tag{7.53}$$

with

$$G_{gr} = \frac{XD}{\left(\frac{\gamma_s}{\gamma}\right)d} \left(\frac{U_*}{V}\right)^n \tag{7.54}$$

where X = the rate of sediment transport in terms of mass flow per unit mass flow rate, that is, concentration by weight of fluid flux.

The generalised dimensionless sediment transport function can also be expressed as

$$G_{gr} = C \left(\frac{F_{gr}}{A} - 1\right)^m \tag{7.55}$$

The values of A, C, m, and n were determined by Ackers and White (1973) based on best-fit curves of laboratory data with sediment size greater than 0.04 mm and a Froude number less than 0.8, as shown in Fig. 7.12.

7.5.2.4 Yang's approach

The rate of potential energy per unit weight of water available for transporting water and sediment in an open channel with reach length X and total drop of Y is

$$\frac{dY}{dt} = \frac{dx}{dt}\frac{dY}{dx} = VS \tag{7.56}$$

Yang (1973) defined the velocity-slope product as the unit stream power. The rate of work being done by a unit weight of water in transporting sediment must be directly related to the rate of work available to a unit weight of water. Thus, total sediment concentration or total bed material load concentration must be directly related to unit stream power. While Bagnold (1966) emphasizes the power per unit bed area, Yang (1973) emphasizes the power per unit weight of fluid transporting sediments.

Yang's (1973) equation for total bed material transport in the sand size range is on page 164.

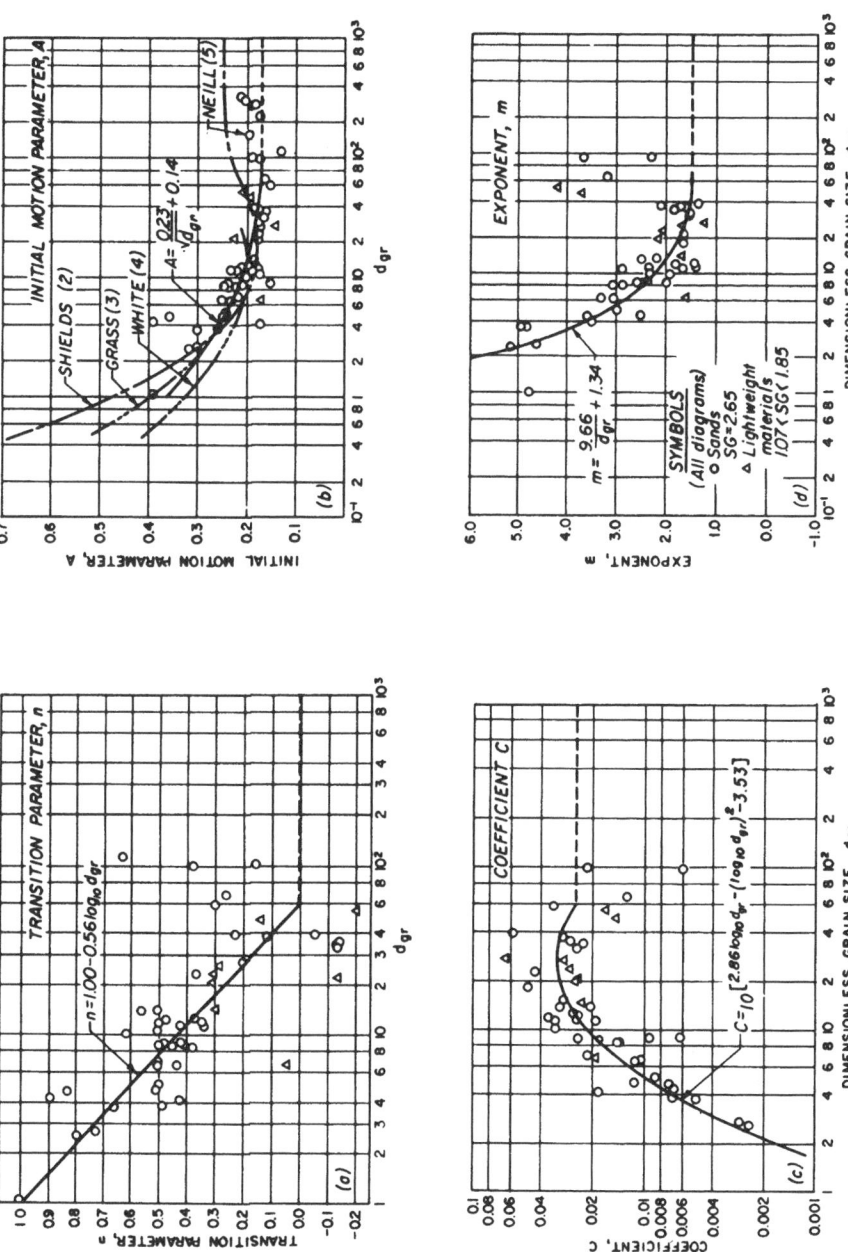

Fig. 7.12. Coefficients in Ackers and White's sediment transport function (Ackers and White, 1973)

$$\log C_{ts} = 5.435 - 0.286 \log \frac{\omega d}{\nu} - 0.457 \log \frac{U_*}{\omega}$$

$$+ \left(1.799 - 0.409 \log \frac{\omega d}{\nu} - 0.314 \log \frac{U_*}{\omega} \right) \log \left(\frac{VS}{\omega} - \frac{V_{cr}S}{\omega} \right) \qquad (7.57)$$

where C_{ts} = total sand concentration in parts per million by weight
  ω = terminal fall velocity of sediment particles in water
  d = median sediment particle diameter
  ν = kinematic viscosity of water
  U_* = shear velocity

Yang's (1984) equation for gravel transport is

$$\log C_{tg} = 6.681 - 0.633 \log \frac{\omega d}{\nu} - 4.816 \log \frac{U_*}{\omega}$$

$$+ \left(2.784 - 0.305 \log \frac{\omega d}{\nu} - 0.282 \log \frac{U_*}{\omega} \right) \log \left(\frac{VS}{\omega} - \frac{V_{cr}S}{\omega} \right) \qquad (7.58)$$

where C_{tg} = total gravel concentration in parts per million by weight.

The dimensionless critical average flow velocity at incipient motion, V_{cr}/ω, in equations (7.57) and (7.58) can be computed from equation (7.11) or (7.12). The basic form of equations (7.57) and (7.58) can be derived directly from basic theories in fluid mechanics and turbulence (Yang and Molinas, 1982) in conjunction with the use of dimensional analysis (Yang, 1973).

7.6 Comparison of transport equations

Vanoni (1975) compared the computed sediment discharges from different equations with the measured results from natural rivers. His comparisons were replotted by Yang (1977) to include Yang's (1973) unit stream power equation. Figure 7.13 shows an example comparison between computed and measured results by Colby and Hembree (1955) from Niobrara River near Cody, Nebraska.

It is apparent from Fig. 7.13 that computed results from different equations differ from each other. One of the basic reasons for this disagreement is the lack of generality of assumptions used in the development of equations. Critical reviews of the assumptions (Yang, 1977, 1983) indicate that the rate of sediment transport can be better determined by the power approach, especially the unit stream power approach. A comparison between Figs. 7.14 and 7.15 made by Vanoni (1978) also indicates that significant improvement can be achieved by using unit stream power for the determination of sediment transport rate or concentration.

Direct comparisons between measured and computed results from different sediment transport equations were made by different investigators. They include, but are not limited to, those made by Ackers and White (1973),

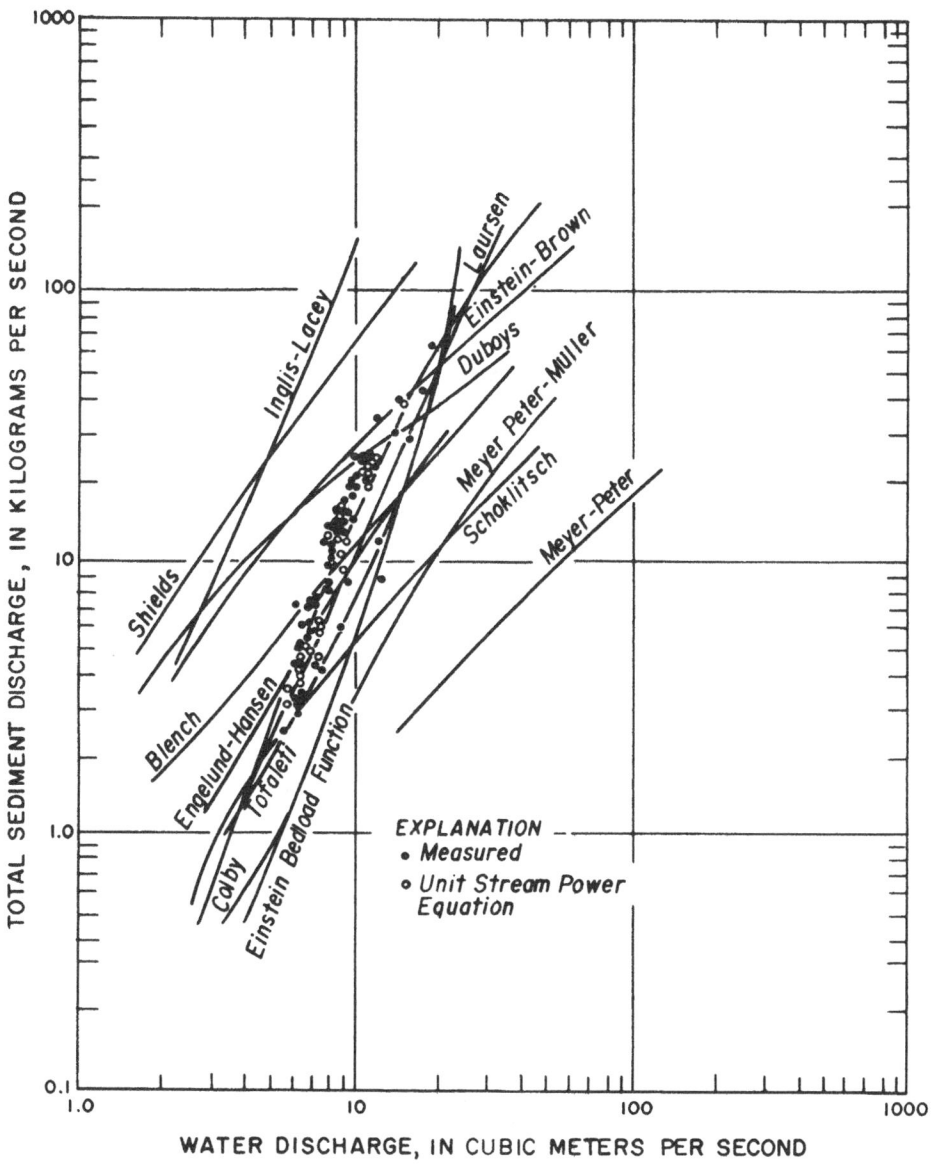

Fig. 7.13. Comparison between measured total sediment discharge of the Niobrara River near Cody, Nebraska, and computed results of various equations (Yang, 1977)

166

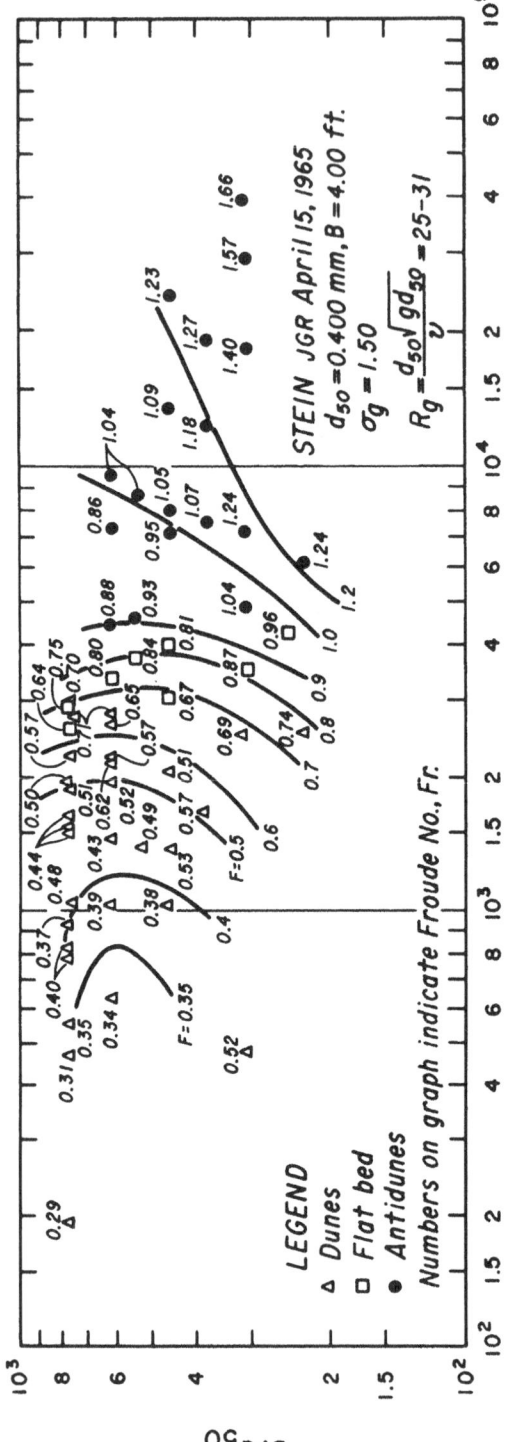

Fig. 7.14. Plot of Stein (1965) data as sediment discharge concentration against Froude number, F_r, and the ratio of flow depth, D, to bed-sediment size d_{50} (Vanoni, 1978)

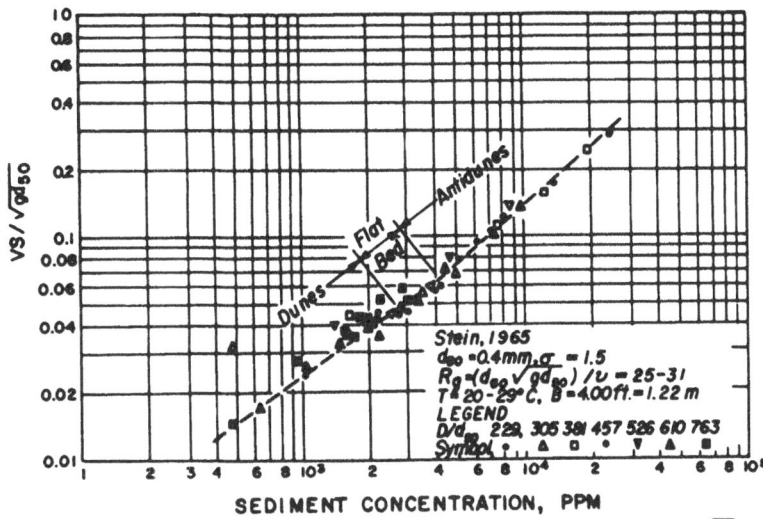

Fig. 7.15. Plot of Stein (1965) data as sediment discharge concentration against $VS/\sqrt{gd_{50}}$ (Vanoni, 1978)

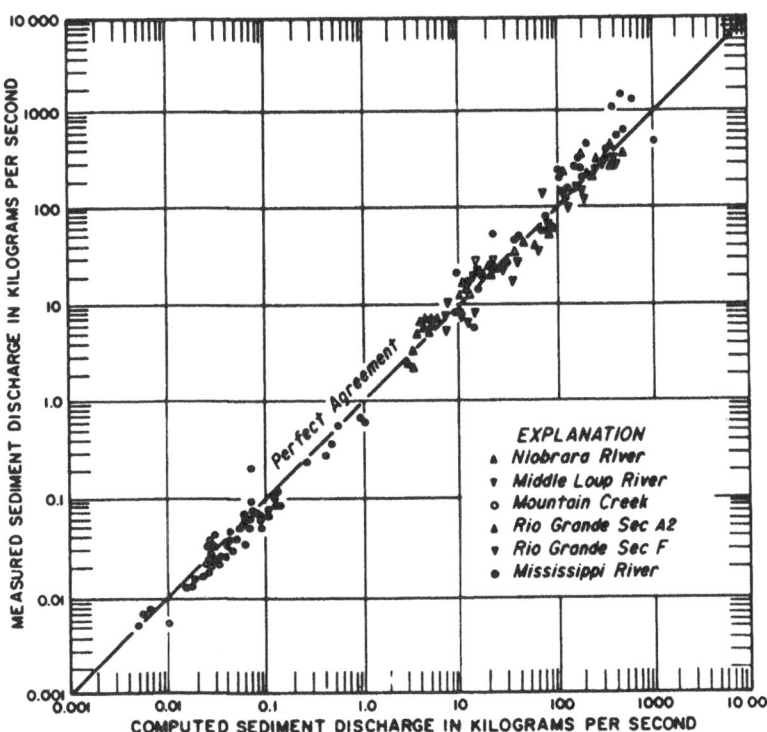

Fig. 7.16. Comparison between measured total bed material discharge from six river stations and computed results from Yang's 1973 equation (Yang, 1979, 1980)

Alonso (1980), Alonso *et al.* (1982), and the ASCE Task Committee (1982). Figure 7.16 shows an example of a comparison made by Yang (1977, 1980). Although Fig. 7.16 shows good agreements between computed and measured results, it does not mean there is a universal equation which can be applied with accuracy under all conditions. Procedures for selecting sediment transport equations under different flow and sediment conditions were suggested by Yang (1977).

REFERENCES

Ackers, P. and White, W. R. (1973): Sediment transport: new approach and analysis, *Journal of the Hydraulics Division, ASCE* 99, No. HY 11, Proceeding Paper 10167, pp. 2041–2060, November.

Alonso, C.V. (1980): Selecting a Formula to Estimate Sediment Transport Capacity in Non-vegetated Channels, chapter 5, CREAMS (A Field Scale Model for Chemicals, Runoff, and Erosion from Agricultural Management System), W.B. Knisel, ed., U.S. Department of Agriculture, Conservation Research Report No. 26, pp. 426–439, May.

Alonso, C.V., Neibling W.H. and Foster G.R. (1982): Estimating Sediment Transport Capacity in Watershed Modeling, *Transaction, ASAE*, 24 (5): 1211-1120 and 1226.

ASCE Task Committee on Relations between Morphology of Small Streams and Sediment Yield of the Committee on Sedimentation of the Hydraulics Division. Relationships between morphology of small streams and sediment yields, *Journal of the Hydraulics Division*, vol. 108, No. HY 11, Proceeding Paper 17450, pp. 1328–1365, November.

Bagnold, R.A. (1966) *An approach to the sediment transport problem from general physics*, U.S. Geological Survey Proffesional Paper 422-J.

Colby, B.R. and Hembree C.H. (1955): *Computation of Total Sediment Discharge, Niobrara River near Cody, Nebraska*, U.S. Geological Survey, Water Supply Paper 1357.

Donat, J. (1929): *Über Sohlangriff and Geschiebetrieb*, Wasserwirtschaft, Heft 26, 27.

DuBoys, M.P. (1879): *Le Rhone et les Rivieres a Lit affouillable*, Annales de Ponts et Chausses, sec. 5, vol. 18, pp. 141–195.

Einstein, H.A. Formula for the transportation of bed-load, *Transactions*, American Society of Civil Engineers, 107.

Einstein, H.A. (1950): The bed-load function for sediment transportation in open channel flows, U.S. Department of Agriculture, Soil Conservation Service, *Technical Bulletin* No. 1026.

Engelund, F. and Hansen, E. (1972): *A Monograph on Sediment Transport in Alluvial Streams*, Teknisk Forlag, Copenhagen.

Fortier, S. and Scobey. F.C. (1926): Permissible canal velocities, *Transactions*, American Society of Civil Engineers, 89.

Hjulstrom, F. (1935): The morphological activity of rivers as illustrated by river fyris, *Bulletin of Geological Institute*, 25: Chapter 3, Uppsala.

Lane, E.W. and Kalinske. A.A. (1941): "Engineering Calculations of Suspended Sediment," *Transactions, American Geophysical Union*, 20: 603–607.

Meyer-Peter, E., Favre H. and Einstein. A. (1934): "Neuere Versuchsresultate über den Geschiebe-trieb," Schweiz Bauzeitung, 103–13.

Rouse, H. (1937): Modern conceptions of the mechanics of turbulence, *Transactions, ASCE*, 102.

Schlichting, H. (1962): *Boundary Layer Theory*, Fourth Edition, Translated by J. Kestin, McGraw-Hill Book Company, New York, NY, pp. 509–526.

Schoklitsch, A. (1934): Der Geschiebetrieb und die Geschiebefracht, *Wasserkraft und Wasserwirtschaft*, 29 (4): 37–43.

Schoklitsch, A. Berechnung der Geschiebefracht, *Wasser und Energiewirtschaft*, No. 1.

Shields, A. (1936): *Application of Similarity Principles and Turbulence Research to Bed-Load*

Movement, Translated from German to English by W.P. Ott and J.C. Van Uchelen, California Institute of Technology, Pasedena, California.

Straub, L.G. (1935): Missouri River Report, In-House Document 238, 73rd Congress, 2nd Session, U.S. Government Printing Office, Washington, D.C., pp. 1135.

Talapatra, S.C. and Ghosh, S.N. (1983): Incipient motion criteria for flow over a mobile bed sill, *Proceedings of the Second International Symposium on River Sedimentation*, Nanjing, China, 459–471, October.

Vanoni, V.A. (1946): Transportation of suspended sediment by water, *Transactions, ASCE*, III.

Vanoni, V.A. ed. (1975): Sedimentation Engineering, ASCE Task Committee for the Preparation of the Manual on Sedimentation of the Sedimentation Committee, Committee of the Hydraulics Division, (Reprinted 1977).

Vanoni, V.A. (1978): *Predicting Sediment Discharge in Alluvial Channels*, Water Supply and Management, Pergamon Press, pp. 399–417.

Yang, C.T. (1973): Incipient motion and sediment transport, *Journal of the Hydraulics Division, ASCE*, 99, No. HY 10, Proceeding Paper 10067, pp. 1679–1704, October.

Yang, C.T. (1977): *The Movement of Sediment in Rivers*, Geophysical Survey 3, D. Reidel Publishing Company, pp. 39–68, Dordrecht, Holland.

Yang, C.T. (1980): Sediment transport and river engineering, *Proceedings of the International Symposium on River Sedimentation*, Beijing, China, 1: 350·386, March.

Yang, C.T. (1983): Rate of energy dissipation and river sedimentation, *Proceedings of the Second International Symposium on River Sedimentation*, Nanjing, China, pp. 575–585, October.

Yang, C.T. (1984): Unit stream power equation for gravel, *Journal of the Hydraulic Engineering, ASCE*, vol. 110, No. HY 12, December.

Yang, C.T. and Molinas. A. (1982): Sediment transport and unit stream power function, *Journal of the Hydraulics Division, ASCE*, vol. 108, No. HY 6, Proceeding Paper 17161, pp. 776–793, June.

CHAPTER 8

River Protection Works

8.1 Introduction

Engineering works related to fish farm construction and other ancilliary works are carried out near river banks. In order to protect the fish farms from destruction due to fluctuating river flows, especially when the river is very active and erosion by meandering occurs, it is necessary to provide various types of river protective devices. This chapter briefly deals with various causes of river banks failures and then discusses the protective measures to be adopted for their prevention.

A river is a dynamic system. It goes through a progressive life cycle from youth to maturity during the geological time circle. During the engineering time span, a river also adjusts its geometry, slope, profile, and pattern to changing natural and man-made conditions. A river is relatively stable if the bed elevation remains relatively constant. Some typical signs of an unstable river are: rushing water in an otherwise tranquil reach, active headcutting, channel width expansion or contraction, aggradation or degradation of river bed. Successful river protection works require a thorough understanding of the causes of bank failure, methods available for bank protection and stabilisation, and the development of a locally acceptable plan of action. The U.S. Army Corps of Engineers did extensive studies on streambank erosion control methods under the Streambank Erosion Control Evaluation and Demonstration Act of 1974, Section 32, Public Law 93-251. Valuable information has been published by the U.S. Army Corps of Engineers (1981). Specific streambank protection guidelines were also developed by the U.S. Army Corps of Engineers (1983).

8.2 Streambank erosion and failure

The first step in developing a successful streambank protection method is to find out the causes of bank erosion and failure. Once the causes are identified, proper correction or protection measures can be developed.

8.2.1 STREAMBANK EROSION

Streambank erosion occurs when individual soil particles at the bank surface are carried away. The major causes of bank erosion are river current, rainfall, seepage, overbank drainage, obstacles in the river, wave action, freeze-thaw and wet-dry cycles, ice and debris, and changes in land use (U.S. Army Corps of Engineers, 1983).

Soil particles along a riverbank can be eroded by river current at a velocity or tractive force exceeding the incipient motion requirement. Because velocity and tractive force usually increase with increasing water discharge and river stage, erosion is more likely to occur during floods or high flows than during normal or low flows. The ability of a soil particle to resist erosion by river current depends on the size and cohesive properties of the particle. Cohesive materials, such as clay, are more resistant to current erosion than non-cohesive materials, such as silt and sand. For non-cohesive materials, the resistance to erosion increases with particle size and specific weight.

Due to uneven distribution of velocity and shear, erosion usually first occurs along the outside of a bend. The secondary current and the slower velocity along the inside of a bend can cause sediment deposition and formation of a point bar during low flows. However, during high flows, the thalweg which means line of maximum current will shift from the outer to inner bend and cut through the point bar which was formed at low flows. Figure 8.1 illustrates this process.

Raindrops can strike and loosen soil particles as water runs down the sloped bank. Loosened soil will be carried away in thin layers as sheet erosion. As sheet erosion continues, the runoff will form small channels and the erosion becomes rill erosion. Erosion due to rainfall can be severe if the bank slope is not protected by vegetation or other protections and overbank drainage is developed.

Part of the rainfall that strikes the soil surface becomes surface runoff and part infiltrates into the soil. Figure 8.2 illustrates that the infiltrated water can combine with ground water to form seepage across the face of a riverbank. The seepage can loosen bank materials and cause bank failure.

The potential for bank erosion can be increased or decreased due to man-made or natural obstacles in a river. These obstacles can cause localised increase or decrease of velocity. The localised increase or decrease of velocity will cause local scour or deposition. Figure 8.3 shows some typical obstacles in a stream and their effect on local current, scour, and deposition. Usually, the existence of obstacles can cause eddies to form. Depending on the intensity of eddies, they may cause severe local scour which may lead to bank failure.

Wind or boat-generated waves can also cause bank erosion. This can be an important factor for consideration in the bank protection design if the river or canal is intended for navigation.

When water particles are frozen in the soil, their volume increases and loosens or separates soil particles. When the ice is thawed, the bank materials are more easily eroded. The wet-dry cycle has a similar effect on bank

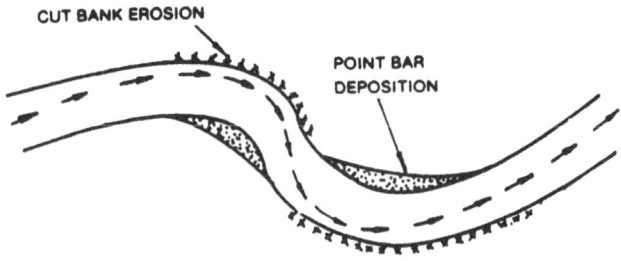

A. Sharply curved bends.

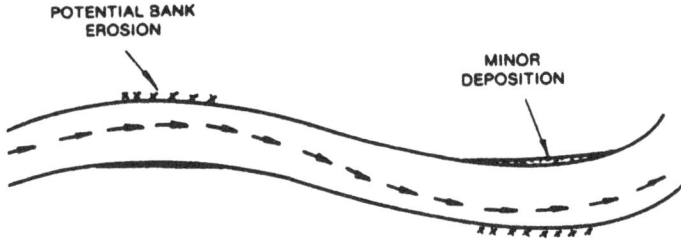

B. Gradually curved bends.

Sharply and gradually curved bends under normal flow conditions. Arrows indicate the path of the maximum stream velocity. During floodflow this path moves across the channel against the point bar.

Fig. 8.1. River current and erosion (from U.S. Army Corps of Engineers, 1983)

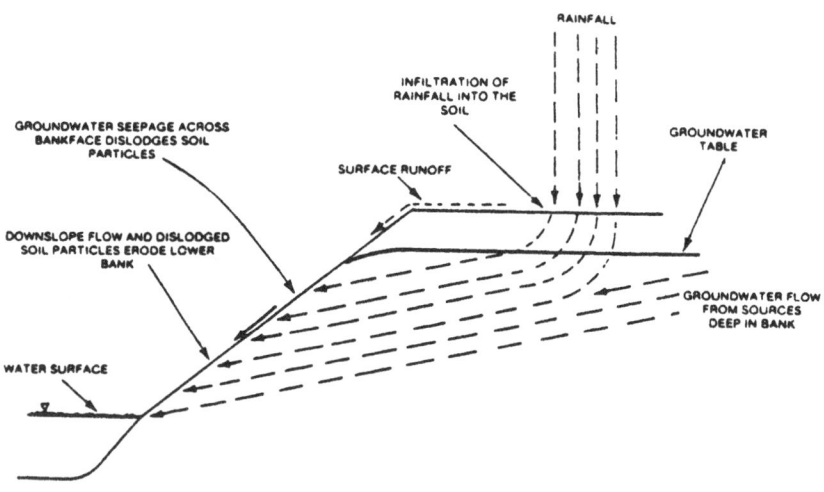

Streambank erosion caused by groundwater seepage across bankface.

Fig. 8.2. Riverbank erosion due to seepage (from U.S. Army Corps of Engineers, 1983)

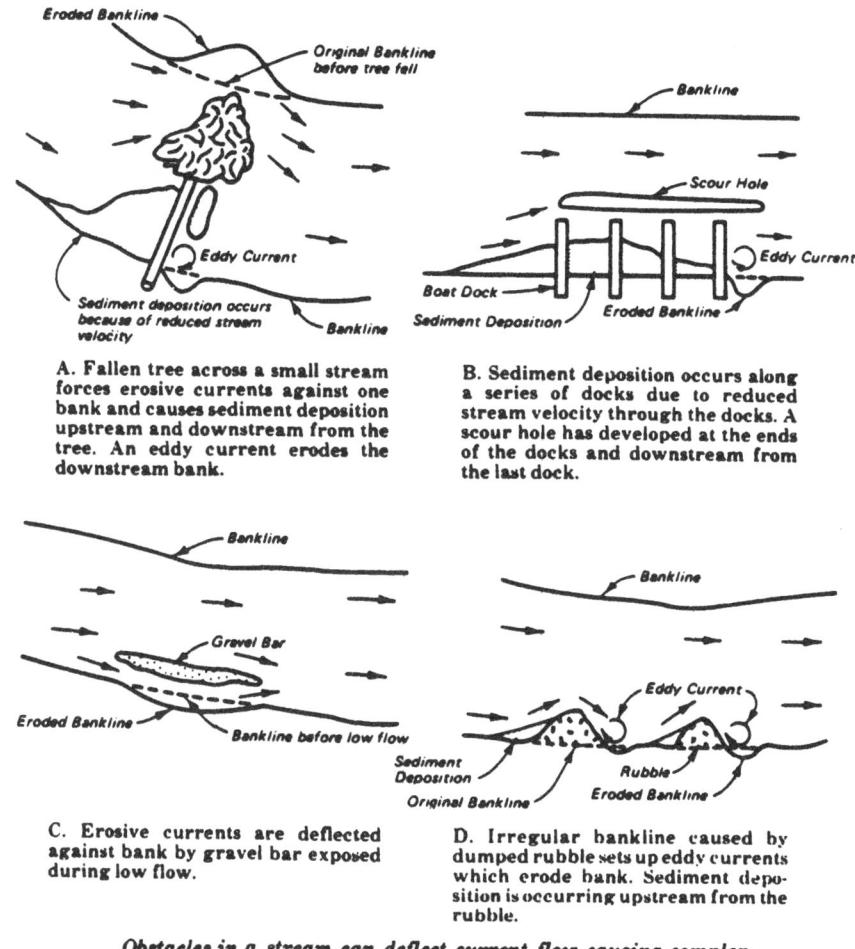

A. Fallen tree across a small stream forces erosive currents against one bank and causes sediment deposition upstream and downstream from the tree. An eddy current erodes the downstream bank.

B. Sediment deposition occurs along a series of docks due to reduced stream velocity through the docks. A scour hole has developed at the ends of the docks and downstream from the last dock.

C. Erosive currents are deflected against bank by gravel bar exposed during low flow.

D. Irregular bankline caused by dumped rubble sets up eddy currents which erode bank. Sediment deposition is occurring upstream from the rubble.

Obstacles in a stream can deflect current flow causing complex erosion and sediment deposition patterns.

Fig. 8.3. Effects of obstacles on local current, scour, and deposition (U.S. Army Corps of Engineers, 1983)

materials. Floating ice blocks and debris can partially block the flow passage and increase local velocity and erosion potential. The debris or floating ice can also cause bank erosion by abrasion or impact.

Urbanisation or change of land use from grassland or forest to farmland can dramatically increase soil erosion if the land is not properly managed or protected. The removal of vegetation can reduce infiltration rate and increase flood peak and soil erosion. These all have negative impacts on the stability of streambanks.

The U.S. Army Corps of Engineers (1981, 1983) made detailed studies on the mechanism which can contribute or cause bank failure. Table 8.1 summarises the surficial bank deterioration mechanisms. Table 8.2 summarises the mechanisms related to sediment transport. Table 8.3 summarises streambank failure mechanics. This information should be useful to design engineers in developing methods of protection.

Table 8.1
Surficial bank deterioration mechanisms (from U.S. Army Corps of Engineers, 1983)

Mechanism	Description
Abrasion	Solid materials carried by wind or flowing water collide with and dislodge surface soil particles. Abrasion also occurs during shifting of winter ice covers.
Biological (animals)	Examples are bank surface destruction during overgrazing and by animal burrows and trails.
Biological (vegetation)	Vegetation normally is conductive to surficial stability; exceptions occur during decay of root material and by tree falls or vegetation patterns that concentrate or cause turbulence in overbank flows or streamflows.
Chemical	Water and acids in water affect cohesive and other types of particle-to-particle bonding; bank material is removed by dissolution.
Debris	Debris gouges, or scrapes material from, bank surfaces as well as causing turbulence and flow concentration.
Flow (water)	Soil particle removal by overbank flows and streamflows is a major cause of bank surface deterioration. Quantity of flow, transport capacity, turbulence, secondary currents, and wave action contribute to the rate and location of surficial particle removal. Seepage flows remove surface particles as well as contributing to mass bank failures.
Freeze-Thaw	Cyclic temperature changes cause fractures due to excessive contraction and expansion and spalling due to successive freezing and thawing of moisture within the bank.
Gravity	The stable slope of a cohesionless bank corresponds to gravitational stability; for steeper slopes, surface particles roll downslope (raveling).
Human actions (on bank)	Certain human actions attack the bank—loosening the bank surface material by farming or other mechanised operation is one example. Other actions may influence natural mechanisms—the destruction of a protective vegetation cover by livestock overgrazing is one example. Many actions are possible.
Human action (stream channel)	Examples of direct actions are dredging and sand or gravel mining of channel sediments. Examples of indirect actions are structures and vessel propeller motion that cause turbulence in the streamflow. Many actions are possible.
Ice	Ice contributes to abrasion and debris. Ice jams restrict a channel and affect stream and overbank flows.
Precipitation	Surficial destruction occurs due to impact by rain or hail as well as during periods of high stream flows and overbank flows.
Waves	Waves due to wind or stream vessel traffic cause surficial deterioration of the bank near the stream surface.
Wet-dry	Alternate wetting and drying cause stress and chemical effects that result in surface soil particle loosening.
Wind	Surface deterioration by wind is normally small as compared with water flow; however, waves due to wind contribute to surficial deterioration.

Table 8.2
Sediment transport mechanisms (from U.S. Army Corps of Engineers, 1983)

Mechanism	Description
Gravity	Gravity is an intermediate means of transport because either materials are removed from the site by other mechanisms or transport ceases due to accumulation.
Human action	Direct transport, such as occurs during local dredging or during mining for sand or gravel, is a site-specific event. Indirect actions are those that either enhance or inhibit natural transport and may either be site-specific or, whenever the action significantly affects streamflow transport, influence erosion at numerous sites along the stream.
Waterflow	Transport by flowing water (hydraulic transport) is the most effective natural transport mechanism as far as streambank erosion is concerned. Hydraulic transport is categorised as stream, overbank, and seepage transport.
	i) Transport by streamflow is enhanced by high flow rates and velocities and by low entering sediment concentrations. Whenever the transport capacity is exceeded, the excess material is deposited and aggradation occurs; conversely, whenever transport is below capacity, material tends to be eroded from the streambed and banks and degradation occurs. Streamflow is determined by drainage basin, rather than local, hydrologic events.
	ii) Overbank transport, which is a major factor in sheet and gully erosion, is hydraulically similar to stream transport. However, local rather than basin-wide hydrologic events are of dominant concern in overbank transport.
	iii) Seepage is dissimilar from streamflow and overbank flow because of the confined flow circumstance. Higher seepage flows occur in more porous material and are commonly enhanced by local precipitation and low stream stages. Piping is an example of sediment transport by seepage flows.
Wind	Exposed fine-grained cohesionless sediments are, to a small extent as far as streambank erosion is concerned, transported by wind.

Table 8.3
Streambank failure mechanisms (from U.S. Army Corps of Engineers, 1983)

Mechanism	Description
Surficial	Stresses within a streambank are changed by particular actions at the bank surface. Examples of surficial actions that effect bank stability are:
	i) Severe surface deterioration may result in an unstable bank configuration. Erosion at the toe of the bank slope due to stream flow, erosion at the water surface due to waves, and erosion along the bank surface due to overbank and seepage flows are three common occurrences.
	ii) Deep tension cracks due to excessive drying of a cohesive soil or

Mechanism	Description
	similar structural change may cause the streambank to weaken and become unstable. Slaking may occur if excessive drying is followed by submergence. iii) Overburden placed along top of bank may cause an otherwise stable streambank configuration to become unstable.
Moisture variation	Stresses and the ability of the bank material to withstand stress without failing are both affected by moisture variation within the bank. Examples of these moisture-induced effects are: i) The slope of a cohesionless bank may be temporarily steeper than the angle of repose of the bank material due to capillarity or other non-permanent stabilising effect; when the non-permanent effect is removed (usually by submergence and saturation of the bank material), the bank becomes unstable. ii) During piping, cohesionless material is eroded from a location on the bank surface by seepage flow; a cavity develops and extends rapidly into the bank along a dominant seepage path. iii) Liquefaction relates to fine-grained and loosely structured materials subject to a rapid increase in pore pressure (such as occurs during rapid drawdown or earthquake loading) and results in a large segment of bank material flowing downslope as a fluid-like mixture. iv) During periods of high water table and low stream levels an added hydraulic loading is placed on the bank structure; this added load may directly cause failure unless relieved otherwise (say by seepage or piping). v) Swelling and shrinking during wetting and drying, respectively, affect the stability of clay soils. Substantial hydraulic pressures may result from water flowing freely into deep tension cracks (see Surficial, above) and into openings between different bank materials. vi) The shear strength of clay soils is highly dependent on pore pressure (slow versus quick shear) and by degree of saturation.
Miscellaneous	Because of the non-homogeneous (heterogeneous, interbedded stratified) character of most streambanks, combinations of failure mechanisms are common; examples are: i) Artesian or gravity flow within a cohesionless or porous layer that evacuates sediment particles by piping can result in shear failures of layers higher in the bank. ii) A thin clay layer that weakens and compresses during saturated bank conditions can also cause shear failures in the upper bank. iii) Lubrication by water and high hydrostatic pressures along interfaces between bank materials that cause low resistance to sliding may result in a massive bank failure. iv) Many other site-specific combinations of mechanisms occur.

8.2.2 STREAMBANK FAILURE

A streambank may fail when and where the active force exceeds the resistance to soil movement. Figure 8.4 shows two basic types of failure, that is, a relatively thin soil layer sliding down the bank, or a large mass of soil

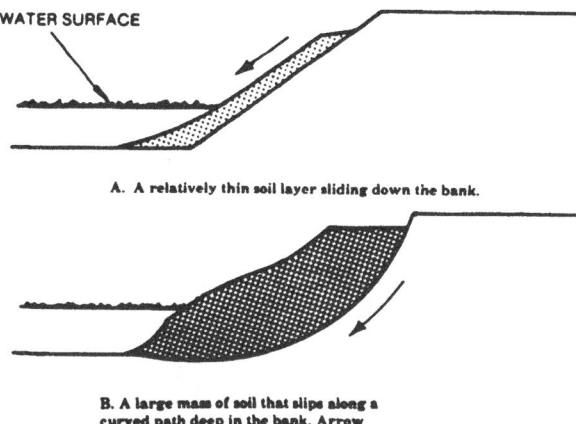

A. A relatively thin soil layer sliding down the bank.

B. A large mass of soil that slips along a curved path deep in the bank. Arrow indicates direction of soil movement.

Two basic types of bank failure.

Fig. 8.4. Basic types of bank failure (from U.S. Army Corps of Engineers, 1983)

slips along a curved path. The basic cause of failure is either due to the decrease in shear resistance of the soil, or an increase of active shear stress, or both. The decrease in shear resistance can be caused by (a) swelling of soil due to absorption of water, (b) increase of ground water or seepage force within the bank, and (c) minor movement of the soil or creep. The increase of active shear stress can be caused by (a) change of channel shape due to local scour or erosion, (b) increase of loading on top of the bank, and (c) rapid drawdown of water surface elevation. Figure 8.5 illustrates the progress of bank failure due to under-cutting at the toe. Figure 8.6 explains the cause and process of bank failure due to rapid drawdown.

In summary, streambank erosion and failure can be caused by one or a combination of the following:

1) Attack at the toe of the underwater slope, leading to bank erosion and failure.
2) Erosion of soil along the bank due to high velocity.
3) Sloughing of saturated banks due to rapid drawdown.
4) Liquefaction of streambank due to saturation of soil.
5) Erosion of soil by seepage out of the bank.
6) Erosion of streambank due to wave action.
7) Erosion due to rainfall or overland runoff.

8.3 Streambank protection approaches

The general approaches in protecting a streambank can be classified as (i) relocate endangered assets, (ii) implement effective land use management practices, (iii) reroute the stream channel away from the problem area, (iv) remove stream flow obstructions, and (v) provide structural protec-

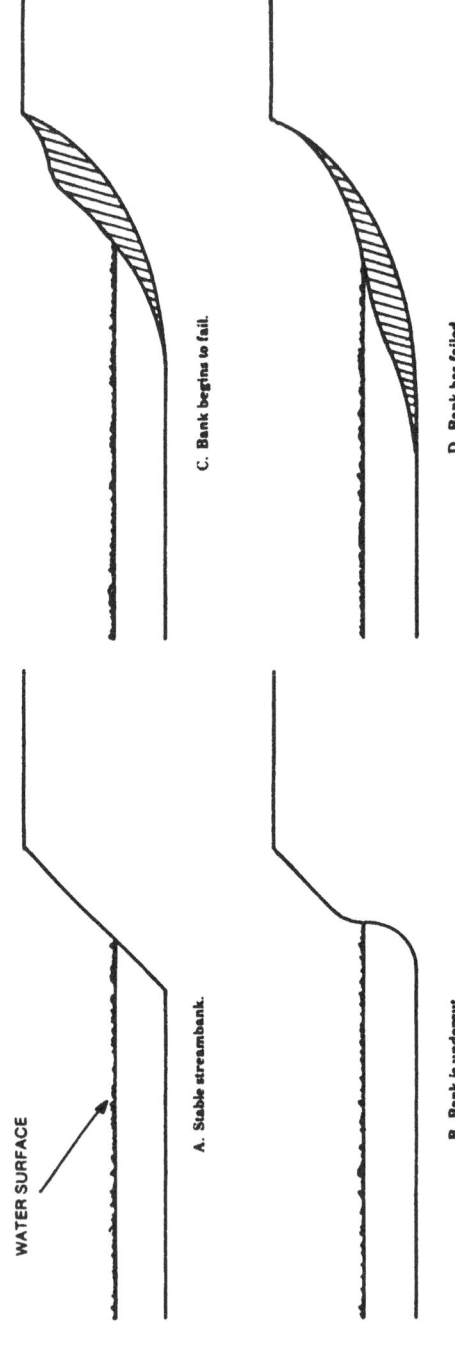

Fig. 8.5. Bank failure due to undercutting at the toe (from U.S. Army Corps of Engineers, 1983)

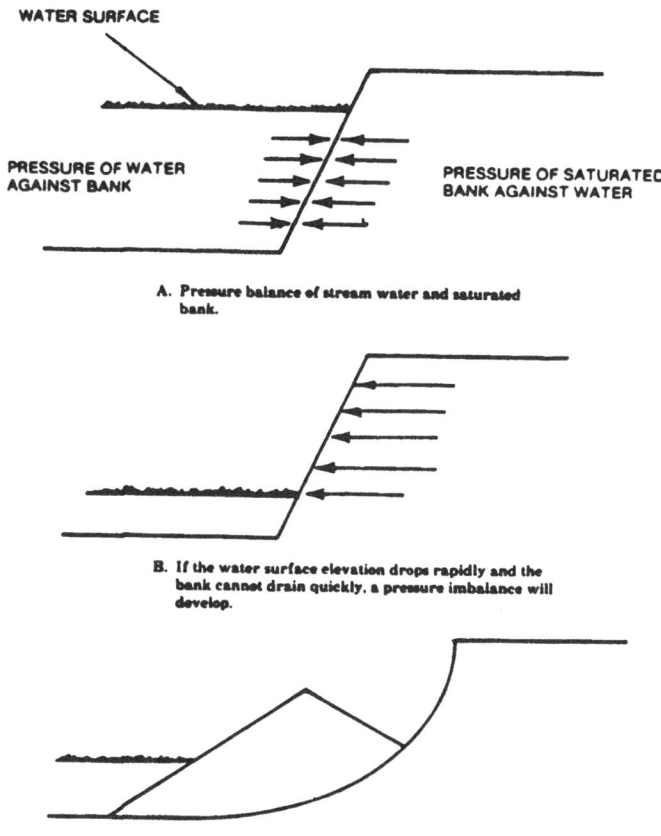

A. Pressure balance of stream water and saturated bank.

B. If the water surface elevation drops rapidly and the bank cannot drain quickly, a pressure imbalance will develop.

C. Bank may fail because of pressure imbalance.

Bank failure due to the water against a saturated bank being lowered faster than the bank can drain (rapid drawdown).

Fig. 8.6. Bank failure due to rapid drawdown (from U.S. Army Corps of Engineers, 1983)

tions such as riprap and gabion. These approaches can be used individually or combined for the development of the most economic and feasible plan of protection based on the site-specific conditions.

Depending on the social and economic value of the land or property, it may be a viable solution to let the bank erode and fail. In many cases, this approach requires relocation of roads, structures, utility lines, etc. Key factors to be considered are the cost of relocation, load value, the rate of bank erosion, and the distance of relocation. This information and the consequences of relocation should be carefully studied before the implementation of a relocation plan.

Effective land use management and maintenance can significantly reduce soil erosion and the chance of streambank failure. These practices include (i) the use of vegetation along the streambank, (ii) regulation of irrigation to enhance the growth of vegetation, (iii) improve or reroute overbank drainage to reduce seepage through the bank surface, (iv) control of runoff to reduce surface or rill erosion on streambank slopes, (v) improve soil conservation practice to reduce rate of runoff or overland flow from rainfall, and (vi) minimising load on top of the streambank by limiting the maximum load and traffic allowed on the streambank. Figure 8.7 illustrates the use of ditch and dike to control overbank drainage.

Sometimes it is more cost effective to reroute the stream away from the point of attack than trying to protect the streambank. Figure 8.8 illustrates how this can be done. Channel relocation is usually accomplished by cutting a pilot channel as shown in Fig. 8.8 (B). Due to the steeper slope, the pilot channel will capture the flow and cause sediment deposition in the original natural channel as shown in Fig. 8.8. (C). Eventually the natural channel is blocked and the flow follows the new course as shown in Fig. 8.8 (D) Although the use of pilot channels may be economical, the cutoff can reduce the length of flow and increase the local channel slope. This could cause a chain reaction in channel adjustments both upstream and downstream from the cutoff. Careful consideration should be given to the dynamic response of a natural stream to man-made cutoffs. A short-term economic plan of action may prove to be an expensive one to maintain in the long run.

Fallen trees, log jams, and mid-channel bars are common obstructions in a stream which can alter the flow characteristics. As a result, localised concentrated flow may cause bank erosion and bank failure. These obstructions may need to be removed to improve flow pattern. The use of a pilot channel and current deflectors can facilitate the removal of bars.

8.4 Streambank protection methods

Different types of streambank protection methods have been developed based on one or a combination of the following principles:

1) Protect the bank and toe from direct contact with high velocity and wave.
2) Divert high velocity away from the area to be protected.
3) Reduce velocity or magnitude of wave.
4) Improve drainage to avoid soil saturation and liquefaction.
5) Trap sediment to protect the toe and bank.
6) Reduce the impact of rain and overland flow.
7) Reduce the channel slope or bank slope or both.
8) Provide surface coverage.
9) Streamline the low path.

The reduction of channel slope can be achieved by building drop structures or grade control structures. Channel realignment and removal of obstructions may be needed to streamline the flow path. The methods which can be used to stabilise streambanks include stone riprap, concrete pavement, articulated concrete mattresses, transverse stone dike, transverse timber

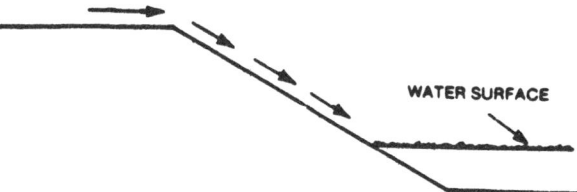

A. Uncontrolled overbank drainage can cause surface erosion and the formation of gullies on an improperly protected bank.

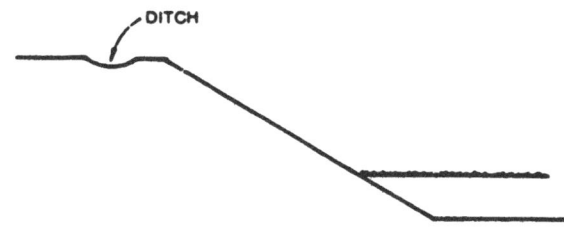

B. A ditch dug along the top of a bank can intercept and reroute drainage.

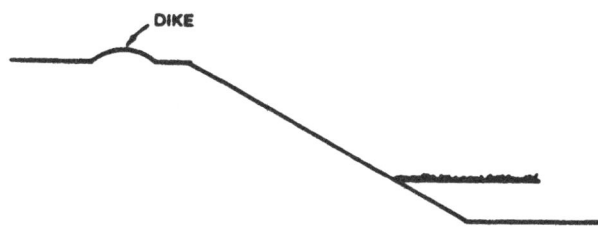

C. A diversion dike can also be used to intercept and reroute drainage.

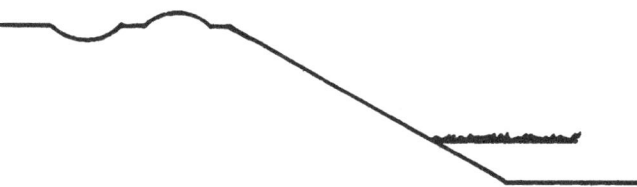

D. Combination ditch and dike (constructed from material excavated from ditch).

Methods to control overbank drainage.

Fig. 8.7. Method to control overbank drainage (from U.S. Army Corps of Engineers, 1983)

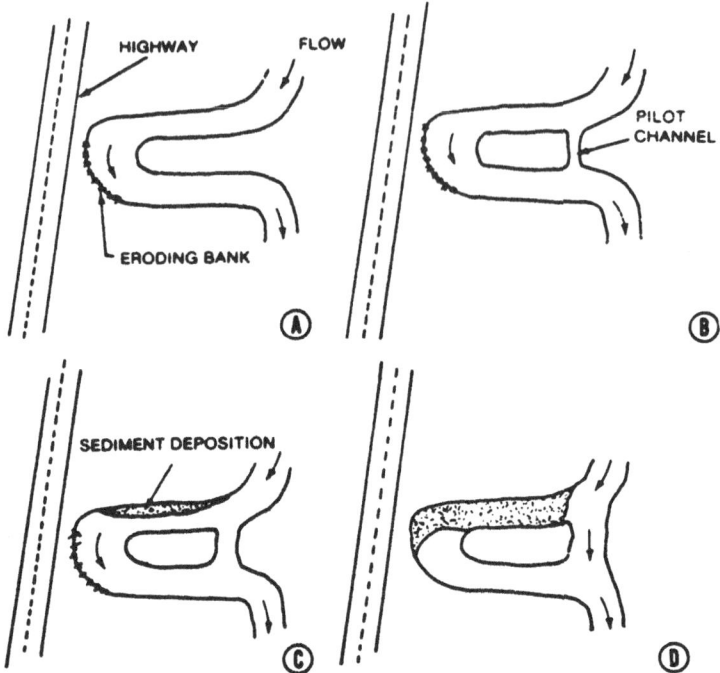

Protection of eroding bank by rerouting the stream. A. Eroding bank threatens highway. B. Pilot channel is excavated. C. Pilot channel begins to capture streamflow; some sediment deposition occurs in the natural channel. D. Pilot channel has captured the streamflow; water movement through the natural channel is now blocked by sediment deposition and bank erosion is stopped.

Fig. 8.8. Protection of eroding bank by rerouting the stream. (from U.S. Army Corps of Engineers, 1983)

piles, wood or wire fence, kellner jack field, vegetation, gabions, erosion control matting, and bulkheads.

With the exception of stone riprap, no specific guidelines can be used by engineers to assist them in developing design specifications for any particular bank protection problems. A comparison of the advantages and disadvantages of different protection methods are presented here to assist engineers in their selection of method of protection.

1) Stone Riprap:
Advantages—Flexible to settlement or other minor adjustments, easy to repair, easy to construct, appearance is natural, riprap is recoverable for future use.
Disadvantages—Expensive.
Remarks—Stone selection and design criteria for riprap are discussed in detail in the appendix based on materials published by the U.S. Army Corps of Engineers (1970, 1971).

2) Concrete Pavement:
Advantages—High degree of reliability over a long lifetime with a minimum of maintenance.
Disadvantages—Very expensive, not flexible, need subsurface drainage to avoid scour under concrete slabs.

3) Articulated Concrete Mattress:
Advantages—Due to its weight and flexibility it can be used on the bank subject to excessive hydraulic flow conditions.
Disadvantages—Expensive and needs specialised construction equipments.

4) Transverse Stone Dike:
Advantages—Because it is impermeable, it does not require the deposition of sediment in the dike field.
Disadvantages—Expensive
Remarks—It is used as a permanent structure to move the thalweg from its position along an eroding bank to an alignment controlled by the location of the structure.

5) Transverse Timber Piles:
Advantages—Effective in slowing the current so sediment can be deposited between timber piles to maintain a greater depth in the main channel.
Disadvantages—It is not effective where sediment load is low.

6) Wood or Wire Fence:
Advantages—No special techniques are required for construction and there is a wide availability of materials suitable for assembly of the fences.
Disadvantages—It is not the most effective means of bank protection, especially where sediment load is low.

7) Kellner Jack Field:
Advantages—A kellner jack field is extremely flexible, and readily conforms to channel geometry. It is suitable to wide, shallow, silt-laden streams which are subject to severe scour during high velocity flows.
Disadvantages—Kellner jack fields are not aesthetically harmonious with a flood plain landscape. They cannot be used in areas with a corrosive atmosphere, where extremely high velocity flows are experienced, or where the banks to be protected are higher than the jacks. They can be used only where debris and high sediment load exist in the river.

8) Vegetation:
Advantages—Vegetation can reduce the stream velocity up to 90 per cent at the boundary layer between the water and soil. It is the only self-renewable method, and in many cases, the most economical and aesthetically pleasing.
Disadvantages—It needs time for the vegetation to be established, and it cannot be used by itself where velocity is high.

9) Gabions:
Advantages—Gabion works are somewhat flexible and are able to accommo-

date minor changes in bank geometry. The stability is high.
Disadvantages—Gabion works should not be used in a corrosive atmosphere unless the steel wires are specially coated. A total failure of gabion works would occur after the steel wires are broken by corrosion or by moving coarse material in the river.

10) Erosion Control Matting:
Advantages—Many of these mats are biodegradable and are the byproducts of other manufacturing processes. The mats allow vegetation to grow through and allow vegetation adequate time to become established.
Disadvantages—It is a short-term bank protection measure.

11) Bulkheads:
Advantages—Bulkheads can be used to protect streambanks where the bank slope is unstable or where additional waterfront area is required by filling behind the bulkhead.
Disadvantages—Depending on the materials used (concrete, timber, etc.), the construction and maintenance costs could vary signifiantly.

In addition to the aforesaid methods, automobile bodies, ceramic material, concrete blocks, rock-and-wire mattresses, sack revetment, soil cements, automobile tires, timber-and-brush matresses, cribs, etc., can all be used for bank protection. The applicability of these methods depend on the availability of materials. Because of the lack of generality of their applications, there is no general design criterion developed for these methods.

More detailed reviews of different streambank protection methods were made by the U.S. Army Corps of Engineers (1981a).

8.5 Plan of action
The chance of success of protecting a streambank can be increased by following a well-organised plan of action. The following plan of action suggested by the U.S. Army Corps of Engineers (1983) can help engineers to reduce the possibility of having a fruitless investment of time and money.

Step 1—Determine why the streambank is in a distressed condition.
Streambank should be checked on a regular basis, especially after floods. The history of deterioration should be documented to find out possible causes of failure. Signs to look for during field inspection are:

—Exposed soil
—Loss of vegetation and fenceposts
—Sheet or rill erosion resulting from rainfall, overbank drainage, or seepage
—Cracks in the bank or in the area immediately behind top bank
—Overhanging banks
—Undermined trees with exposed roots
—Scour along the bank toe
—Changes in channel bed elevation
—Wave action

—Rapid drawdown
—Increased load on the top of the bank
—Higher flood stages than have occurred in past years
—Logs, debris, and sandbars in the channel that could deflect eroding
   currents into the bank

These types of problems should be recorded during an inspection visit on a dated map sketch of the site. This sketch should show the location of the streambanks (top and toe), eroded areas, vegetation, fence lines, buildings, sandbars, logs, debris, and any streambank protection works already in place. Side views of the bank should also be made at various locations. Be sure to reference all side views to some fixed point such as a tree or fencepost. Photographs should be taken (and dated) and the locations from which the photographs are made shown on the sketch. Additional visits should be made on a regular basis, including once during low water season and once immediately after any high water period.

Step 2—Decide if the bank is worth protecting.

There is no step-by-step guideline which can be followed for all cases in determining whether a distressed streambank is worth protecting. Tangible and intangible factors should both be weighted against estimated costs of proposed solutions.

Step 3 —Inventory of available resources.

Once the decision is made to protect a distressed streambank, inventory of available financial, technical, manpower, equipment, and construction material resources should be made. The results of this inventory may determine to a large extent the type of protection most likely to be successful under the given site-specific conditions.

Step 4—Select a bank protection method.

Based on the causes of streambank failure, resources available, and the advantages and disadvantages of different protection methods, feasible methods can be identified. Detailed comparison of feasible methods should be made before the final selection of protection method.

Step 5—Develop a project plan.

Plan should be laid out in an orderly manner. These include preparation of drawings, financial arrangements, legal contracts, environmental impact statements, permits, road, construction equipment and materials.

Step 6—Construct the project.

Safety precautions should be followed during the construction. Seasonal variations of natural conditions and effective arrangement of manpower, materials, and equipment should all be taken into consideration for project construction.

Step 7—Inspect and maintain the project.

A streambank protection project generally requires some maintenance after the construction is completed. Proper maintenance can prolong the life of the project and avoid needless expense and property damage. Each inspection or maintenance activity should be well documented for record and future reference.

REFERENCES

U.S. Army Corps of Engineers. (1970): Hydraulic design of flood control Channel, *Engineering Manual* EM1110-2-1601, July.

U.S. Army Corps of Engineers. (1971): Additional Guidance for Riprap Channel Protection, *Engineering Technical Letter* ETL110-2-120, May.

U.S. Army Corps of Engineers. (1981): Final Report to Congress, the Streambank Erosion Control Evaluation and Demonstration Act of 1974, Section 32, Public Law 93-251, Main Report, December.

U.S. Army Corps of Engineers. (1981a): Final Report to Congress, the Streambank Erosion Control Evaluation and Demonstration Act of 1974, Section 32, Public Law 93-251, Appendix A— Literature Survey, December.

U.S. Army Corps of Engineers. (1983): *Streambank Protection Guidelines,* October.

CHAPTER 9

Design of Brackish Water Pond System

9.1 Introduction

The design of a brackish water fish pond system requires careful considerations in order to make it functionally efficient and economically viable. The engineering design has to be attuned, to satisfy the biological requirements. Depending upon the particular stage of the fish's life cycle these requirements are volume and exchange of water, availability of food organisms, optimum oxygen content, turbidity, salinity, temperature and chemical qualities. The design of a brackish water pond system essentially consists of the following steps:

i) site selection, engineering investigations, and layout design;
ii) design of water management system;
iii) design of sluice gates as structures for water control; and
iv) design of dikes and internal bunds.

9.2 Site selection, engineering investigations and layout design

9.2.1 SITE SELECTION

For preliminary assessment of the suitability of a site, it is suggested that existing survey maps and the Admiralty chart be consulted. The study is primarily intended to help identify the nature of area, that is, whether it is bare, sandy or forest covered. This information can be used to delineate the tidal creeks, to identify the areas covered by swamps, lagoons, as well as to find out whether there is flooding of the land from the nearby river system. Other factors that need to be looked into by an actual site visit are:

i) the location of the site with reference to sea or brackish water source;
ii) infrastructural facilities such as approach road, electricity, fresh water source;
iii) topography of the area with reference to tide levels and its general gradient;

iv) types of vegetation, such as mangrove forests in and around the area;
v) nature of soil conditions at the sites;
vi) the highest high water level at the site;
vii) cultural practices to be followed;
viii) availability of labour, and construction materials;
ix) existing pattern of land and water use near the site;
x) discharge of any pollutants from industry; and
x) availability of marketing outlets and general price levels.

For further evaluations of the site, it is necessary to examine some of these essential factors in detail:

i) range of tide and general elevation of the land;
ii) characteristics of soils;
iii) topography; and
iv) fresh water supply.

9.2.1.1 Range of tide and general elevation of the land

In the case of gravity fed fish ponds, the depth of water to be maintained is determined by the height of the incoming tide and the elevation of the pond bed with respect to zero tidal datum as it is essential that the existing tidal range should be able to fill up the pond. With regard to ground elevation with respect to tide, the design pond depth should allow the most economical construction. The optimum pond bottom elevation should be the one which permits the pond to be drained at any day of the year and filled with sea water to the desired depth within five days or less during the critical spring tides. Areas accessible only to high spring tides usually are not selected because they require large excavation, with the associated problem of disposal of excess soil. Similarly, low areas that may require higher and wider dikes should be avoided to reduce costs.

9.2.1.2 Characteristics of soils

One soil property, *texture*, can usually be adopted as a criterion to determine the suitability of the site for fish pond construction. Soil texture is the relative proportion of sand, silt, and clay content of the soil. According to Dureza (1982) soils belonging to any one of the following textural classifications is suitable and desirable for fish pond. They are clay, clay loam, silty clay loam, silty loam and sandy clay loam as shown in Fig. 9.1. Clayey soils are preferred for dike construction because they have good water retention property as well as good compaction characteristics. Loamy soils are also suitable because they have a good organic content which favours the culture and growth of natural fish food.

9.2.1.3 Topography

It is necessary to examine closely the topography of the site because fish pond design, and layout are made largely in accordance with land

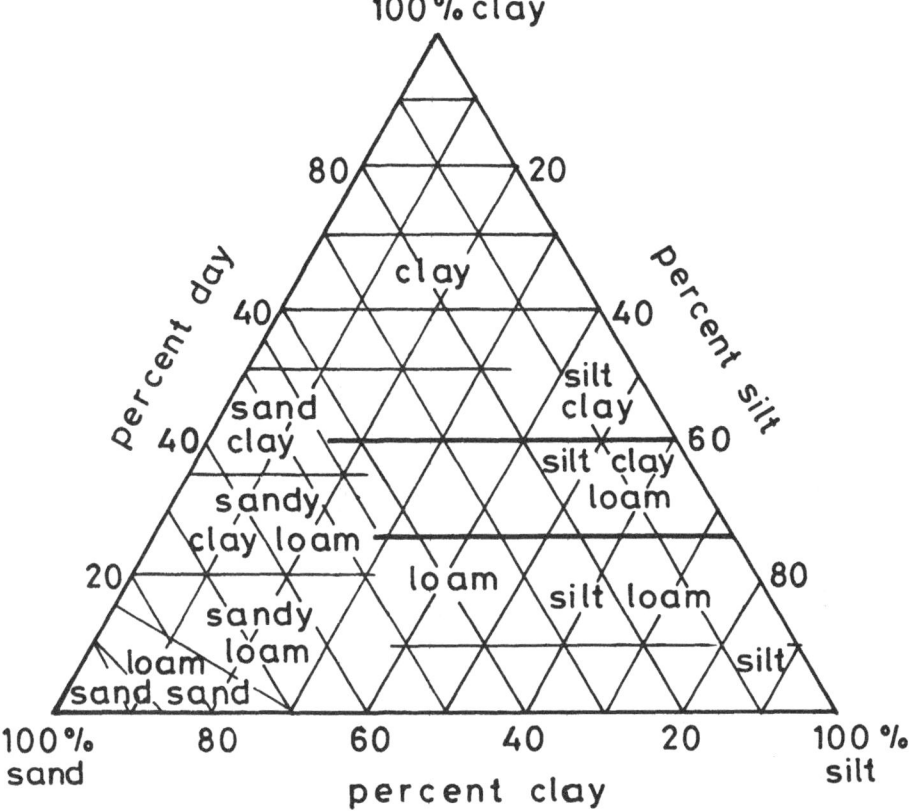

Fig. 9.1. Soil classification based on soil texture

topography. Flat coastal swamplands whose elevations are within the range of ideal pond bottom elevation are preferred. Sites with undulating topography or with elevations varying from lower than mean lower low water (00 tidal datum) to 4 or 5 m in height should be avoided. Figure 4.4 shows four zones, on the coastal edge as probable sites for a fish pond according to Denila (1980) in Chapter 4.

9.2.4 FRESH WATER SUPPLY

A suitable fish site should have an adequate source of fresh water supply. The sources to be examined are mainly ground water, springs, irrigational canals and creeks. The available flow must be sufficient to meet the demand of fish farms as well as the daily needs of people working in the farm.

9.2.5 ENGINEERING INVESTIGATIONS

For the final assessment of the suitability of the site as well as for preparation of a project report, it is necessary to carry out various surveys for collection of necessary input data. Engineering investigations or surveys essentially consist of three parts:

i) Engineering surveys
ii) Soil survey
iii) Hydrologic and hydraulic survey

9.2.2.1 Engineering survey

An engineering survey will prepare topographic maps to show the nature of ground relief or its characteristics, such as difference in elevation, location of boundaries, and physical facilities such as existing buildings, roads, rivers, canals, bridges and other features. The map is of vital importance in the final design of fish farms as it provides the relationship of the site with the tidal fluctuation, determines the general gradient through which the water movement occurs, helps in locating water control structures, in estimating quantities of earth work etc. For details on how to conduct a field survey, readers are referred to standard text books on surveying like those by Kanetkar (1959) and Clark (1963). Only the basic minimum outlines will be provided here. To carry out the survey, one has to use various survey equipments such as theodolite, levels, compass, tape, levelling rods, chains, plane table etc. Levels are used mainly for preparation of contour maps or the ground elevation maps. Other instruments such as compass, plane tables, etc. are used for traversing and locating the details on the map.

In general, survey work involves measurements of distances, angles and relative heights. Angles are measured both along the horizontal and vertical planes. Angles and directions are usually expressed in terms of their bearing with respect to the north and south whichever is nearest with the added designation of east or west wherever it applies. Figure 9.2 shows that a bearing can never be greater than 90°. After the traversing is done, a check is made by using the interior angles. In a closed polygon the angles inside the figure between adjacent lines are called interior angles. The sum of the interior angles in a closed polygon is $(N-2)$ 180°, where $N =$ number of sides. The actual value of every interior angle is computed from the field data, namely, bearings. An error occurs when the total interior angles obtained from survey data do not satisfy the value determined from the formula.

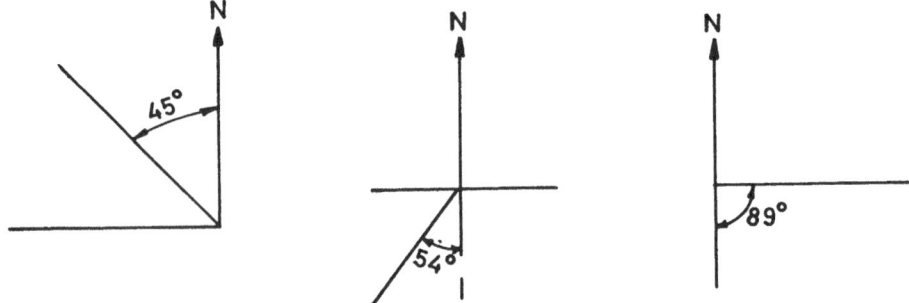

Fig. 9.2(a). Bearing of survey lines in a traverse

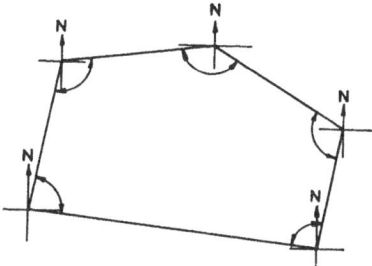

Fig. 9.2 (b). Interior angles in a traverse

Another basic survey operation is levelling for preparation of relief maps. Direct levelling is commonly used when the elevation is determined with the help of a level together with level staff. Differential and profile levelling are two types of direct levelling. The first method determines the difference in elevation between two points which is necessary in the preparation of a contour map. In the second method the difference in elevation of points along a prescribed line and at measured distances is determined, such as that along the central line of water supply canal in a fish farm.

9.2.2.1.1 Contour mapping—Contour lines show the configuration of the ground on a map. Each contour line represents points of the same elevations. A contour survey for a fish farm is commonly done by laying-out-square method and by sounding method. In the laying-out-square method, the area is criss-crossed with a square network. At each point of intersection, the level is taken and the elevation is determined with reference to a bench mark and thereafter the contour map is prepared as shown in Fig. 9.3.

In case the coastal fish farm site is under shallow water, the usual survey method adopted is sounding. The procedure of carrying out the survey is described herein as it is not usually available in the references cited earlier. The instruments needed for the survey are magnetic compass, levelling staff, tape and two small boats. One boat is needed for carrying the instrument the other for the staff man.

Step I: The predicted tide curve on the day of the survey is prepared in advance. The curve provides tide level during any time of the day.

Step II: The survey team establishes the boundaries of the survey under reference. It also fixes some landmark locations with reference to corner points of boundary lines as shown in Fig. 9.4.

Step III: The staff rod man at point *X* locates his position on the map by taking down the bearings of at least two of his tie lines which are the corner points of the boundary. The bearings obtained when plotted will intersect at one point which defines the position of

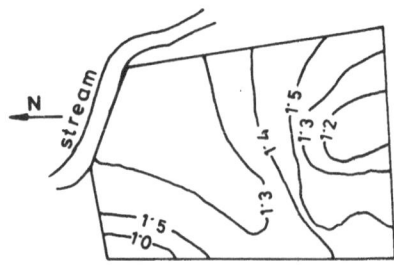

Fig. 9.3. Contour map.

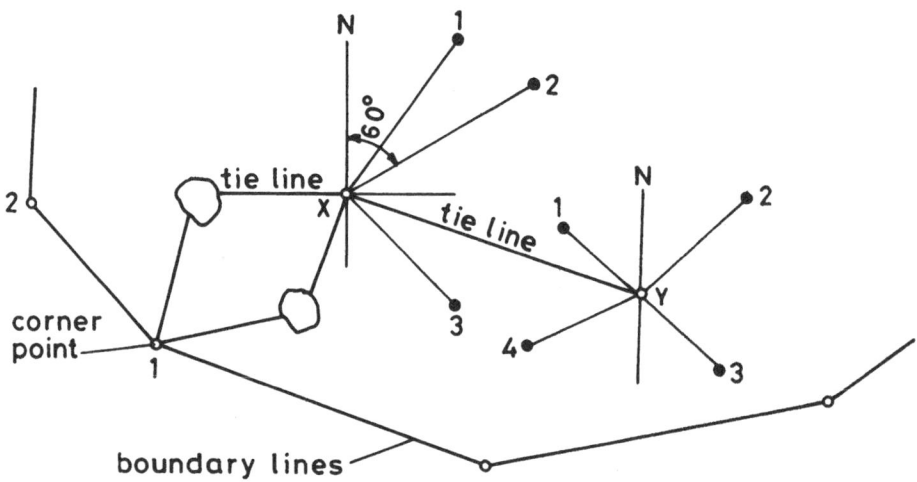

Fig. 9.4. Instrument station in an area with random ground points

the instrument man. The actual distance of the boat from the tie line may be measured by a tape for checking purposes.

Step IV: The instrument signals the staff man to hold the staff rod down to the ground surface to get the bearing of the ground point. The staff man notes the depth of water at the point and the corresponding time. Then the distance between point X and ground point is measured.

Step V: The ground point elevation can then be computed from the relationship:

Elevation of survey point = Height of tidal elevation – staff level readings at the point

The process is then repeated for various ground points.

9.2.2.1.2 Profile levelling—The steps involved in profile levelling are illustrated in Fig. 9.5. It begins by setting up the instrument·at a convenient location from where a number of stations are visible and readings are taken in the form of back and foresight. The instrument is then shifted before a foresight is taken on the transfer point. At the new location of the level a backsight is taken from the turning point and then additional foresights are taken. The whole process is repeated till the work is done. After the necessary computations the ground profile can be drawn on a paper with suitable scales as shown in Fig. 9.6. This provides necessary information with regard to channel slope, cuts, or fills.

9.2.2.1.3 Measurement of areas—After the survey it is necessary to determine the area which can be done by using a planimeter or by employing Simpson's or Trapezoidal rule (Clark, 1963). Apart from the above other methods also exist. The principle and procedure for determining areas by trapezoidal rule is illustrated herein. Consider the area of a fish farm which is bounded by an irregular curve such as a tidal creek and line *AB* as shown in Fig. 9.7. Draw offsets along line *AB* at equal intervals. The area bounded by *ABCD* is

$$\text{Area} = \left(\sum h + \frac{h_0 + h_n}{2} \right) d \qquad (9.1)$$

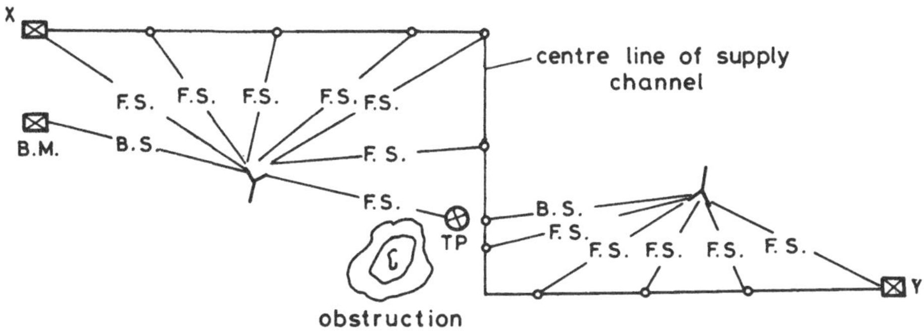

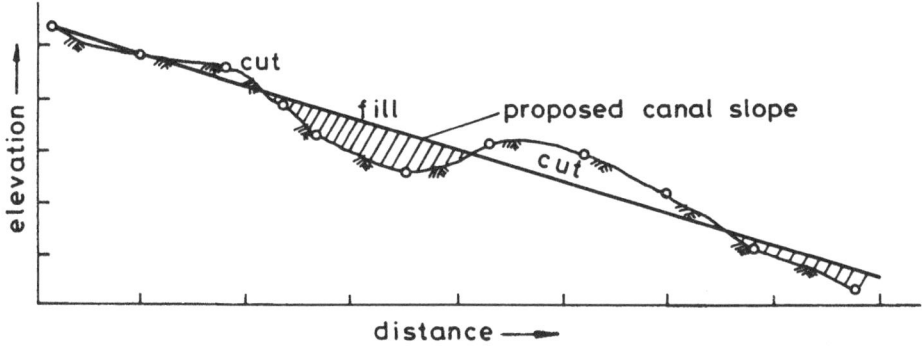

Fig. 9.5. Illustration of profile levelling procedure.

Fig. 9.6. Typical profile of a supply canal

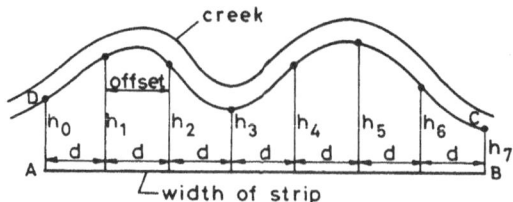

Fig. 9.7. Illustration for trapezoidal rule

where h_o and h_n = distances of the two offsets,
  d = distance between offsets; and
  Σh = sum of the other remaining offsets.

9.2.2.2 Soil survey

Subsoil investigations at farm sites are necessary to obtain information to enable the design of a brackish water fish farm. For this purpose, field investigations are necessary in order to collect undisturbed samples for testing of engineering properties of soils. The methods are probing and sampling. Probing is a method of determining soil resistance to penetration. The results of probing should always be confirmed by conducting soil tests on the samples collected at the representative planes. The probings at the site can be made with the help of steel or wooden rods fitted with sharp metal shoes and by driving the rods with a sledge hammer. The number of blows required for a 20 cm penetration is an indication of the soils resistance to penetration.

For sampling, a 50 mm diameter, 600 mm long PVC pipe can be slowly pushed into the soil, and the soil surrounding the pipe should be carefully removed without disturbing the pipe. At the bottom of the pipe the soil can be cut by a thin wire to release the pipe with the soil intact inside the tube. The ends of the PVC pipe should be sealed with plastic sheeting to prevent loss of moisture.

As a standard practice, several samples should be obtained from predetermined locations of a given site. Usually soil samples should be drawn from about 10 random locations per hectare. The samples from each location should be marked with information on the date of boring, location, water table elevation, and thickness of each soil layer. It is adviseable to keep rough sketches on the location and depth of boring from each site. As a general rule the depth of soil layer to be investigated should reach the hard or impermeable stratum. In case the hard soil is at a greater depth, the depth of investigation should cover at least 3 m. The soil samples are then analysed in a soil testing laboratory equipped with relevant instruments to find out (a) physical properties such as textural class, grain size analysis, permeability, cohesion, angle of internal friction, and load bearing capacity; (b) chemical properties such as pH values and presence of acid forming substances such as pyrites and sulphides. The most important physical property of the soil for fish pond engineering is the texture of the soil, because it provides information regarding its ability to resist the flow of water, attainable compaction and its load bearing capacity.

Load bearing capacity is important for determining foundation nature for sluice gates, perimeter dikes, and the necessity of piling. Usually pond soils have poor load bearing capacity, and in most cases pile driving becomes necessary to strengthen the foundation.

The pH value of soil provides an idea of the chemical nature of soil which has a direct relationship to fish production. The desirable pH range of soil and water for aquaculture varies from 6.5 to 8.5. Lower than six is too acidic for most fish species and higher than nine is unfavourable for growth. pH measurement can be done on site or in a laboratory by using a pH meter. A less accurate value can be obtained by a litmus paper. Extensive determination of soil pH values are necessary up to a depth of 1 m of soil. For details of soil testing equipments, procedure etc. reference may be made to standard books on soil mechanics such as those by Terzaghi and Peck (1960), Taylor, (1958) and Singh (1967).

9.2.3 LAYOUT DESIGN OF POND SYSTEM

9.2.3.1 General

The fish farm is an establishment which consists of a system of ponds, water supply and drainage channels, control structures such as gates, and dikes or embankments, and other support facilities. Each of the components should be properly located and fitted into the system so as to maintain a balance between economy, aesthetics, and functionality. In the layout plan, the basic principle is to have a minimum number of sluice gates and minimum length and size of canals and dikes. However, these criteria should match the biological requirements for creating a suitable environment for the species to develop. The support facilities consist of farm buildings, roads, bridges, storage shed for feed and equipment, chilling tanks and a watch tower.

9.2.3.2 Types of pond, their layout

A fish pond, may have single or multiple compartments. A nursery pond, transition pond, production or rearing pond, and catching pond are some of the essential compartments. A nursery pond is small in size, having a square or rectangular shape with about 1 to 4 per cent of total production area. A nursery pond is used for rearing the fry before transferring into a larger pond. It is located in an elevated portion of a farm or near the corner of a rearing pond compartment (Djajadiredja and Danley, 1982). According to Alcantara (1982) the most suitable location of the nursery pond is where it can always be supplied with water as well as draining during ordinary flow. Location of the nursery pond near a perimeter dike has to be avoided to prevent ingress of predators and unwanted species into the nursery pond. The transition pond or holding pond is intended for efficient and quick transfer of fingerlings. About 10 per cent of the production area is usually allocated for the transition pond. The fingerlings are reared in this pond for varying periods before stocking them in a rearing pond.

The rearing pond or growing pond occupies nearly 80 per cent of the total farm area. The bottom elevation of the rearing pond is about 20 cm lower than the transition pond. The pond bottom usually has a slope towards a catching pond or supply channel for easy harvesting. The usual manageable size of a catching pond is around 1 to 5 ha per compartment. A catching pond, as the name suggests, is intended for catching the fish and it is constructed adjacent to the gate inside a big pond compartment. The simplest form of pond layout is a single compartment. In recent times, an improved layout consisting of multiple compartments has come into general use which fits into the general system of an appropriate production management scheme. Some typical pond layout systems for milk fish and shrimp culture are shown in Figs. 9.8 and 9.9 (Alcantara, 1982; Chalyandya et al. 1982). The orientation of fish ponds, should be planned in such a way that the longitudinal direction of the pond is parallel to the prevailing wind direction. The wind direction as experienced by South Asian countries is shown in Fig. 9.10. The suggested orientation of pond compartments is shown in Fig. 9.11. Such a position takes advantage of the wind energy in having good water aeration through mixing and circulation.

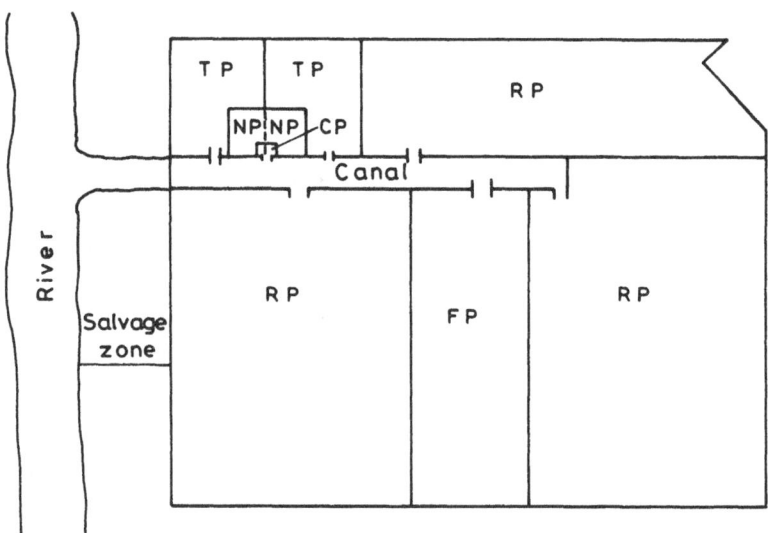

T.P.—Transition pond; N.P.—Nursery pond;
R.P. Rearing pond; C.P.—Catching ponds;
F.P.—Feeding pond

Fig. 9.8. A conventional pond system (Alcantara, 1982)

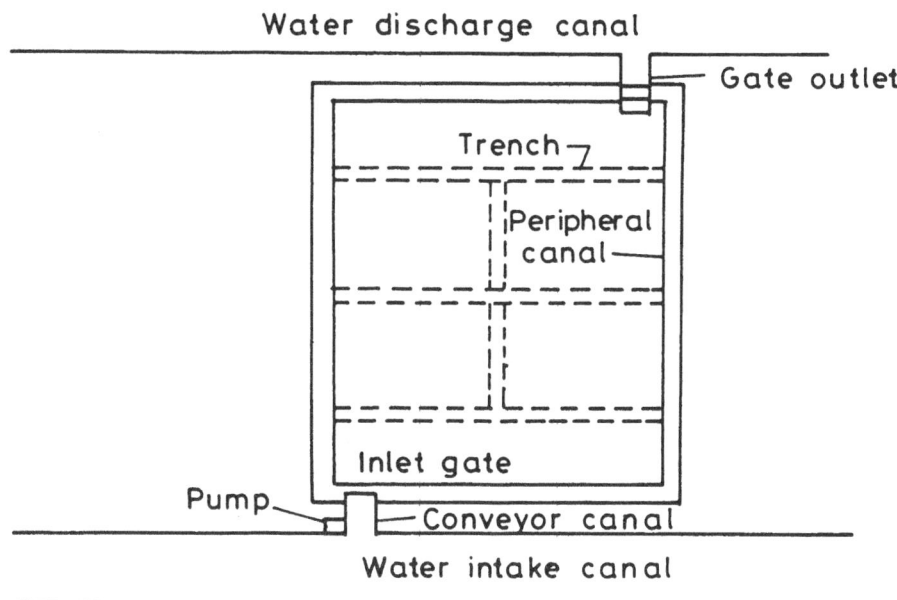

Fig. 9.9. Traditional pond layout in Thailand (Chalayondeja et al., 1982)

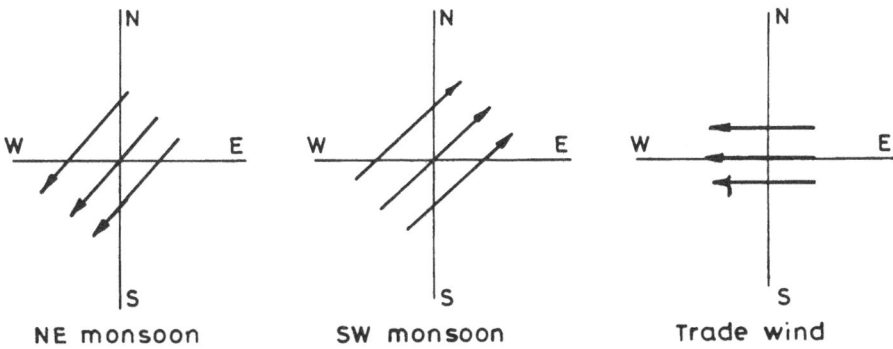

NE monsoon   SW monsoon   Trade wind

Fig. 9.10. Wind direction in South Asia

9.2.3.3 Location of supply, drainage channels and gates

In the case of intensive shrimp monoculture farms, each compartment should have individual water supply and drainage outlets to make them independent of each other. The location of water control gates depends primarily on the water management scheme. In general, main gates and secondary gates are located where entrance and circulation of water can be made efficient. Ordinarily, a single gate per pond connected to a canal provides passage for tidal inflow and outflow. In a flow-through system, two gates located at opposite ends of the ponds are required. Separate canals that accommodate inflow and outflow from the gates are provided. Canals should be located where they can serve the largest number of ponds. The canal

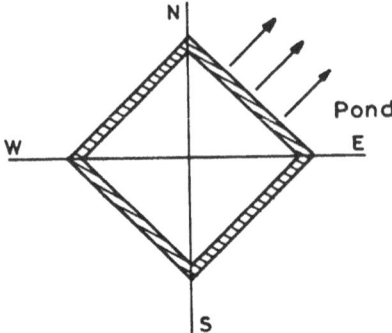

Fig. 9.11. Pond compartment orientation with respect to wind direction

lengths should be minimised as far as practicable. Apart from the preceding, a peripheral canal or diagonal ditch deeper than the pond bottom is provided to allow a hiding place for the fish during the dry season when temperatures rise from 27 to 32°C. The depth of ditch ranges from 0.5 to 1 m as shown in Fig. 9.12.

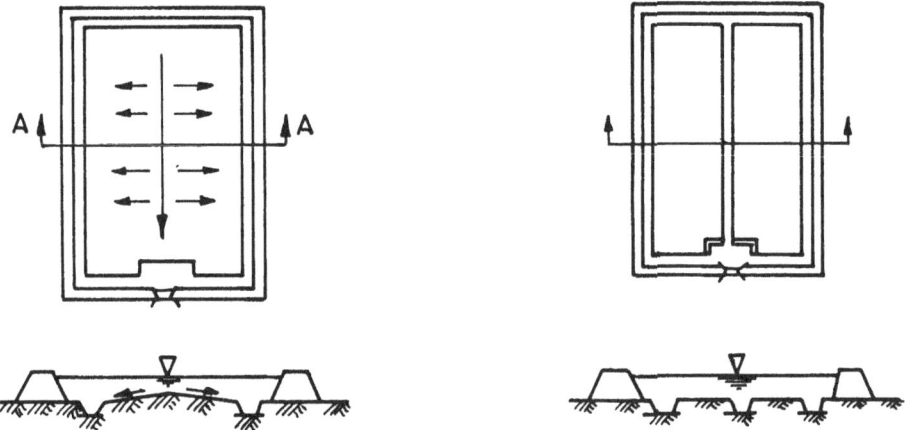

Fig. 9.12. Pond with peripheral and central ditch

9.3 Design of water management system

9.3.1 GENERAL

In the design of a water management system for culture ponds, the following are important considerations:

    i) Design tide curve or selection of appropriate tide curve at site.
    ii) Determination of the elevation of the pond bottom.
    iii) Consideration of the pond management criteria.
    iv) Determination of sluice gate size.
    v) Determination of the size of the supply or drainage channel.

9.3.2. DESIGN TIDE CURVE OR SELECTION OF APPROPRIATE TIDE CURVE
AND DATUM AT SITE

The tide curve must be determined at the time when the environmental factors of production such as water levels, water quality, or hydraulic environment become critical to fish life. Generally, the tide curve at the site of the pond is not known and has to be observed during the investigation stage through continuous recording of water levels. Thereafter, an appropriate datum has to be selected for plotting purpose as well as for fixing the relative levels of all farm structures. This datum selection is the most important aspect of aquacultural engineering because all water, pond, and channel bed levels are to be determined with reference to the datum for efficient water management and design of an economical fish farm. The datum and the tide levels for a site can be arrived at by the transfer of data from the nearest standard port. In order to correlate the tide levels with that of the nearest standard port, the spring high water near new or full moon at the site should be observed and computed with reference to the high tide on that day for the standard port by referring to appropriate tide tables. For illustration, consider a standard port P at a distance of 200 km south of a site surveyed at a place Q. The spring high tide levels on say December 4, 1980 at the standard port and the corresponding high tide levels at the site on that day with respect to an arbitrary datum chosen were as follows.

	Standard Port P			Farm Site at Q		
Date	Time in hrs.	Tide level (m)	Tidal range (m)	Time in hrs	Tide level (m)	Tidal range (m)
	0213	00.07		0250	00.13	
12.4.80	0825	01.04	0.97	0900	00.97	00.79

The datum correction equal to the difference in high water levels between P and Q for that is applied so that the high water levels at both place are the same; this means that the datum at Q is reduced by $(1.04-0.97)=0.07$ m. Now the mean high water spring tide level at Q with respect to corrected datum can be taken as that of P. Because the datum has been fixed by comparing high water levels of the standard port and the site, there is no need to apply any correction for mean high water spring. For arriving at other tide levels at the site Q a range correction will have to be applied. The tide levels at P and the corrected levels at Q along with the correction calculations are given below.

Tide levels	For Port P (m)	Corrections		For Site Q (m)
MHWS	1.15	$(1.15-0.84)$	$\frac{0.79}{0.97}=0.25$	$1.15-0$ $=1.15$
MHWN	0.84			$1.15-0.25=0.90$
MLWN	0.43	$(1.15-0.43)$	$\frac{0.79}{0.97}=0.59$	$1.15-0.59=0.56$
MLWS	0.14	$(1.15-0.14)$	$\frac{0.79}{0.97}=0.82$	$1.15-0.82=0.33$

This method of transferring the datum should be done based on the high water levels during the spring tide of a fair weather day to minimise the errors due to abnormal weather conditions. Further, it should be ensured that the nature of tides, that is, whether semi-diurnal or diurnal, be the same at both the site and the standard port. Having fixed the datum, the design tide curve can be drawn along the zero datum level which is usually the Mean Lower Low Water (MLLW). In the absence of full tidal record or observation at the site, a tide curve can be prepared based on a simplified approach. In this approach, it is necessary to know the difference between mean or, low water level and the mean sea level. It is assumed that tide can be depicted by a sine curve having a constant periodicity. Accordingly, the standard high and low water tide curves are drawn as shown in Fig. 9.13. A horizontal line representing MSL is drawn on a sheet of paper, first, then a circle is drawn from a point on this line with a radius equal to half of the difference between the mean high and low water levels. The circumference of the circle thus drawn is divided into 12 parts or 12 houses and numbers are assigned from 1 to 12 corresponds to lunar hour. Note the difference in solar and lunar time; the latter corresponds to about 62 minutes of the solar calendar. Thereafter, the perpendicular lines passing through the points are drawn. It is then followed by drawing a horizontal line from each hour on the circumference of the circle. The numbering of the intersection of each pair of vertical and horizontal lines represents the same hour. By connecting these intersections with a curve, the needed tide curve can be prepared. For the preparation of a tide curve subjected to diurnal inequality, the steps indicated in Fig. 9.14 may be followed.

In case the low water level observation data is not available, the following correction relationship should be used (Kato, 1980).

$$H = \frac{2h}{1 - \cos\dfrac{\pi t}{t'}} \tag{9.2}$$

where H = tidal range in m; h = difference in elevation between high water and the limit of observation, in m; t = time required to reach h from high water level, and t' = time required to reach low water level from high water.

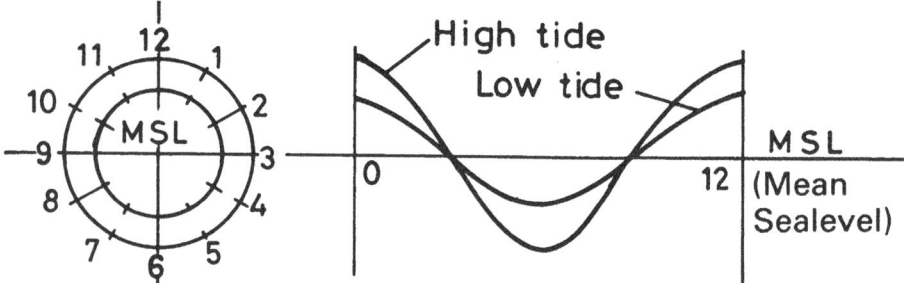

Fig. 9.13. Drafting of tide curve without diurnal inequality

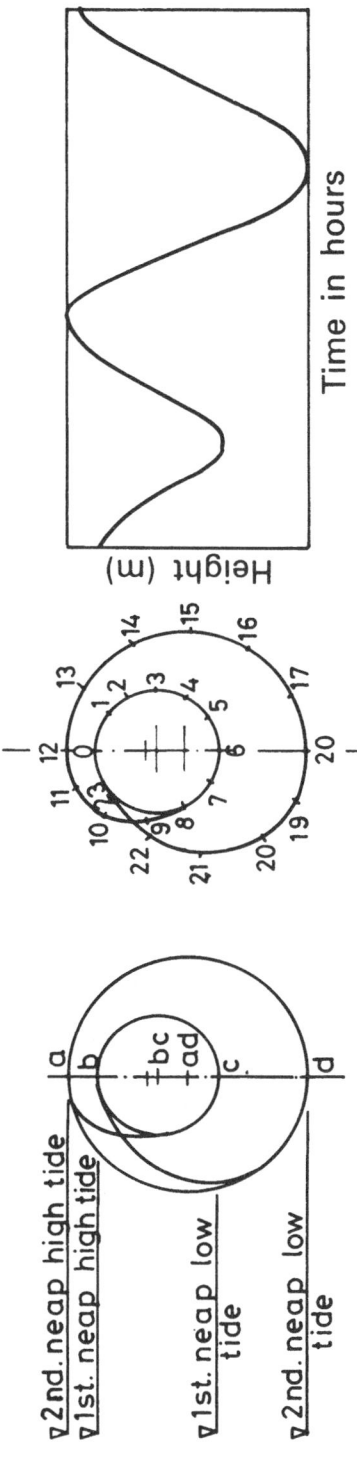

Fig. 9.14. Darfting of tide curve with diurnal inequality

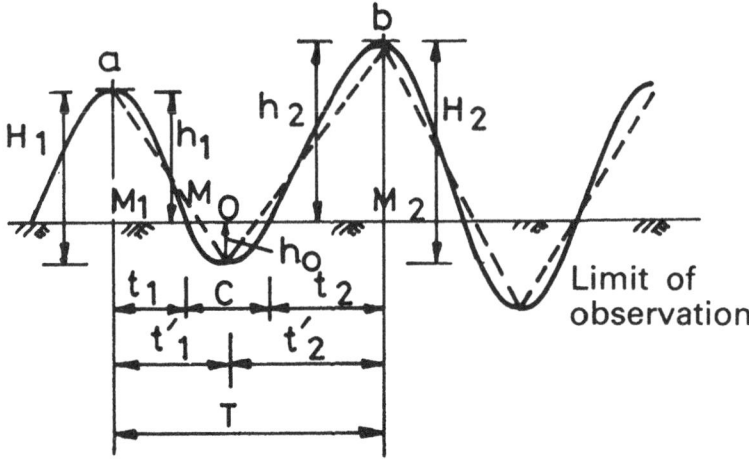

Fig. 9.15. Correction when low level is not available

Referring to Fig. 9.15 by substituting h_1, t_1, $t_1' = t_1 + 1/2\,[T - (t_1 + t_2)]$, h_2, t_2 and $t_2' = (T_2 - t_1')$ in the equation (9.2) one obtains H_1 and H_2. The low water height h_0 can be expressed as

$$h_0 = H_1 - h_1 = H_2 + h_2$$
If $H_1 - h_1 = H_2 - h_2$
$$h_0 = 1/2\,\{(H_1 - h_1) + (H_2 - h_2)\}$$

After plotting h_0, the point C is located which is then connected by a straight line with points a and b. The tidal curve corresponding to the observed portion can then be drawn.

A more rigorous analysis for the design tide curve can be obtained by performing harmonic analysis. As is well known, the relative motions of the earth, moon, and sun cause a number of tide producing forces. Even though the amplitude and phase of the tidal response to each constituent are difficult to express in terms of the forces causing them, the period of each constituent is determined from astronomical studies. Tide predictions can be performed accurately from a harmonic analysis of previously recorded tidal observations. The basic premises for harmonic analysis are as follows:

1) The resultant tide at any location is composed of a finite number of constituents, each with its own periodicity, amplitude and phase angle.

2) The constituents are each simple harmonic in time and are mutually independent.

Each constituent is the result of uniform simple harmonic motion and has an oscillation about the mean tide level. Tidal constituents are denoted by letter symbols which are abbreviations of the main characteristics, subscript numbers indicating whether they are diurnal, semi-diurnal, quarter-diurnal etc; namely, O_1, M_2, M_4, etc.

The vertical ordinates of each component curve above mean tidal level are in sinusoidal relationship to the angular moments of a point relating above mean tide level at semi-amplitude radius, the angle of rotation being measured from a point on the mean tidal level. Figure 9.16 shows the sinusoidal pattern for a single component. In this case the ordinate at any point of the curve, that is, tidal height is a $\sin 2\frac{\pi t}{T}$ where a = amplitude of the wave and $2\frac{\pi t}{T}$ = the angle of rotation respectively.

The resultant tidal variation h at a particular locality composed of K constituents may be formulated as

$$h = \sum_{r=1}^{K} a_r \sin(\sigma_r t + \delta_r) \tag{9.3}$$

$$H = a_0 + \sum_{r=1}^{K} a_r \sin(\sigma_r t + \delta_r) \tag{9.4}$$

where $\sigma_r = 2\pi/T_r$;
  H = height of water surface with respect to a datum;
  a_0 = height of mean water level with respect to the datum;
  h = height of the water level with respect to mean water level; and
  t = time at any instant

a_r, δ_r and T_r being the amplitude, phase angle and period of the rth constituent respectively. Equation (9.4) may be expanded and recast as

$$H = a_0 + \sum_{r=1}^{K} E_r \cos \sigma_r t + \sum_{r=1}^{K} F_r \sin \sigma_r t \tag{9.5}$$

where $E_r = a_r \sin \delta_r$ and $F_r = a_r \cos \delta_r$

If the tidal computation is done by considering only the tidal component M_2,

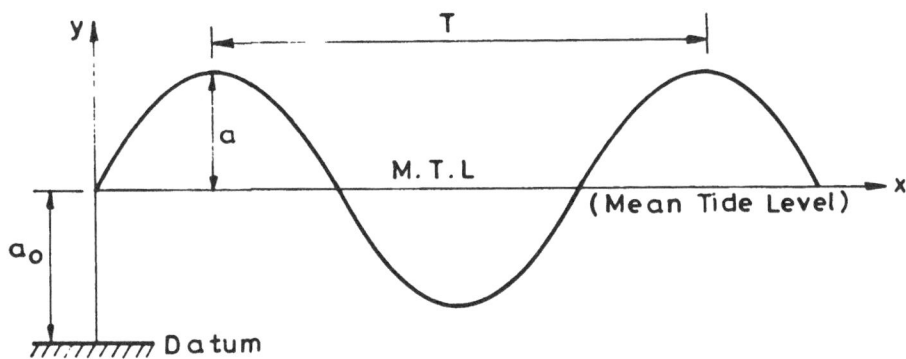

Fig. 9.16. Sinusoidal form of a tide curve

which is usually the major component of a tide, the equations (9.3) and (9.4) become

$$h = a \, \sin\frac{2\pi t}{T} \qquad (9.6)$$

$$H = a_o + a \, \sin\frac{2\pi t}{T} \qquad (9.7)$$

With the help of above equation the tidal height due to single component can be predicted easily.

Tidal predictions by harmonic analysis procedures can be separated into two steps:

Step I: Determination of the unknown phase and amplitude characteristic of each constituent.

In this step, the unknown phase and amplitude characteristics of each constituent at a given locality are determined from measured tidal records at the particular site. The records must include long period tidal fluctuations, for example, the effects of wind-generated waves must be eliminated, because they would decrease the accuracy of the analysis. With sufficient data, it is possible to determine the characteristics of the various constituents comprising the resultant tide. In most cases, the measured tide can be reproduced with reasonable accuracy if about ten constituents are considered. The periods of the important constituents are determined from knowledge of the earth, moon, and sun system, which causes the tide-producing forces. Tables listing important tidal constituents and their respective periods are available in handbooks and manuals. One such reference is given under item (16) at the end of the chapter.

Step II: Prediction of future tidal variation.

The preceding information of tidal constituents can be employed to predict future tidal variations at that locality. Taking the characteristics of all the tidal constituents as known, the tidal variation is given by equation (9.4) or equation (9.5). This summation could easily be performed on a computer or using a mechanical tide prediction machine.

9.3.2.1 Harmonic constituents by the Fourier method

Consider a tidal record from which the amplitude a_r and phase angle δ_r are to be determined for each of K important constituents. Multiplying both sides of equation (9.5) by $\sin(\sigma_p t)$, integrating over a long time interval $\triangle t$ (that is, ideally $\triangle t$ would approach infinity) and then dividing by $\triangle t$,

$$\frac{1}{\Delta t}\int_0^{\Delta t} H \sin(\sigma_p t)dt = \frac{1}{\Delta t}\left\{ a_0 \int_0^{\Delta t} \sin(\sigma_p t)dt \right.$$

$$+ \sum_{t=1}^{k} E_r \int_0^{\Delta t} \cos(\sigma_r t) \sin(\sigma_p t)dt$$

$$+ \sum_{r=1}^{k} F_r \int_0^{\Delta t} \sin(\sigma_r t) \sin(\sigma_p t)dt \qquad (9.8)$$

By integration, it can be shown that

$$\lim_{\Delta t \to \infty} \frac{1}{\Delta t} \int_0^{\Delta t} H \sin(\sigma_p t)\, dt = \begin{cases} 0, & \text{if } \sigma_p \neq \sigma_r \\ F_r/2, & \text{if } \sigma_p = \sigma_r \end{cases}$$

or

$$F_r = a_r \cos \delta_r = \lim_{\Delta t \to \infty} \frac{2}{\Delta t} \int_0^{\Delta t} H \sin(\sigma_r t)\, dt \tag{9.9a}$$

similarly, it can be shown that

$$\lim_{\Delta t \to \infty} \frac{1}{\Delta t} \int_0^{\Delta t} H \cos(\sigma_p t)\, dt = \begin{cases} 0, & \text{if } \sigma_p \neq \sigma_r \\ E_r/2, & \text{if } \sigma_p = \sigma_r \end{cases}$$

or

$$E_r = a_r \sin \delta_r = \lim_{\Delta t \to \infty} \frac{2}{\Delta t} \int_0^{\Delta t} H \cos(\sigma_r t)\, dt \tag{9.9b}$$

Equations (9.9a) and (9.9b) define all a_r and δ_r and the constant a_0 is given by

$$a_0 = \lim_{\Delta t \to \infty} \frac{1}{\Delta t} \int_0^{\Delta t} H\, dt \tag{9.10}$$

The characteristics of the constituents are now, at least in principle, determined. The indicated operations to calculate these characteristics is defined if H is a continuous time function. It is more likely that H would be digitalised as a discrete time series, in which case the integrations would be replaced by their approximate equivalent numerical summations.

9.3.2.2 Harmonic constituents by the method of least squares

Harmonic analysis of tidal observations over long periods for determination of tidal constants at a certain location can be performed on computer using the least square method. The mathematical procedure for the method is described below.

Let H_v $(v = -s, -s+1, ... -1, 0, 1, ... s-1, s)$ be the hourly observations taken by a tidal gauge, with respect to central time $t = 0$ (that is, $v = 0$ is the middle observation at $t = 0$). The harmonic series for k constituents may be written in the form

$$H(v) = a_0 + \sum_{r=1}^{k} E_r \cos(\sigma_r v) + \sum_{r=1}^{k} F_r \sin(\sigma_r v) \tag{9.11}$$

This series will approximate the observed series as closely as possible if

$$\phi = \sum_{v=-s}^{s} [H(v) - H_v]^2 = \text{minimum} \tag{9.12}$$

The conditions under which ϕ is a minimum are

$$\frac{\partial \phi}{\partial a_0} = 0$$

$$\frac{\partial \phi}{\partial E_r} = 0, \quad r = 1, 2, \ldots, K \qquad (9.13)$$

$$\frac{\partial \phi}{\partial F_r} = 0, \quad r = 1, 2, \ldots, K$$

Substitution of equations (9.11) and (9.12) into equation (9.13), yields the following simultaneous linear algebraic equations

If $p = 0$; $\displaystyle a_0 N + \sum_{r=1}^{K} E_r s(\sigma_r) = \sum_{v=-s}^{s} H_v$

If $p \neq 0$, $(p = 1, 2, \ldots, K)$

$$a_0 s(\sigma_p) + \tfrac{1}{2} \sum_{r=1}^{K} E_r [s(\sigma_r - \sigma_p) + s(\sigma_r + \sigma_p)] = \sum_{v=-s}^{s} H_v \cos(\sigma_p v)$$

$$\tfrac{1}{2} \times \sum_{r=1}^{K} F_r [s(\sigma_r - \sigma_p) - s(\sigma_r + \sigma_p)] = \sum_{v=-s}^{s} H_v \sin(\sigma_p v) \qquad (9.14)$$

For simplicity let

$$N = 2s + 1$$

and

$$s(\sigma) = \frac{\sin(s + \tfrac{1}{2}\sigma)}{\sin(\sigma/2)} = \sum_{v=-s}^{s} \cos(\sigma v)$$

From these $(2K+1)$ equations, that is, equation (9.14), the $(2K+1)$ unknowns a_0, E_r $(r=1, \ldots, K)$ and F_r $(r=1, \ldots, K)$ can be computed,

9.3.3 DETERMINATION OF THE BOTTOM ELEVATION OF THE POND

Pond bottom elevation is a main factor of consideration in the use of design tide curve as shown in Fig. 9.17. In using tide curve for design the consideration should be given mainly to satisfying the biological needs of the cultured species, that is, adequate flow of water into the ponds, and the minimum and maximum water level to be maintained in the ponds. For economic construction of a pond, pond bottom elevation should strike a balance between the excavation or filling work and the tidal range. If a site has a relatively high ground elevation, pumping water to the ponds may prove more economical than excavating the soil to the desired elevations. The elevation of other structures such as sluice gate, supply channel, drainage channel, and dikes are also based on the design tide curve. It is essential that both sea and fresh water be taken or discharged completely within the shortest possible period. This requirement dictates the cross-section of the

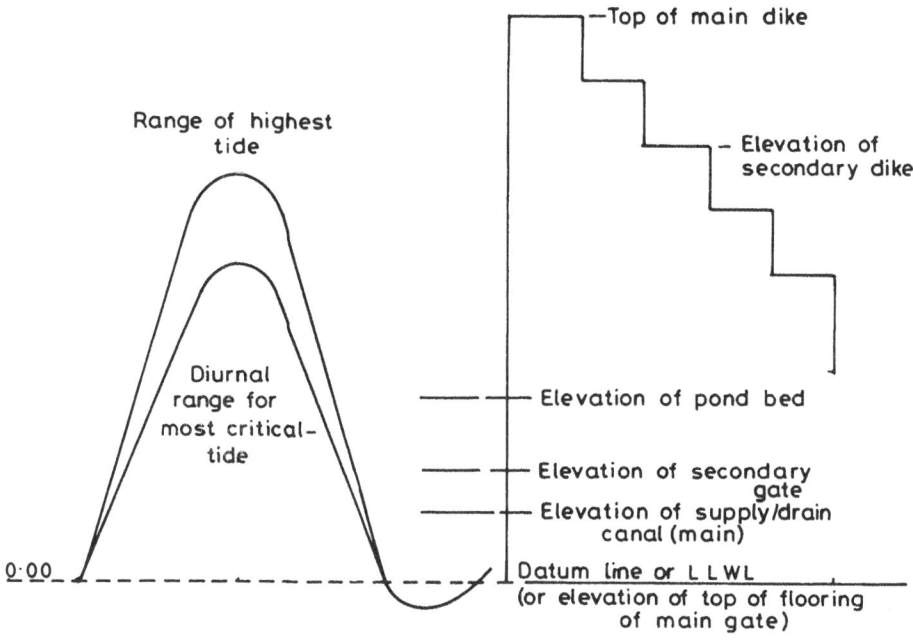

Fig. 9.17. Use of design tide for finding pond bed elevation

sluice. In the design of water supply and drainage systems, careful study should be made the site specific relationship between tide and elevation of the foundation as well as topography and flood hydrograph to estimate the amount of labour and cost and to determine whether pumping will be necessary. Installation of pumps will be required if the foundation is higher than the low water level. Also, reclamation work is necessary if the required elevation of the foundation is lower than the existing site elevation. The final determination of foundation level must be preceded by a careful study to see whether the installation of a pumping system can compensate the cost of excavation.

9.3.4 DETERMINATION OF SLUICE GATE SIZE

After determining the design tide and pond elevations, the next step involves the design of the sluice opening, that is, the determination of cross-sectional dimensions through hydraulic computations. To do this, a provisional width is assumed first. The inflow outflow volume of water in a given unit time is then measured with reference to the established tide curve and the elevation of pond bottom as shown in Fig. 9.18 to produce a time dependent curve in the pond. The time-elevation curves in and out of the pond are compared to examine whether the designed water level can be obtained within the specified time. If water cannot be filled to the desired level in the specified time, this process is repeated by selecting another width. The computational -procedures are: (1) the tide curve is drawn in the first quadrant, and (2) the water level in the pond is drawn in the second quadrant with storage along the x-axis and the corresponding water depth along the y-axis (Fig. 9.18). To perform the calculations as outlined below, the discharge through the sluice is assumed to be the same as that over a broad crested weir. The relevant expressions for supercritical and subcritical flow are given as follows (Kato, 1980).

For supercritical flow

$$Q = 2/3\,C_d\,\sqrt{\frac{2g}{3}}\,BH^{3/2} \text{ for } h < 2/3H \qquad (9.15)$$

For subcritical flow

$$Q = C_d'h\,\sqrt{2g(H-h)}\,B \text{ for } h > 2/3H \qquad (9.16)$$

where C_d, $C_d'h$ = coefficient of discharge;
    h = water level in the pond;
    H = total head, and
    B = width of sluice.

The values of C_d and C_d' are usually of the order of 0.9 and 1.0 respectively.

As an example of computation pond water depth is assumed to be 0 when the outer sea level is at an elevation 0.2 m, and the gate is fully open at the beginning of the calculation. The calculations are then made at a selected interval, say one hour, assuming that the difference between pond depth and outer sea level is free from fluctuation during this interval. During the time interval 0 to 1 hour, the head causing flow = ½ $(H_0 + H_1) - h_0$, where H_0 and H_1 = tidal levels corresponding to $t = 0$ and $t = 1$ hour respectively and h_0 = initial water level in the pond. Depending on the interval whether $h < 2/3\,H$ or $h > 2/3\,H$, equation (9.15) or equation (9.16) will be employed to find Q.

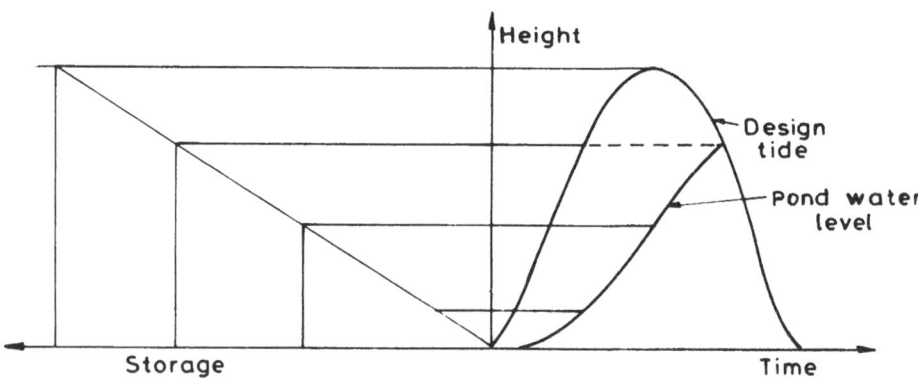

Fig. 9.18. Variation of water level inside pond with time against design tide

The total assumed volume of inflow is $3,600\,Q\,m^3$. The water level in the pond can be calculated by dividing the inflow volume with the corresponding farm area. In this way, calculation will proceed till the required sluice size is determined which is commensurate with the filling time. Generally the designed sluice opening is arrived at by considering the maximum desired rise in the farm water level in three consecutive tidal high waters.

The adequacy of sluice opening for draining the farm during a period of heavy rainfall also has to be checked. For this purpose, it will be necessary to examine the maximum rainfall intensity on the farm and the outside tidal condition. Usually, all available rainfall data for as large a period as available near the site have to be analysed. From the record, maximum 24 hour rainfall intensities, for example, for the monsoon period June to September, have to be ascertained. Assuming the maximum water depth to be maintained in the pond is 1.20 m from the culture point of view and the maximum rainfall depth is 25 cm during a 24 hour period, the depth of water in the pond to be drained will be $1.20 + 0.25 = 1.45$ m. Thereafter, one has to refer to the expected high water, above pond water elevation when tidal blockage will occur during monsoon months. In determining the highest tidal high water above the reference datum, one can find the duration of tidal blockage. The dimension of the sluice is determined by considering the remaining period of tidal cycle to see if the sluice is able to drain the water.

Numerical computations for the determination of required sluice size to satisfy appropriate filling and draining criteria of a brackish water fish farm are quite tedious. These computations can be performed on a computer using a suitable programme. Relevant programmes for filling and draining considering various hydraulic flow conditions and appropriate coefficients have been developed by Kulasekharan (1985) under the guidance of Ghosh. Typical computer outputs showing the pond filling and draining curve for a farm of 10 ha with constant and variable amplitude tide having a time period of 12.25 hours is shown in Figs. 9.19 and 9.20.

In the event of a main gate being installed on a main canal from which water is supplied to a number of ponds, Nakamura (1976) suggests that due consideration be given to the transformations of sea water fluctuations to the water level inside the main gate. This is reflected on the passage of water to the pond through the subgate by a factor. This factor is the ratio of water level fluctuation in the pond to the fluctuation of main sea water level. Nakamura (1976) gives the following relationships for the determination of sluice gate size in terms of effective pond area.

$$Q = A_e h \tag{9.17}$$

$$\bar{V} = \frac{2Q}{AT} \tag{9.18}$$

$$V_{\max} = \bar{V}/0.76 \tag{9.19}$$

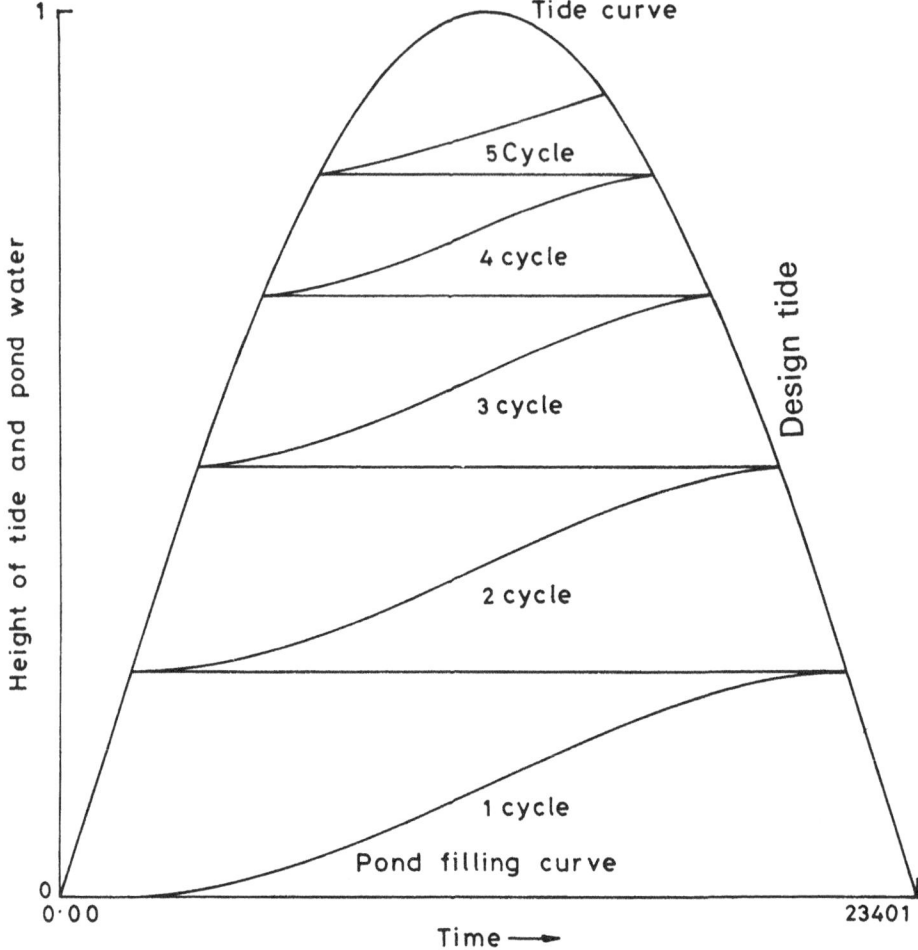

Fig 9.19. Pond filling curve with time during various tidal cycles for sluice opening of 1.15 m and pond area of 1 ha

where, Q = volume of water flowing in or out of the pond at each tidal flux and reflux,

A_e = effective pond area = $A_p\ (1 - \pi t_L/T)$

A_P = pond area,

$t_L = L/(gh)^{1/2}$,

L = length or distance from the intake to the further point of the pond.

h = initial depth of water in the pond,

g = acceleration due to gravity,

h = total rise of water in the pond during one tidal cycle,

$\underline{A}$ = area of sluice opening,

$\overline{V}$ = mean flow velocity, and

V_{max} = maximum flow velocity, through sluice opening.

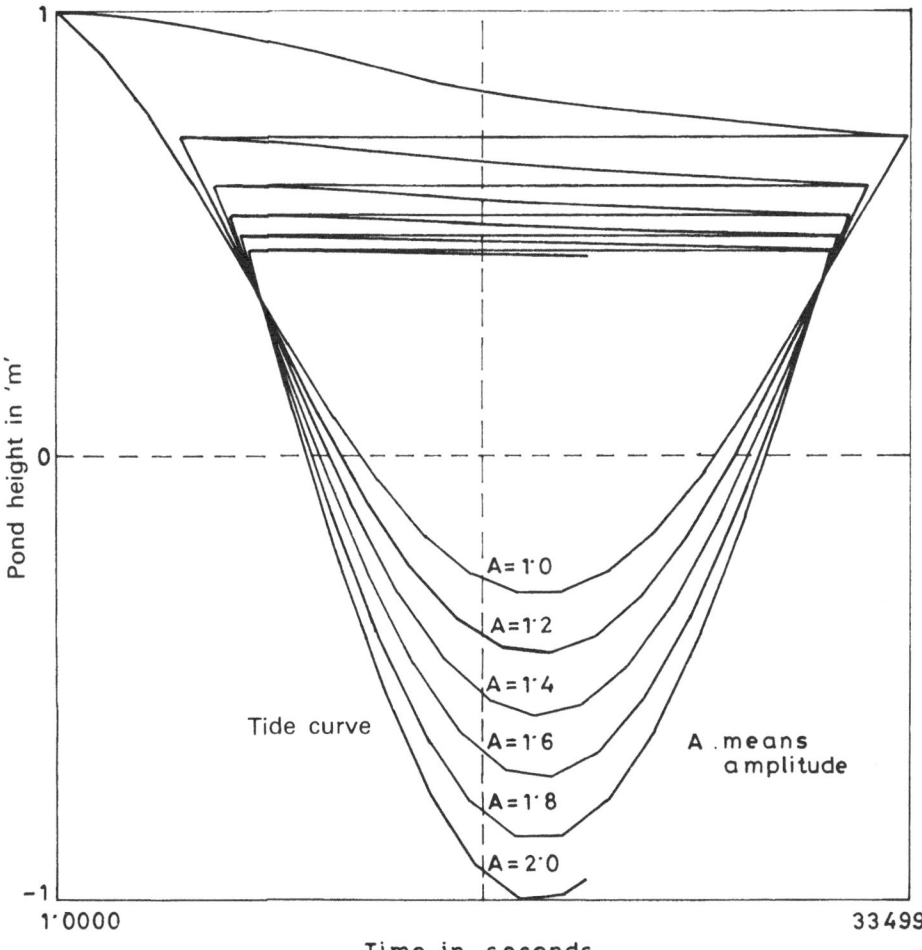

Fig. 9.20. Pond drainage curves with time for various tidal amplitudes for pond area of 1 ha and 1.15m sluice gate opening

For subgates, the volume of water considered is the total quantity of water rising above the pond level within the pond during a tide. For the farm as a whole, the total amount of water rising in all ponds or channels above the existing pond water level should be considered. When the pond level is kept slightly higher than mean tide level, a 10 per cent per tide usually provides a satisfactory requirement for design of pond water exchange.

9.3.5 Determination of the size of the supply drainage channels
    Brackish water from the outer sea or tidal creek is drawn into the fish pond at a specified rate and time through a channel and discharged into the sea or creek, through the same canal. The channel should be designed to carry the required flow within the provided cross-section efficiently without siltation or scour. The canals in fish farms are usually made of alluvial soil.

They are designated the main supply canal, secondary supply canal, drainage canal, and diversion canal.

The main canal starts from the main sluice gate and usually runs through the central portion of the fish farm. In designing the size of the main canal, consideration should be given to the emergency discharge of the entire farm water and water from the surrounding area, during a period of heavy rainfall. A secondary canal is the portion between the main canal and the sluice gate of the farthest fish pond. It is usually constructed in large farm areas. A supply channel is used as a drainage channel in many designs. However, a separate drainage channel may be provided in intensive shrimp culture which if provided, is usually located at the other end of the pond, opposite and parallel to the supply canal.

The purpose of a diversion canal is to protect a farm from being flooded by runoff from its watershed. The capacity of the diversion canal should be able to drain the peak runoff from the contributing watershed for a 10 to 15 year frequency storm. Its gradient should be such that the water flows toward the drainage area or around the fish farm to a convenient outlet. The main channel design discharge shall be the maximum required discharge passing through the sluice opening to fill up the farm area. This discharge can be determined from the computations of sluice openings. Similar computations for the secondary sluice will determine its required design discharge. The capacity of supply-cum-drainage channels can also be obtained from considerations of the influx of tide into the pond or the farm as a tidal prism. For an individual pond, the tidal prism is considered the total quantity of water rising above the pond level within the pond during the tide. For the farm as a whole, the total quantity of water rising in all the ponds and channels above the pond water level, should be considered.

Generally, a tidal prism is defined as the flow entering the estuary from the sea or tidal creek over a flood cycle which does not include the fresh water from the opposite direction. If the tidal velocity is given, as assumed by harmonic analysis, the average velocity over the period of flood tide is $(2/\pi)$ times the maximum velocity V_{max}. The tidal prism can be computed from

$$Q = \frac{2}{\pi} V_{max} \frac{T}{2} A \qquad (9.20)$$

or,

$$A = \frac{Q\pi}{V_{max}} T \qquad (9.21)$$

For practical purpose, equation (9.21) can be rewritten as

$$A = \frac{Q\pi C}{V_{max} T} \qquad (9.22)$$

where A = cross-sectional area of channel in square metres,

Q = quantity of water required for exchange in cubic metres,

C = a coefficient with a value between 0.8 and 0.9,

T = duration of tidal influx above minimum pond water level in seconds, and

V_{max} = mean maximum velocity of the tidal flow.

The canal cross-section usually is of trapezoidal shape with a side slope of 1:1 for the alluvial soil as shown in Fig. 9.21. The depth of main canal ranges from the mean higher high water excluding the height of free board for the mixed tide to the datum plane. In the case of diurnal tide, the depth will be from mean high water to the mean lower low water or datum plane. For the secondary canal, the depth will be from the designed pond water level to the mean tide level. The lower limit of the water depth depends on tidal range. Once the discharge is known in a design the required cross-sectional area can be found either with the help of Lacey's (1930) regime equations or on the basis of tidal prism concept as has been explained previously. The expression for velocity in regime channel is given, Lacey (1930).

$$ v = \left[\frac{Qf^2}{140} \right]^{1/6} \tag{9.23} $$

where v = velocity (m/s)

Q = discharge (m^3/s),

f = silt factor defined as $f = 1.76 \sqrt{d_m}$; and

d_m = silt diameter in millimetres = d_{50}.

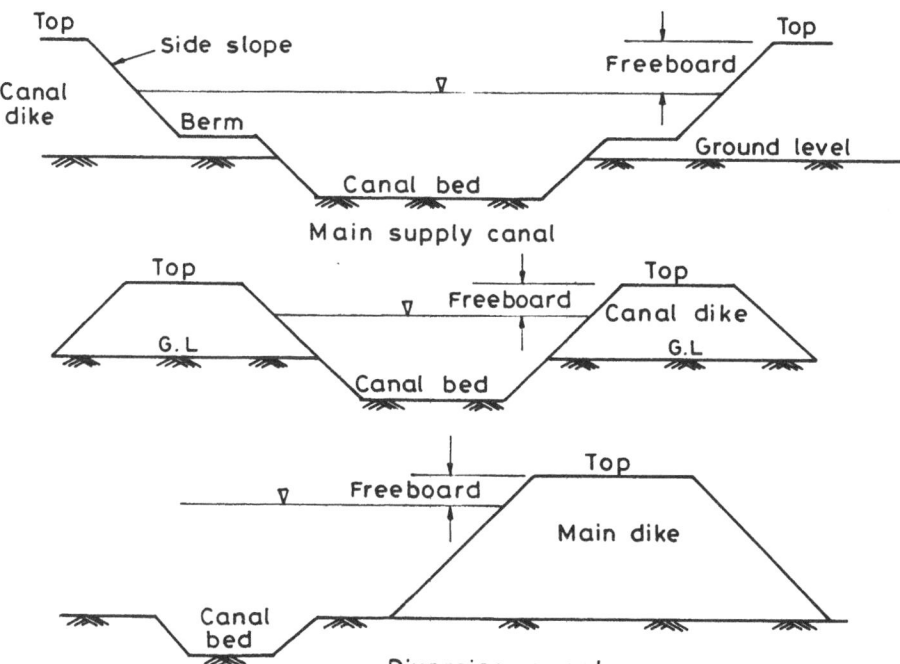

Fig. 9.21. Typical canal sections

The values of f can be calculated from the grain size analysis curve of the soil sample. Dividing the discharge by velocity, one obtains the cross-sectional area of the channel

$$A = \frac{(140)^{1/6}Q^{5/6}}{f^{1/3}}$$ (9.24)

where A = cross-sectional area in square metres.

For the assumed trapezoidal cross section one can write

$$A = Bd + nd^2, \quad \text{and}$$

$$p = B + 2\,(\sqrt{n^2 + 1})d$$

where B = channel bed width,
    d = depth of flow, and
    n = side slope.

Thus, the value of B and d can be obtained by solving the preceding two equations. The calculated depth should conform to the range as mentioned previsouly. Knowing the depth of flow to be maintained, one can also calculate B using either one of the equations.

Similar efforts will result in the estimation of the channel cross-section based on equation (9.22). In this connection, it should be remembered that the velocity of water in canals should be non-silting and non-scouring. For most soils, design velocity should not be lower than 0.3 m/s to avoid silting nor greater than 0.7 m/s to avoid scouring. The usual design velocity ranges from 0.5 to 0.7 m/s. The velocity computed by equation (9.23) usually lies within this range. For drainage purposes, the channel is given a slope which can be calculated using Manning's formula

$$v = \frac{1}{n}R^{2/3}S^{1/2}$$ (9.25)

where,
    n = roughness coefficient of the canal wall

$$= \left(\frac{d_{50}^{1/6}}{26}\right)$$

d_{50} = grain size in millimetres,

R = hydraulic radius = A/P, and

S = slope.

9.4 Design of water control structures viz, sluice gates

9.4.1 MAIN SLUICE GATE

The main gate connects the pond system to the source of water and regulates the exchange of water between the tidal creek or sea inlet. It is located at a convenient place in the proposed fish pond, faces the source of water and depends upon the availability of a suitable foundation. Some of the information needed for its hydraulic design in addition to the width of sluice opening is listed as follows:

1) The elevation of the floor of the main gate should be lower than the lowest pond bottom elevation inside the pond system of the farm. It should be as low as, or slightly lower than, the lowest tide shown in Fig. 9.17. In case the installation of a pump is required, the creek and elevation of the floor or apron may be made lower.

2) The height of the main gate depends on the highest tide in combination with the river flood and is generally at the same elevation as that of the main dike.

3) Main gates may have one or more openings. Usually a 1 to 1.2 m width per opening is appropriate for easy handling. A wider gate will require a lifting mechanism for the operation of the gate such as that adopted in major civil engineering structures like diversion dams.

Before discussing other aspects of the main gate, attention will be focussed on the various structural components of a control gate. These are briefly mentioned with reference to Fig. 9.22.

i) Floor: The gate rests on a main floor which serves as its foundation. The elevation at which it should be placed has already been mentioned. The gate foundation must be rigid and stable, and should be able to carry the whole weight when the gate is fully constructed. Usually there are two design practices for gate foundations. The first type lets the floor and apron gate rest on a combination of piles and layers of boulders and gravels. The second type as shown in Fig. 9.23 uses the piles alone to strengthen the apron.

ii) Apron: The extended and clear part of the floor upstream and downstream of the gate is known as the apron. The main function of the apron is to serve as a protective device against scouring action of flow.

iii) Cut off walls: For safety of the structure against seepage and consequent piping action, cut off walls are necessary at both ends of the gate floor if it lies over alluvial soil. The depth to which the cut off walls should be extended can be estimated from the probable scour depth by Lacey's (1930) equation

$$R = 0.47 \left(\frac{Q}{f}\right)^{0.333}$$

(9.26)

where R = maximum scour depth with respect to flood level (m),
   Q = discharge (m³/s), and
   f = silt factor.

For safety against piping, the exit gradient has to be calculated based on

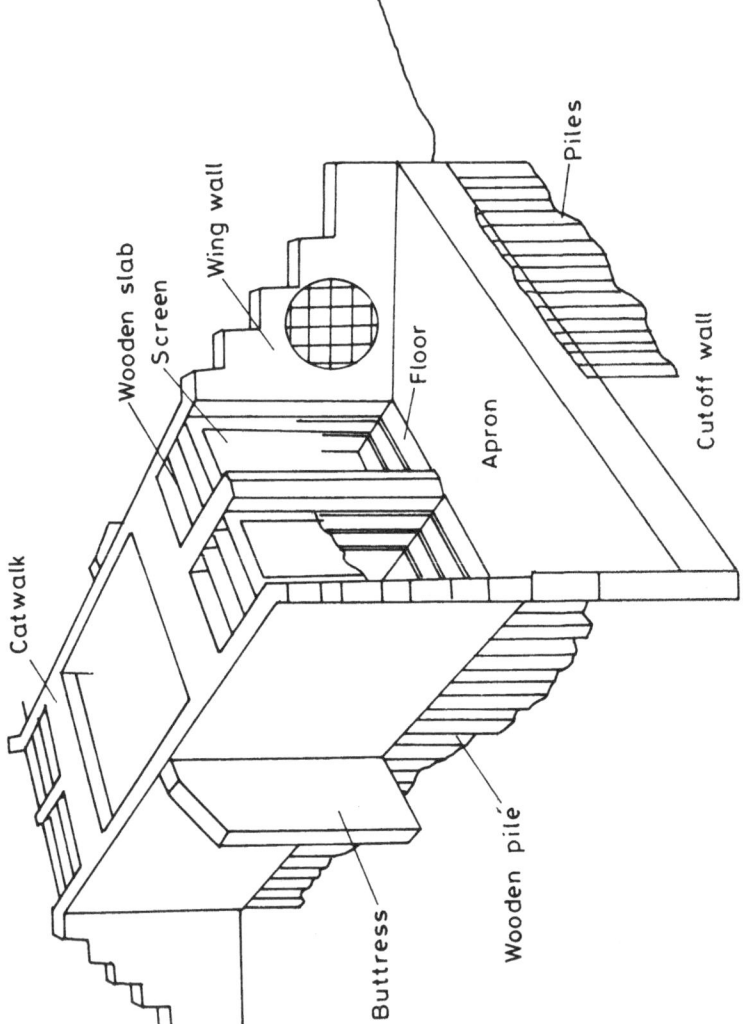

Fig. 9.22 Diagram showing the various structural components of a control gate

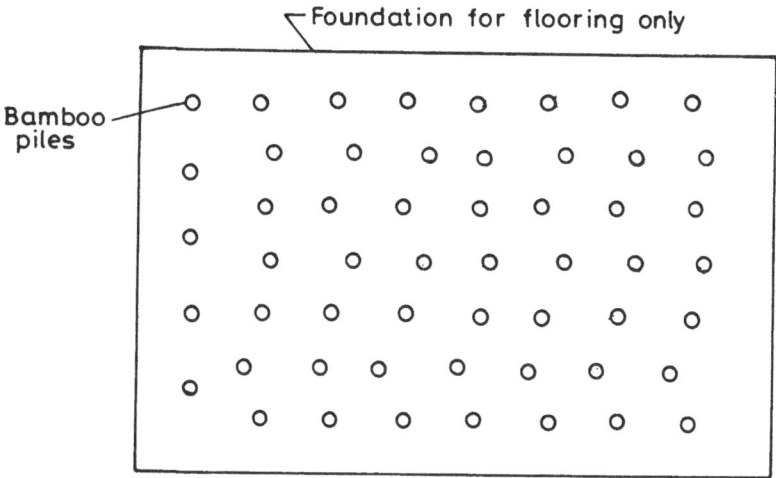

Foundation for flooring only

Bamboo piles

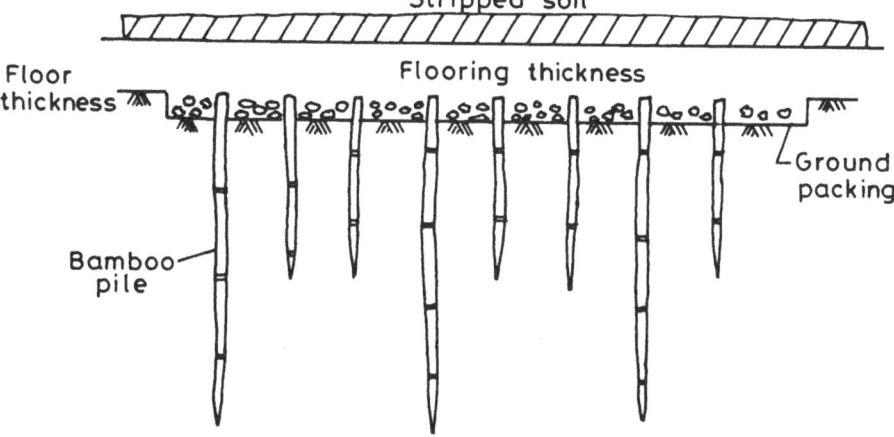

Stripped soil

Floor thickness

Flooring thickness

Ground packing

Bamboo pile

Fig. 9.23 Sluice gate foundation resting on piles

Khosla's (1936) relationship, $H/\pi d\sqrt{\lambda}$, where $d=$ depth of cut off wall; $\lambda = 1 + \frac{\sqrt{1+\alpha^2}}{2}$; $\alpha = b/d$ and $b=$ the width of the floor. Normally, a minimum of 0.6 m is provided as the thickness of cutoff wall. The value of the safe exit gradient is usually 1/5.

iv) Side or breast walls and buttress: These walls are built across the embankment to define the boundary of a sluice way. They are used as a retaining wall for the dike fill. Sometimes buttress supports are also built for strengthening the walls. The buttress helps in reducing seepage flow along the side walls. In the side walls, grooves are made for fixing up screens or stop gates. Usually the top of the walls are made the same height as that of the dike.

v) Wing walls: The functions of the wing walls are to provide the transition of the flow from the creek to the sluice and then from the sluice to the canal. The transition helps to guide the flow gradually to reduce the velocity from a higher to a lower value and vice versa. The wing walls also help in retaining the earth on both sides of the gate. For the structural design of wing walls and their stability analysis, one has to know the earth pressure exerted by the soil as well as various other loading criteria. For details, reference may be made to the book by Teng (1965).

vi) Walking platform: A walking platform, either made of a reinforced concrete slab or thick wooden planks, is provided between side walls. Generally a minimum of three catwalks are provided, two at each end and one at the centre near the stop log grooves.

vii) Flashboards, screens and pillars: Flashboards are generally made of wooden planks about 50 mm thick and 300 mm wide inserted into grooves to control the amount of water flowing through the gate. Screens made of polythene meshes are attached to a wooden rectangular frame which fits into the grooves. They are used to prevent the exit of the cultured fish and entry of predators into the ponds. Pillars are vertical supports provided in wooden gates where wooden walls are needed. They are placed at regular intervals to form a framework for the gate.

9.4.2 DESIGN OF SECONDARY GATES

Secondary gates are provided for the control of water to and from a canal to the pond compartments. The considerations for their planning and design are the same as for those of a main gate except that their respective elevations depend on the canal bed where they are being constructed. Common practice is to provide an elevation of about 150 mm of the flooring of these gates above the canal. The floor elevation of the farthest gate from the main gate should be checked against the design tide curve in order to ensure its capability to fill the pond within the prescribed time limit. Normally the width of the opening with wing walls varies from about 0.6 to 1 m, (Fig. 9.24).

Sometimes less expensive hume pipe culvert gates made of hollow concrete blocks or with slabs and screens are also provided.

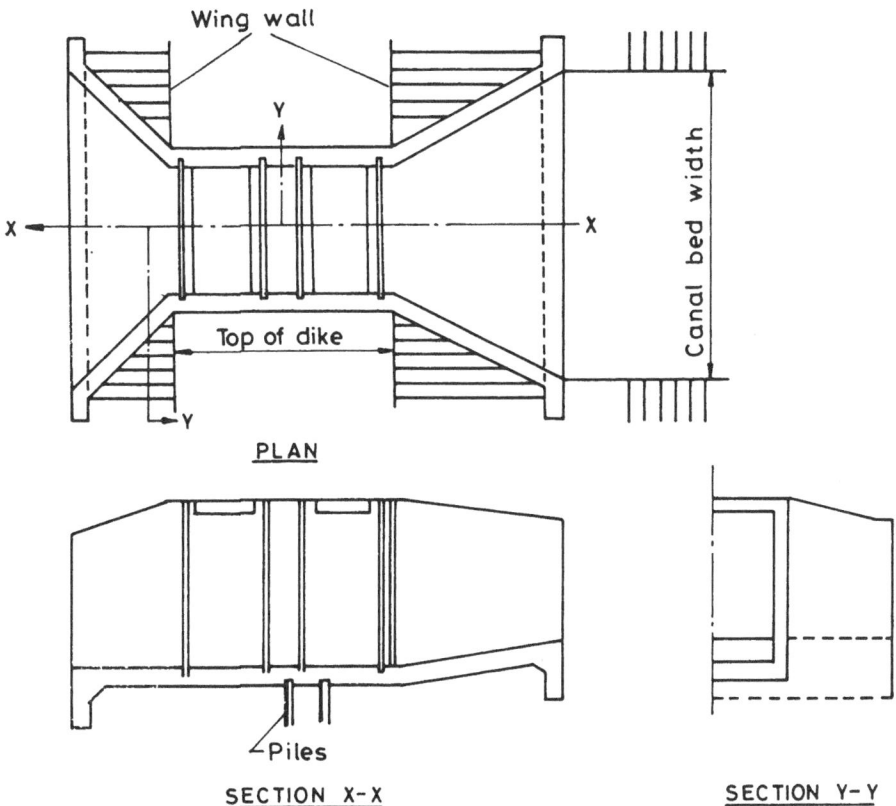

Fig. 9.24. A typical layout of a secondary gate

9.5 Design of peripheral dikes and internal bunds

9.5.1 MAIN BUNDS OR PERIPHERAL DIKES

The function of a peripheral dike is to retain water for use in the fish farming operation as well as to protect the fish farms from destruction due to floods, storm surge, astronomical tides, etc. Its design and strength depend on prevailing conditions at site. Some sites may be free from floods and storm surges. In this case, the peripheral dike should be designed only to withstand currents during an astronomical tide. These dikes are located at what is known as protected areas. The height of this type of peripheral dike should be based on astronomical tide level and water depth being retained on the pond site in addition to allowances for wave action and settlement of the bunds. In the case of exposed areas its height should be fixed based on the highest flood or storm surge level plus adequate freeboard as the case may be. In deciding the freeboard, due allowance should be made for the settlement of the periphery bund. A minimum of 0.6 m freeboard is necessary in the case of well-compacted periphery bunds. Most often the periphery bunds may have vehicles on them for transporting the fish caught. In this case it should have a width of 3 m at the top. The outer slope of the bund must be at least 1:2.5 and preferably 1:3 (vertical to horizontal) if it is exposed to strong floods and storm surges. Depending on the soil conditions the inner slope can be the same or steeper than the outer slope. Figure 9.25 shows a typical cross-section of perimeter dikes. A perimeter dike is usually built on the seaward side, along the river banks or in certain spots that are vulnerable to flooding. Some countries such as the Philippines and Indonesia, specify a belt of mangrove vegetation on a river bank and sea shore as an addition to the protection of dikes against attack due to waves and currents. This belt also can be used for absorption of wave energy, and to some extent for the conservation of the environment. The alignment of a dike is determined by survey to avoid (i) crossing of streams or creeks, (ii) areas having very poor soil conditions, and (iii) locating it where there is erosion of a coast or by a river. As previously mentioned, the cross-section of the perimeter dike should be designed to (i) prevent overtopping at high tide combined with a maximum flood height from the river system, and (ii) prevent failure due to seepage and slippage. A more detailed discussion on this subject is given in the chapter on River Protection Works.

The following factors have to be considered to determine the required height of the dike in the exposed area. These are (i) the higher high water level during spring tide, (ii) the effect of wind on tide levels, and (iii) wave impact, that is, waves in front of the dikes and wave uprush. The effect of wind is the friction exerted by the wind when blowing over a water surface causing a piling up of the water. The set up Z (Fig. 9.26) can be expressed by

$$Z = 0.36 \times 10^{-6} \, u^2 \, (1/d)\cos \phi \qquad (9.27)$$

where Z = set up in m, that is, difference in level between A and B;

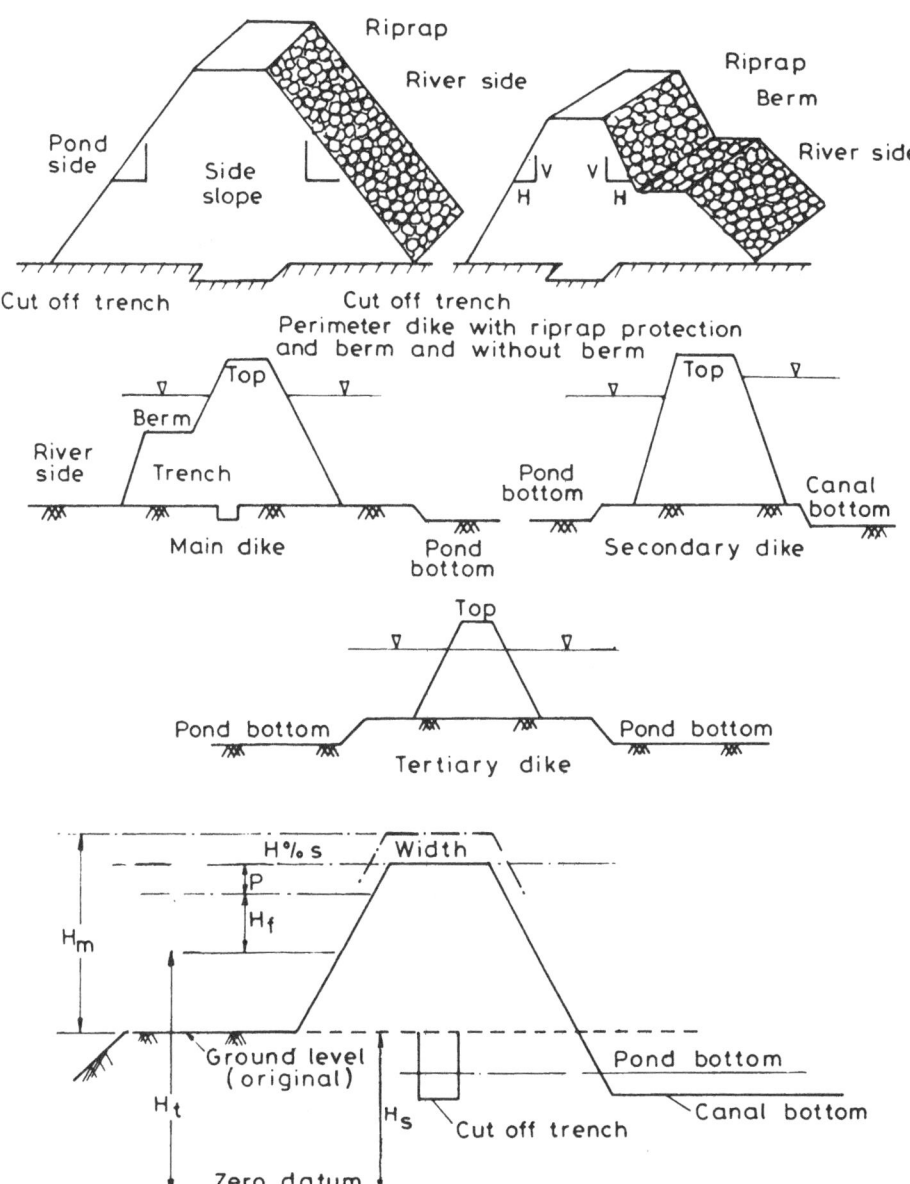

Fig. 9.25. Figure showing typical dike sections and height of main dike

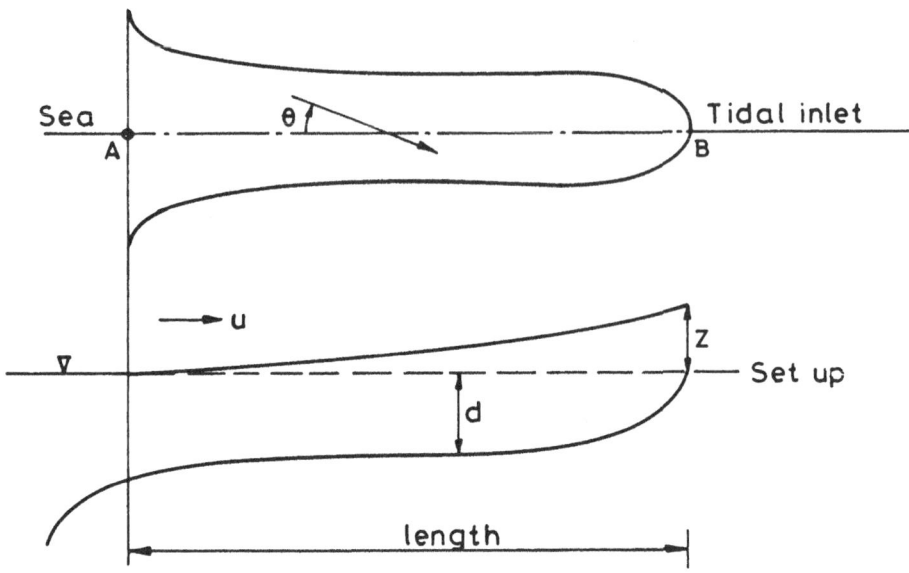

Fig. 9.26. Set up due to wind action in a lake

u = wind velocity (m/s) at 6 m height;
l = distance in m;
d = depth of water in m:
ϕ = angle between AB and the wind direction.

For a simple topographic situation, the wave height H in front of the dike can be computed from wind velocity u depth d, and fetch F, that is, distance over which the wind is blowing. Nomographs (Fig. 9.27) provide the dimensionless parameter of gF/u^2, $g\ d/u^2$, and gH/u^2 Bretschneider *et al*, (ref. 12, 1977).

Wave uprush R is the vertical distance from the undisturbed water level in front of the dike over which the top of the wave ascends along the slope. This undisturbed water level is determined by the astronomical tides and the wind effect as explained before. A relationship between wave height H, in front of the dike where the waves have not yet been affected by the presence of dike, and the wave uprush R (Fig. 9.28) is for dikes with an outer slope roughness equivalent to that of stone-pitching

$$R = 8H \tan \alpha \cos\beta \qquad (9.28)$$

where R = wave uprush; H = wave height;
α = slope of the dike; and β = angle between the direction of the wind and the line at right angle to the dike.

For a slope of 3.5:1 (horizontal to vertical) and $\beta = 0$, the relationship becomes $R = 2.3\ H$. The wave uprush can be reduced by applying a berm on

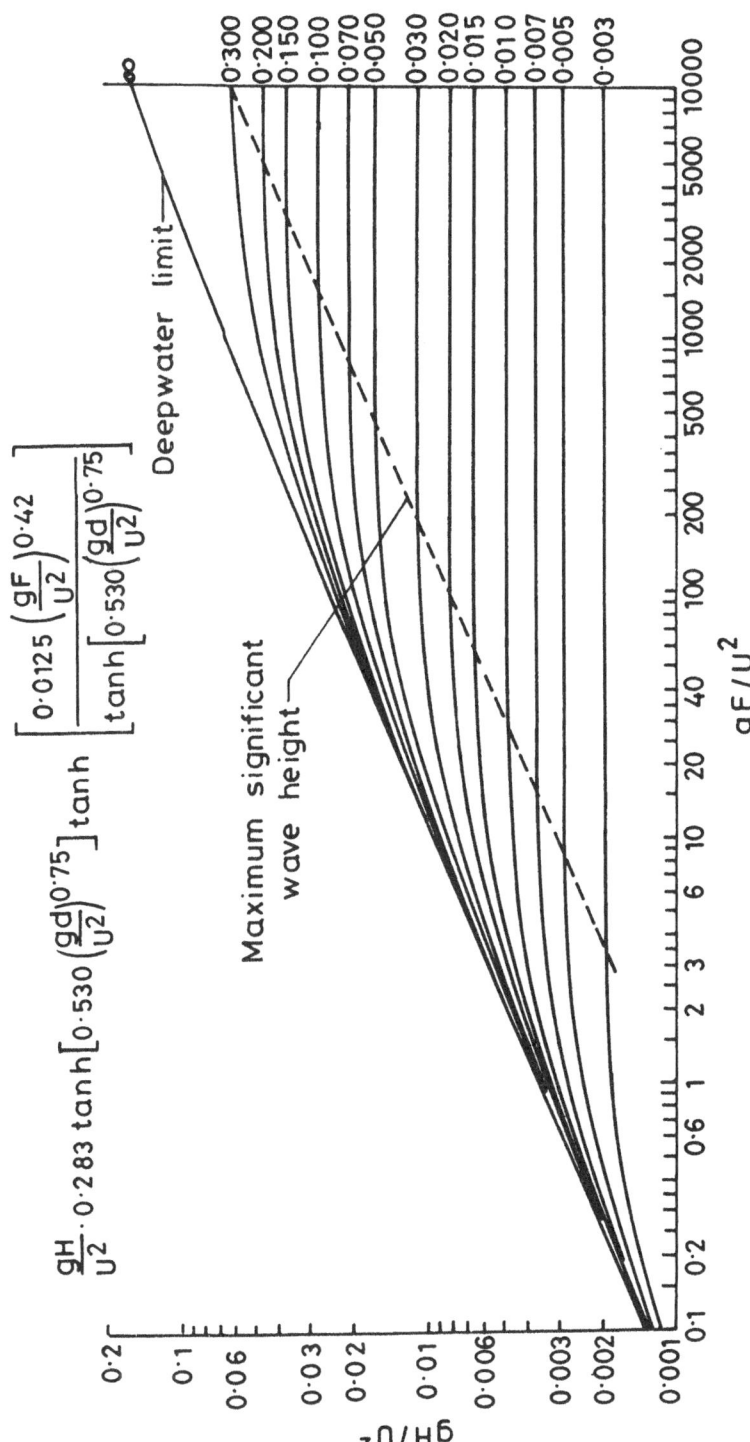

Fig. 9.27. Wave height as a functions of gF/U^2 and gH/U^2

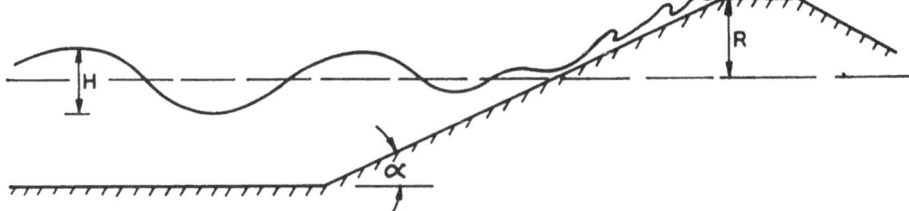

Fig. 9.28. Wave uprush on an embankment

the outer slope near the elevation of the undisturbed levels. The effect of this berm depends on the width B in relation to the wave length λ. In this case, $R = 8H \tan \alpha \, (\cos\beta - B/\lambda)$. The formula is applicable if $1/8 < \tan \alpha < 1/2$, $B < 2/3\lambda$, and $\beta < \pi/3$. For values of $\beta > \pi/3$ and up to $\pi/2$ a value of 0.5 instead of $\cos\beta$ should be used.

The crest elevation of a dike is the most important factor in the design. The conventional method is to consider the highest tide and flood that occur at site. According to the modern approach a certain frequency of excedence of the top of the waves is taken as a starting point. Selection of design frequency depends on the value of the land to be protected, from flooding damage, costs of repair work, and the risk of loss of human lives. An economic appraisal provides a comparison of the cost increase with increasing dike height and decrease in damage as shown in Fig. 9.29. For fish farms located along sea coasts, the maximum flood water based on a frequency of 10 to 15 years may be adopted. This is based on the fact that the economic life of a fish farm is generally taken as 12 to 15 years. The design height is provided with a freeboard after considering shrinkage and settlement of 0.3 to 1.0 m above the highest water level. The table below provides a relationship of soil classes and suitability for like material and recommended allowance for shrinkage and settlement.

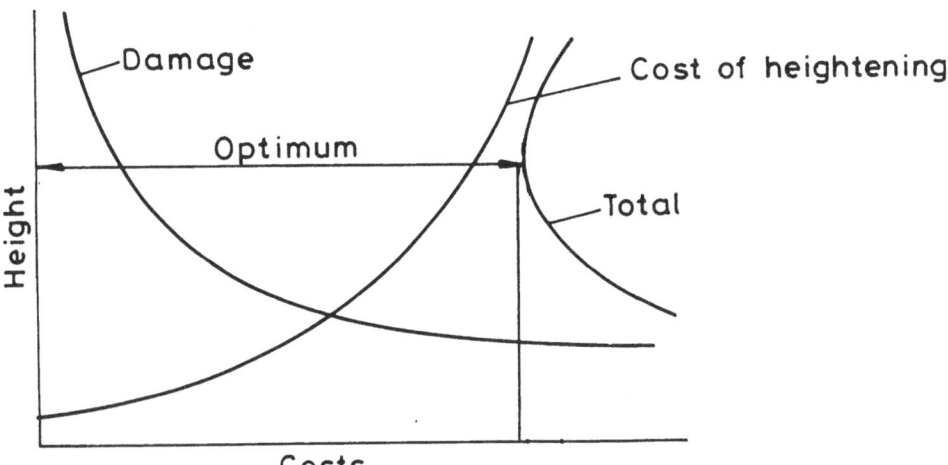

Fig. 9.29. Cost and damage function of embankment

Table 9.1
Soil classification After BFAR—UNDP/FAO 1981 (Tang, 1982)

Class	Permeability	Compressibility	Compaction	Suitability
Clay	Impervious	Medium	Fair to good	Excellent
Sandy clay	Impervious	Low	Good	Good
Loamy	Semi-pervious to impervious	High	Fair to very poor	Fair
Silty Sandy	Semi-pervious to impervious	Medium to high	Good to very poor	Poor
Peaty	Pervious	High	Good	Poor
Peaty		Negligible		Very poor

Table 9.2
Settlement allowance for different soil condition (Tang, 1982)

Condition	Allowance for structure settlement in (per cent)
Poor material and poor methods and practice in construction	15
Soil exceptionally high in organic matter	40 or more
Compacted by construction equipment	5–10

The total height of the main dike above the ground level shown in Fig. 9.25 can be computed by the following formula.

$$H_m = \frac{(H_t - H_s) + H_f + F}{\left(1 - \dfrac{\%S}{100}\right)} \tag{9.29}$$

where H_m = height of the main dike; H_t = highest astronomical tide;
$\quad H_s$ = elevation of the ground surface, with respect to datum;
$\quad H_f$ = flood allowance;
$\quad F$ = freeboard; and
$\quad \% S$ = per cent of shrinkage and settlement.

To prevent leakage and reduce seepage loss, attention should be paid to dike design and its proper construction. After the dike is constructed, leakage usually occurs by crustacean burroughs particularly the species called Thalassina anomala (Tang 1982). They make burroughs in the soft mud under the dike thereby causing piping by which sand and silt particles are moved by seepage flow. The measures for preventing leakage include.

i) minimising destruction by crustaceans by desalinising and drying out the embankment soils,

ii) minimising the amount of seepage flow through proper compaction, core trenching, embedding a vertical plastic membrane inside the dike, covering the dike wall with riprap, etc.

For estimation of loss due to seepage through a dike, the following factors have to be known: (i) permeability coefficient of the dike, (ii) height of water level in the pond, (iii) effective dike width, and (iv) whether the dike foundation is permeable or impermeable. Consider a dike placed on a permeable foundation as shown in Fig. 9.30. The seepage can go through the body of the embankment and its foundation. The expression for the seepage loss is given as

$$q = K \left(\sqrt{h^2 + w^2} - w \right) + K_1 \, (h/w) \, (H) \qquad (9.30)$$

where q = seepage flow/cm length of dike in (cm^3/hour);
K = coefficient of permeability of the dike (cm/hour);
K_1 = permeability coefficient of the permeable foundation (cm/hour);
h = height of water level (cm); and
w = effective width of dike.

In case the dike rests on an impermeable foundation (Fig. 9.31) then

$$q = K \left(\sqrt{h^2 + w^2} - w \right) \qquad (9.31)$$

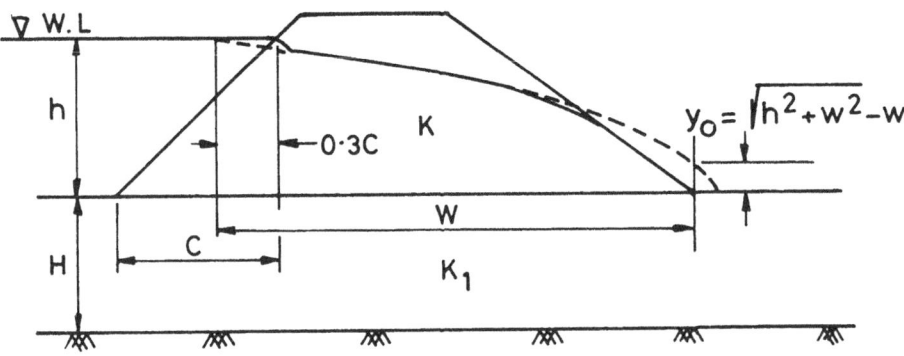

Fig. 9.30. Dike resting on a permeable foundation

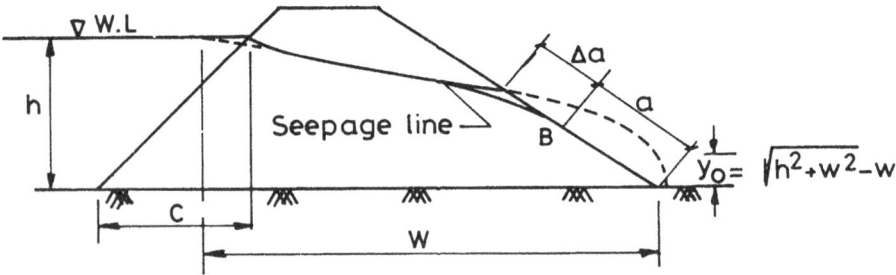

Fig. 9.31. Dike resting on an impermeable foundation

For a brackish water fish farm, the dike foundation is mostly permeable. As the K value of alluvial clay is relatively constant, the amount of seepage depends on tidal fluctuations and effective width of dike as shown in Table 9.3.

Table 9.3
Variation of seepage due to tide in dike (Tang, 1982)

Dimension of dike			Minimum seepage inflow in rising tides		Minimum seepage outflow in ebb tides	
Top width (m)	Height (m)	Slope	Effective width (m)	Rate of seepage flow (cm³/hour)	Effective width (m)	Rate of seepage flow (cm³/hour)
2	2	3:1	11	0.13	14	0.28
2	2	2:1	8	0.18	10	0.32
2	2	1:1	5	0.29	6	0.69

Secondary dikes are smaller than main peripheral dikes, and are usually provided on both sides of the main supply or drainage canal. Secondary dikes should be able to contain the requisite flow depth while filling or draining the farm. Finally, their total height is obtained by providing a suitable freeboard. The width and slope of the dike are usually provided so that the line of seepage shown in Fig. 9.31 is not exposed. Tertiary dikes are partition dikes that separate the ponds to contain the desired water levels in the pond.

A dike's stability has to be checked against failure due to slippage. The usual method of analysis is the slip circle method which is briefly outlined as follows:

The methods for stability analysis are based on the shear strength of the soil and certain assumptions with respect to embankment failure. The Swedish or slip circle method shown in Fig. 9.32 which assumes a cylindrical surface of rupture is a comparatively simple method of analysing embankment stability. The factor of safety against sliding is defined as the ratio of the average shear strength of soil to the average shear stress determined by statics on a potential sliding surface. The force exerted by any segment within the slip circle is equal to the weight of the segment and acts vertically downward through its centre of gravity. The components of this weight acting on a portion of the circle are the force normal to the arc and the force triangle with lines in the radial and tangential directions. Porewater pressures acting on the arc result in an uplift force which reduces the normal component of the weight surface. The factor of safety can be expressed in accordance with porewater pressure, that is,

$$F.S = \frac{CL + \tan \phi \ (N - U)}{T} \qquad (9.32)$$

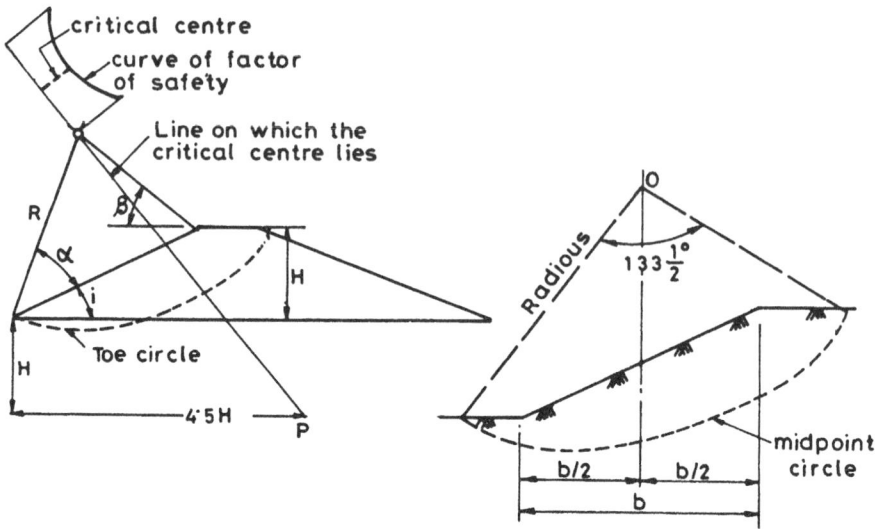

Fig. 9.32. Critical failure surfaces

when there is no porewater pressure,

$$F. S = \frac{CL + N \tan \phi}{T}$$

where N = summation of the normal forces acting along the arc;
  U = summation of the uplift forces due to porewater pressure along the arc;
  T = algebraic sum of the tengential forces along the arc; and
  L = length of the arc slip circle.

In homogeneous material, toe circles generally are the critical circles for steep slopes. For flat slopes, mid-point circles are critical. In non-homogeneous material, the critical circle is located so that the maximum portion of its length passes through the material having the least shear strength. The lowest shear strength may be at the core of the drain or foundation layer. Usually the minimum value of factor of safety should be of the order of 1.25.

Locating critical slip circle by trial and error requires considerable time. To reduce the number of trials, Fellenious (ref. 6, 1967) method is used for locating the line on which the centre of the critical circle is most likely to be in a homogeneous section. Various circles with their centres on this line are tried in order to find one with minimum factor of safety. The position on which the centre of the critical circle lies depends only on the height and slopes of the embankment. The coordinates of P and H downward from toe and $4.5H$ horizontally away are as shown in Fig. 9.32. Point M is located with the help of direction angles α and β as given in Table 9.4.

Table 9.4
Location of direction angles

Dike slope	Angle (degrees)	α (degrees)	β (degrees)
1:1	45	28	37
1.5:1	33.8	26	35
2:1	26.6	25	35
3:1	18.4	25	35

The centre of the critical circle lies on the line *PM*. Trial circles are drawn on the line and the factor of safety corresponding to each centre is calculated. The critical centre is the one where the factor of safety is the least.

For dams and foundations of homogeneous materials the most dangerous failure surface is below the toe of the slope. Fellenious found that the angle intersected at 'O' in this case is about 133.5°. To find the centre for the most dangerous circle below the toe, the following procedure as shown in Fig. 9.32 is suggested. A vertical line at the mid-point of the slope should be erected. The centre 'O' of the first trial circle should be on this vertical line. In locating the trial circle, use an angle 133.5° between the two radii at which the circle intersects the surface of the embankment and the foundation. After the first trial circle has been analysed, the centre is moved somewhat to the left of the radius, shortened, and a new trial circle is drawn for analysis.

REFERENCES

Alam Singh; (1967): Soil Engineering in theory and practice Bombay, Asia.

Alcantara, L. (1982): Variation of Fishpond layouts for different types of brackish water management, Seminar Coastal Fishpond Engineering, Surabaya, Indonesia.

Chalayondija, K.P., Tharnbuppa and Sikga S. (1982): The present design and construction of ponds for rearing penaeid Shrimps in Thailand, Seminar Coastal Fishpond Engineering, Surabaya, Indonesia.

David, Clark. (1963): Plane and Geodetic Surveying for engineers. Constable Co., London.

Djajadiredja, R. and Danley, T. (1982): Aspects of design and construction of Coastal ponds for milk fish seed production Seminar Coastal Fishpond Engineering, Surabaya, Indonesia.

Dureza, V. (1982): Site Selection and pond construction guide for brackish water Milk fish ponds. Workshop Aquaculture Project Development and Management, Philippines.

Kanetkar, T.P. (1959): Surveying and Levelling, Grih Prakaspa in Pune.

Kato, J. (1980): Guide to design and construction of Coastal aquaculture ponds, New Edition, JICA, Tokyo.

Khosla, A.N., Bose, N.K. and Taylor, J. (1936): Design of Weirs on permeable foundation Central Board of Irrigation and Power, Govt. of India Publication No. 12.

Kulasekaran, R. (1985): Computer aided design of sluices for water exchange in tide fed farms. Unpublished M. Tech. Thesis Aquacultural Engineering, I.I.T., Kharagpur.

Lacey, G. (1930): Stable Channels in Alluvium, Proc. Inst. of Civil Engrs, Lond, Vol. 229, pp. 259–384.

Nakamura, N. (1976): Design of an aquaculture pond with a tidal inlet, FAO Technical Conference on Aquaculture, Kyoto, Japan 26th May to 2nd June.

Tang, Y.A. (1982): Planning, Design and Construction of a Coastal Fish Farm. Advances in Aquaculture, FAO, by Fishing News Book Ltd.

Taylor, D.W. (1958): Fundamentals of Soil Mechanics, Wiley Publication.

Teng, W.C. (1965): Foundation Design, Prentice-Hall of India (Private) Ltd., New Delhi.

Terzaghi, Karl and Peck, R.B. (1960): Soil Mechanics in Engineering Practice John Wiley.

Shore Projection Manual (1977): Vol. I, U.S. Army Coastal Engineering Research Centre, Department of the Interior, U.S.A.

Tide Tables for the Hugli River, Calcutta Port Trust, Govt. of India. Published by the Surveyor General, Govt. of India (Every Year).

CHAPTER 10

Sedimentation Problems in Brackish Water Fish Farms and Engineering Measures for Its Mitigation

10.1 Introduction

Coastal fish ponds in the tropical and subtropical zones are constructed in tidal swamps, estuaries, bays and sheltered lagoons (Figs. 4.3 and 4.4). This is because of the necessity of having nutrient-rich sea water with adequate salinity for culture and biological activities of the relevant species of fin fish and shell fish. Water in the coastal areas contains varying amount of suspended materials, of which plankton favours greater fish production by stimulating the growth of fish food organisms, but the forms like sand and excessive amount of clay and silt, are not desirable in the fish pond. Wallen (1951) noticed behavioural changes in fish exposed to clay turbidities greater than 20,000 mg/litre, but individuals of 16 species survived exposure to 100,000 mg/litre of clay turbidity for one week or longer. Appreciable mortality occurred at turbidity values above 175,000 mg/litre. Even though turbidity caused by suspended soil particles will seldom have an immediate direct effect on fish, in the long run it may harm fish populations. Clay turbidity will restrict light penetration, adversely affecting productivity, and some of the particles will settle at the bottom and smother fish eggs and destroy benthic communities. The main concern with regard to protection of the aquatic fauna is not the suspended particles (turbidity) *per se*, but the amount of solids in suspension that can potentially settle out (settleable solids) Boyd (1982). Buck (1956) reported faster growth and greater reproduction of fish in less turbid ponds.

These suspended materials are brought into the estuarine and deltaic reach by the river and its tributaries from the respective catchment areas. As they enter into the estuarine reach where the flow is two way and the environment is saline, they settle down in some specific areas forming sediment bars (Fig. 10.1) as a result of complex interaction of physical, chemical and biological processes depending on the upland flow, tidal influx, temperature variation, pH value, clay minerals, algal growth and season. In case such bars form in the neighbourhood of tidal creeks from which the fish farm draws its supply, then it is likely that in the course of time, the flow in the creek is going to be reduced or stopped completely unless the bar mouth is opened or washed off during a flood season.

In the case of a coastal inlet, the formation of bar mouths as shown in Fig. 10.1, is essentially due to the complex interaction of tidal and wave action. When and where the tidal prism is small, the wave action tends to force the sand into the inlet. This action predominates over the force of the ebb current which tends to keep the channel open and the result is the considerable silting of the inlet channel mouth and beyond. On the other hand if the tidal prism is large, the ebb currents will force the littoral transport offshore in the form of a crescent shaped bar. Usually the combination of a small tidal prism and insufficient stream runoff coupled with high intensity of wave action is the reason for inlet closure. This is because the waves build a beach with a relatively high berm such that the normal stream flow can reach the sea as subsurface flow. Because of high flow during the rainy seasons a flood channel across the beach to the sea is quickly cut. As soon as the runoff has ceased, the waves quickly rebuild the beach berm and the entrance is again completely closed.

In the case of river channels of the Sunderban tidal island of India it has been found to be sinuous to dendritic in pattern. Further, each island is characterised by definite sites of erosion and deposition. Erosion commonly takes place on the convex sides of the islands whenever they are distributed within the meandering distributory channel. Some islands show a terrace-like rise from the shore upwards whereas others exhibit a gradual rise from the

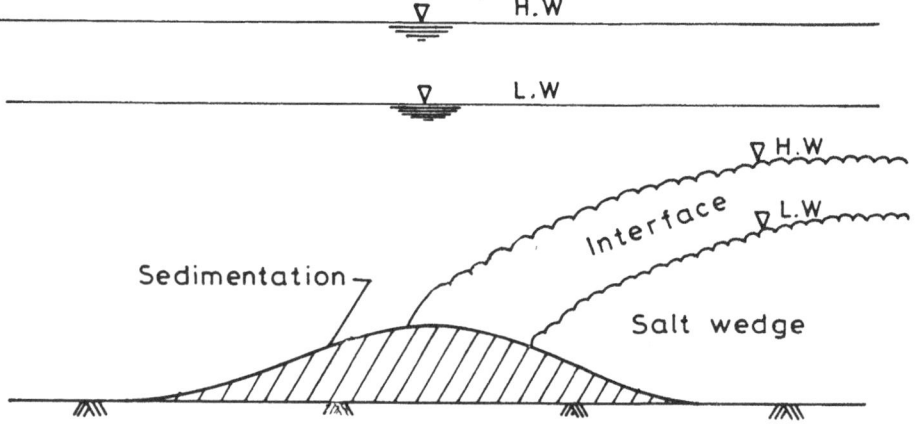

Fig. 10.1.

margin towards the central part. The central parts of the islands are quite elevated with exposure to the sun for most part of the year. The sediment of these zones gets the opportunity of being reworked by wind action. These regions are often characterised by wind-borne structures such as dunes, eolian ripples, wind shadow structures (Chaudhuri, 1984).

It can be seen that successful operation of fish farms located in deltaic and coastal areas depends on understanding the underlying principles governing the erosion and depositional patterns of the silt laden water carried by tidal creeks as it effects its stability, sedimentation processes in the saline environment in the estuary and the fish ponds, interaction of tide, wave and winds in the formation of bar mouths in coastal inlets. In this contribution some aspects of the above problems are discussed along with possible remedial engineering measures to reduce sedimentation in ponds and prevention of closure of inlet mouth.

10.2 Sediment transport in tidal flow

It is generally well known that a tidal wave becomes asymmetrical as it progresses up the estuary. This results in a skewed bed shear cycle (Fig. 10.2). Let us consider now a tidal creek having bed material which is stable when $\tau_0 < \tau_c$ where τ_0 = bed shear stress, τ_c = critical shear for incipient motion for that particular size. Figure 10.2 illustrates a case with scour when $\tau_0 > \tau_{cs}$ on the flood and ebb during the time interval δt_1 and δt_3 respectively and in between there will be deposition during the intervals δt_2 and δt_4 and slack water when $\tau_0 < \tau_{cd}$. If we consider a typical reach, the cross-sectional shape

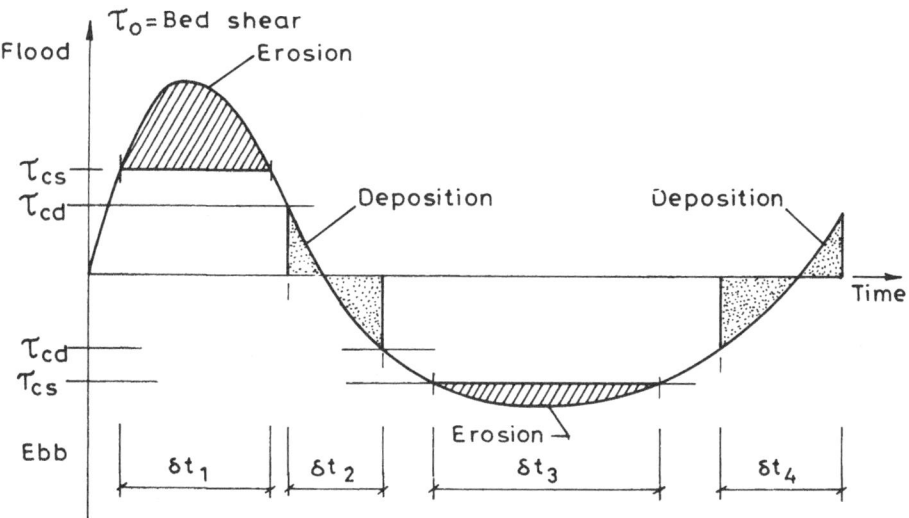

Fig. 10.2. The variation of shear stress during a tidal cycle

at the end of a tidal cycle will remain unchanged if scour and depositional volume are equal and hence cancel each other. The erosion and depositional volume can be expressed in terms of the following relationship after Krone (1962).

$$q_e = K \ (\tau_0/\tau_{cs} - 1) \ \text{and} \tag{10.1}$$
$$q_d = cw \ (1 - \tau_0/\tau_{cd})$$

Where K = erosion flux at $\tau = 2\tau_{cs}$, τ_{cs} being the limiting bed shear for scour and τ_{cs} = limiting shear for deposition, c = volume concentration of particles with fall velocity w. The eroded particle volume during one tidal cycle of period T can be expressed as

$$\int_0^T (q_e - q_d)dt \tag{10.2}$$

In order to evaluate the integral, it is necessary to know both the bed shear τ_0 and the concentration c. the above equations are for fine uniform sediment carried mostly in suspensions.

The bed shear stress can be estimated from velocity measurement following a relationship of the type

$$\tau_0 = \alpha\rho \ V^2 \tag{10.3}$$

Where V = velocity of flow, ρ = density of water,
α = constant

Alternately, it may be obtained from a flow simulation model. The concentration of suspended sediments can be obtained by measurement linking c with the prevailing tide (Fig. 10.3), or it may be obtained from a suspended sediment transport model based on diffusion equation for sediment suspension.

The sediment in the estuarine region essentially consists of silt and clay minerals and they are very fine in size and behave as a cohesive, colloidal particle. These colloidal particles remain suspended in water against gravitational forces and are, as a rule, negatively charged and repel each other. Contact with electrolytes of opposite charge neutralises the charge of the colloids and reduces the strength of repulsion between colloids and they agglomerate. Experiments have shown that due to the physico-chemical properties of the clay suspension in the flow with high salt content, flocculation between clay particles occur. The floc is an aggregate of flocculent grains which according to their structures can be subdivided into catenary, X-type, honey comb and group structures (Fig. 10.4). Experimental results (Zhizhong *et al.*, 1983) show that the diameter of the sediment with which it flocculates fast is around 32 μm and amongst the various factors that influence the amount of flocculation, the clay mineral component of the fine sediments and the electrolyte are very crucial. The diameter of most flocs lie between 16 μm to 32 μm and its largest size may be as high as 700 μm. In view of these size differences, the flocs do not settle down with constant fall

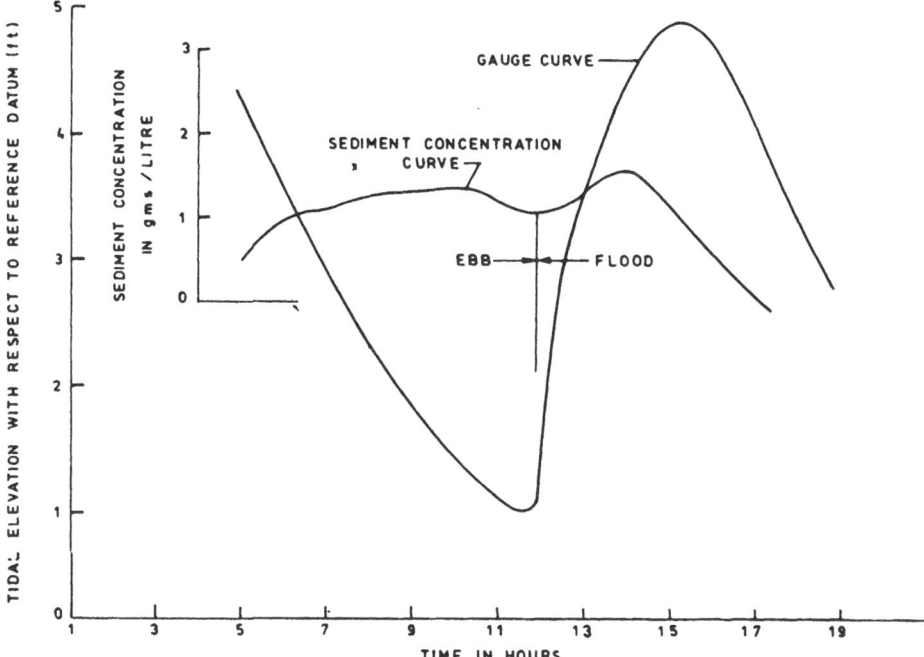

Fig. 10.3. Variation of sediment concentration with tide

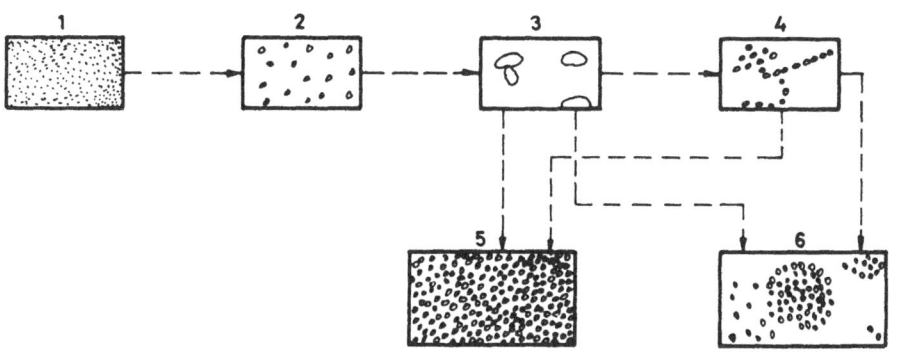

Fig. 10.4. Floc formation process (Zhizhong *et al.*, 1983) 1. FLOCS. 2. FLOC AGGREGATE. 3. MEDIUM AGGREGATE 4. CATENARY, X-TYPE 5. HONEY COMB 6. LARGE NETWORK

velocity. Figure 10.5 shows a result of the average settling velocity of flocs as a function of salinity variation (Zhizhong *et al.*, 1983).

When a floc of particles becomes heavy enough, it precipitates. The effectiveness of electrolytes in coagulating colloids increases with the charge on the electrolyte. For example, calcium is 30 times as effective in coagulating as sodium ion, and aluminium and ferric ions are 1000 times as effective as sodium ions (Boyd, 1982). Organic matters also induce the turbidity to precipitate.

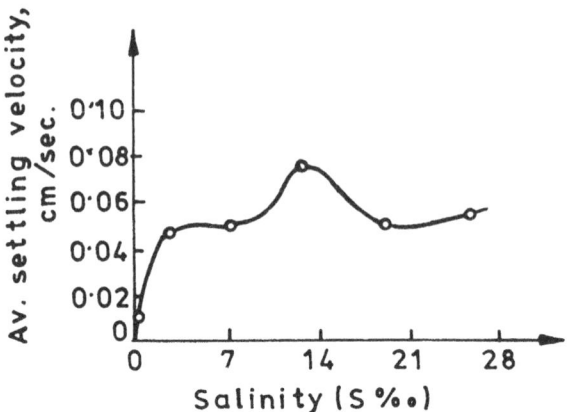

Fig. 10.5. Variation of average settling velocity of flocs with salinity (Zhizhong *et. al.*, 1983)

10.3 Sedimentation processes in marine environment including fish pond

The sedimentation processes in marine environment are generally complex. Here physical sedimentation by simple gravitational deposition usually takes place for coarser size fraction of the river sediment material usually the sand whose sizes are greater than 45 μm. Similarly fine clays in aggregate form as flocs settle down by differential settling.

On the other hand there also occurs chemical sedimentation due to precipitation of materials like calcite $CaCo_3$, apatite $Ca_5 (PO_4)_3$, (OH, F, Cl). Further in these environments there occurs seasonal growth of algae and formation of diatoms under favourable conditions which plays a vital role in the formation of aggregates and their eventual deposition. Investigations carried out in the Loire river in France and reported by Manickam (1983) present very interesting results on the depositional behaviour of various substances in marine environment.

The main intention of discussing the above processes is because of the fact that sedimentation processes in the fish pond will also be of similar nature. In order to estimate how much material will settle down in fish ponds which from the hydraulic point of view functions similar to a settling basin, it is necessary to estimate their settling velocity. If one considers the length of fish pond to be equal to *l*, then a sediment particle will begin to settle when the flow velocity is smaller than the critical velocity for the particle size considered and the relationship governing these can be expressed as

$$l = \frac{hv}{w} \tag{10.4}$$

where $h =$ water depth in the pond, $v =$ mean velocity of flow, $w =$ fall velocity in moving water.

Thus it is possible to estimate the likely amount of sediment to be trapped in the fish pond knowing the quantity of silt entering the pond having different grain size compositions and salinity level in the pond. Culture of fish in cages with the the application of artificial feed, in the protected bays of the coast contributes largely to the sedimentation. Kadowaki *et al.* (1978) investigated the dispersion of suspended solids in relation to the fish feeding programme. Average daily amount of feed supplied to the observed culture unit (18 cages) reached 1.6 tonnes per day in October and 1.9 tonnes per day in November. There exists a relationship between the amount of food offered per day and the dry weight of suspended solids collected in sediment traps underneath the cages. The dry weight of organic matter varied between 35.5 $gm/m^2/day$ in October to 45 $gm/m^2/day$ in November. During feeding and shortly after, the transparency of the water decreased in the vicinity of the cages while the number of particles counted increased.

10.4 Measures to reduce sedimentation in fish ponds

The measures to be adopted for reduction of sedimentation in the fish pond can broadly be outlined as follows. The first case is related to intake of saline water from creeks at the upper layer of tidal curve where the silt concentration will be less. Figure 10.6 shows the simultaneous variation of silt concentration, velocity and salinity during flood and ebb period in a tidal channel (Zhizhong *et al.*, 1983). This will require suitable intake design other than the present practice of drawing water by the sluice gate from the relatively lower strata. In the second case, sedimentation is induced prior to the water being introduced into the fish pond. For achieving the above, the following guide lines may be adopted.

The quiet period preceding and following slack water after high tide is the main period of sedimentation. Extending the duration of the period by diminishing water velocity before and after slack water will promote sedimentation provided the current that supplies the sediments are not hampered too much. Usually part of the freshly deposited sediments are again carried away by the ebb current or the flowing flood current during periods of strong wave action. In such a case measures which consolidate the sediment and decrease wave action will increase the ultimate amount of sediment that is captured. Two measures for achieving the above objective can be employed in combination. The first method consists in promoting vegetation. Vegetation breaks the velocity of water during period of flood and helps lengthen the quiet period. Some kind of mangroves which are strong and tall can also decrease the wave action. There exists an optimum condition for sedimentation between the velocity of current and the density of vegetation. In case the breaking action is strong compared with the current, then in the central part of the vegetation the supply of sediment will decrease due to the low capacity of currents for carrying sediments. On the other hand, scattered obstacles such as isolated vegetation boundary in the water, tend to increase turbulence and produce erosion effects. The more suitable conditions are vegetation with increasing density backward from the borders. In order to prevent fresh deposited sediment being eroded the most suitable vegetation is a dense short

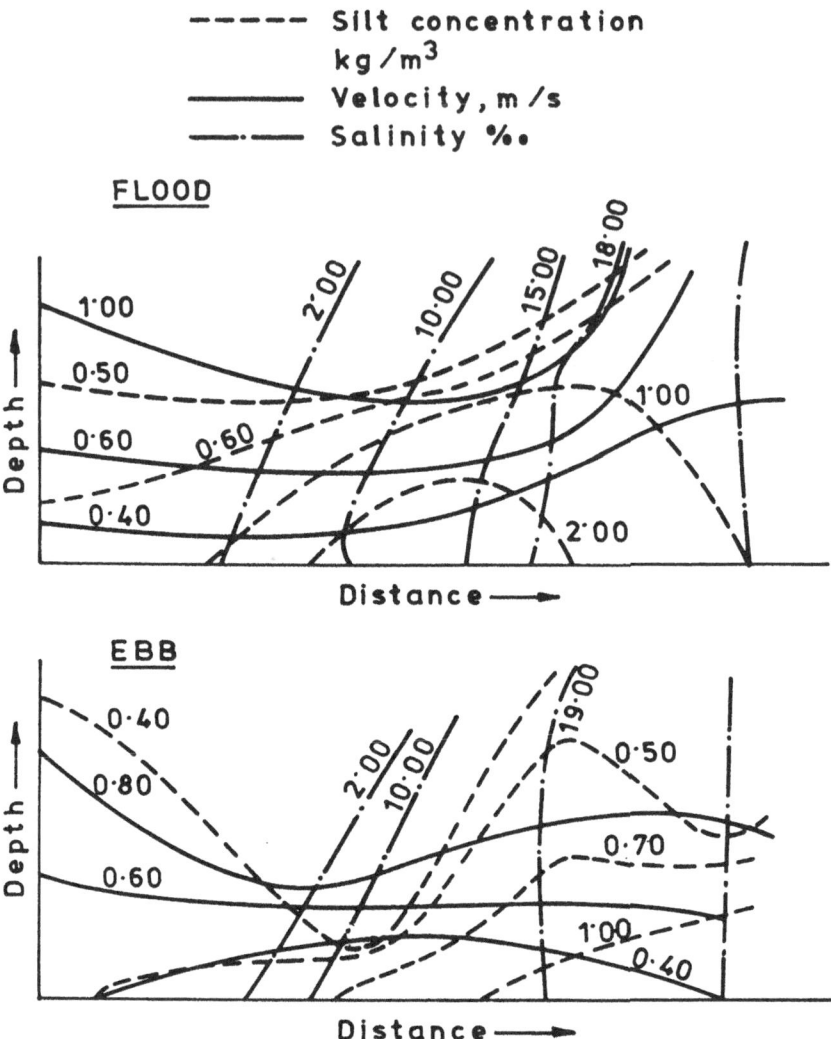

Fig. 10.6. Simultaneous variation of silt concentration, salinity and velocity along a tidal channel (Zhizhong *et al.*, 1983)

permanent grass and weed vegetation. The dense root system gives good mechanical protection against erosion. This vegetation does not hamper introducing flood water which carries the sediment. Immediately after deposition the silt particles are protected in the shallow quiet water zone between the short plants. In the Sunderban area, most of the islands harbour a luxuriant mangrove population. These mangroves by their dense root system enhance mud deposition specially in the upper intertidal zones and provide resistance to bank erosion. These marshy areas with thick mud deposits are characterised by strong biogenic activities.

The third type consists of providing suitable space in the initial pond layout system of the fish farms for dumping the deposited silt from the fish ponds. In this approach it is envisaged that over a period of say five years, great volumes of deposited silt will be removed from the pond and then dumped in the space. For this purpose it is necessary to know the expected volume of sediment of be deposited in the course of say five years or so.

10.5 Closure of coastal inlet and engineering measures for its reopening

It was O'Brien (1931) who first recognised that a relationship exists between tidal prism and the area of entrance. This can be expressed as

$$A = 1000 \, V^{0.85} \tag{10.5}$$

where A = area in ft^2 below mid-tide level,
V = tidal volume in square miles feet between MLLW and MHHW.

Recently, the problem of inlet stability has been treated by Bruun and Gerritsen (1958). They considered that in the equilibrium state, the eroding tendencies balance the tendency of inlet closure under the pressure of sediments moved into it by the ocean littoral currents and bay currents. Accordingly they sought to arrive at the solution by introducing the concept of stability shear stress as related to the maximum tidal discharge at inlet. The stability shear stress is defined as

$$\tau_s = \rho_g \frac{V_m^2}{C^2} \tag{10.6}$$

where V_m = Maximum velocity of inlet flow,
C = Chezy resistance coefficient,
τ_s = Stability shear stress.

They showed the equilibrium values of τ_s lies between 300 to 500 gm/m^2 depending on the size of inlet, type of soil (or sand) and the rate of littoral drift.

In many cases, however, due to inadequate tidal flow and littoral movement of sand, bar mouths are formed in front of coastal inlets. These act as a barrier for drawing of water for fishery development. It is therefore necessary to correct it by suitable engineering measures. These usually consist either of a dredged channel or jetties or a combination of both. A jetty is defined as a structure normal to the shore extending into a body of water. Its purpose is to direct and confine the tidal flow in the inlet channel and to prevent or reduce its shoaling by materials carried by littoral movement. Jetties at the bay entrance also serve to protect it from cross currents and waves due to storm. Furthermore, when it is located at inlets through the barrier beach it also serves to stabilise the inlet location. In this case the jetty acts as a partial or complete barrier to the littoral drift (Fig. 10.7).

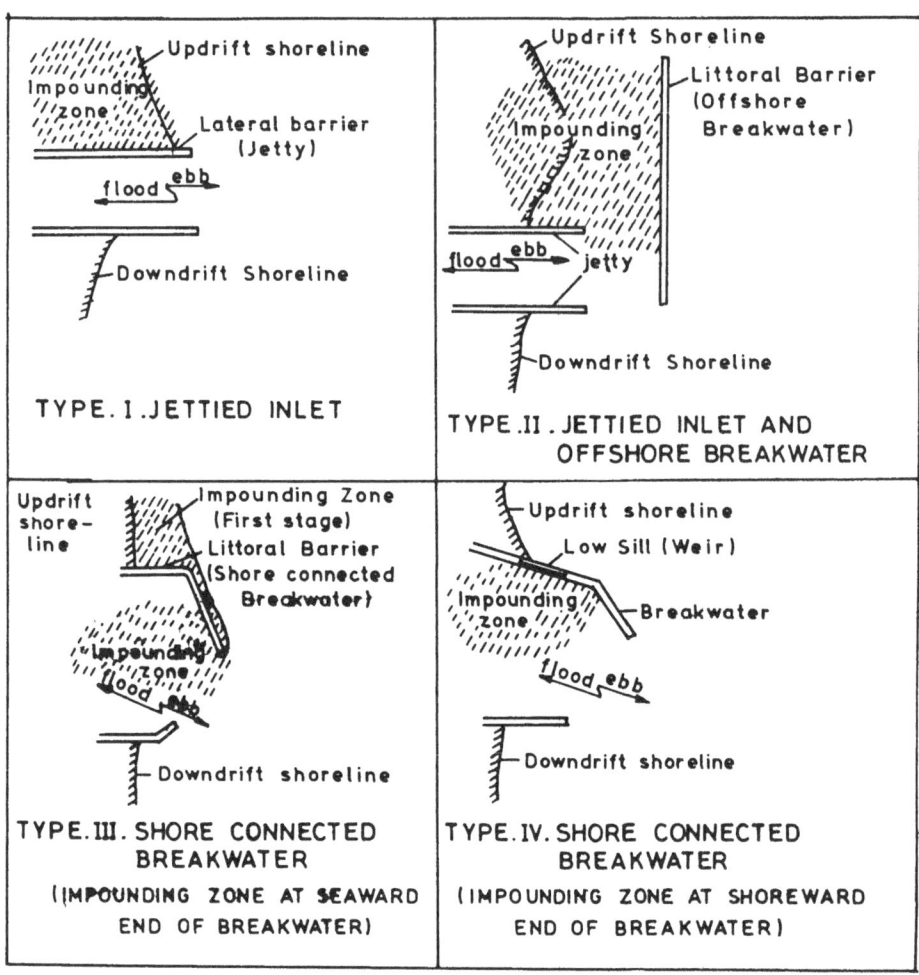

Fig. 10.7. Types of littoral barriers for inlet stability (Ref. Watts, 1965: Shore protection Manual, Vol. II, U.S. Army Coastal Engg. Research centre, 1977)

Accordingly in areas where the rate of littoral drift is significant, accretion occurs in the updrift side and erosion occurs on the downdrift side. Depending on the rate of drift and the length, orientation of the jetty, the updrift accumulation may have such a magnitude so as to cause the littoral material to pass around the jetty into the inlet channel. Hence for most of the jetty system there has to be a provision for passing the littoral drift from the updrift to the downdrift side.

To sum up, it can be said that

i) Sedimentation processes in a marine environment and in fish ponds are relatively complex phenomena requiring the knowledge of hydraulic engineering, colloidal chemistry and biological processes.

ii) Stability of creek cross-section and coastal inlet have to be thoroughly assessed based on existing knowledge of sediment transport mechanics in tidal and coastal environment and fluvial morphology.

iii) As far as practicable, naturally induced siltation processes by encouraging vegetation growth and utilising existing mangrove forest may be incorporated in the planning and layout of fish ponds. These types of solutions apart from being cheap also help preserve natural ecological balance of the areas.

iv) Development of fish farms where engineering measures such as building up separate intake structures to draw water from the higher level and jetties to prevent shoaling of coastal inlet are costly as they require laboratory investigation, design study and construction cost. It is therefore advisable to integrate these engineering structures with the general development cost of such areas such as improved communication facilities through navigation, so as to make the fishery development economically viable. Apart from that, such structures where necessary will require fish farming on a large scale.

10.6 Estimation of littoral transport

Two important concepts which form the basis of sediment transport theories are based on shear stress at the bed and energy flux or stream power. These concepts have been discussed in Chapter 7. The shear stress approach assumes that the bedload transport rate and sediment concentration near the bed are direct function of shear stress. The shear stress considered is the effective shear stress which is the excess quantity over the critical shear stress at incipient motion. The value of shear stress may be determined by vector addition of wave orbital velocities and current close to the boundary which may be derived from momentum flux equation. The energy flux or stream power approach is based on the rate of total work done per unit area or unit weight of water. The total work can be divided in two components, i.e., the work done in keeping material in suspension and in overcoming resistance to bed load movement, by sliding. In addition work is done in overcoming bed resistance not associated with bed movement, e.g., due to boundary texture or bed form. Usually the energy flux concept is more convenient for application to coastal problem, to give the total flux of sediment past a normal to the coast. This is based on the fact that at coastal locations the rate of sediment transport appeared to be directly proportional to wave energy flux. In order to apply the energy flux method it would require that the dissipation of wave energy used to move sediment be a fixed ratio of total wave energy flux. The total wave energy flux also includes energy losses in breaking waves, frictional losses not directly associated with sediment transport, and energy transported in long shore currents or rip currents. It is unlikely that the above items should have the same ratio of energy used in moving sediment to total energy. Variations in grain size, density, shape, and bottom friction during tidal rise and fall would affect the ratio. The formula based on energy flux method has been quite successfully employed to estimate the littoral transport or longshore transport rate.

Gross transport rate Q_g is the sum of littoral drift transported to the right and to the left past a point on the shoreline in a given period of time, i.e.,

$$Q_g = Q_{rt} + Q_{lt} \tag{10.7}$$

The net transport rate is $Q_n = Q_{rt} - Q_{lt}$. Usually Q_g is used to predict shoaling rate in uncontrolled inlets and Q_n is used for design of protected inlets. One needs information regarding both Q_{rt} and Q_{lt}, for the design of jetties.

10.6.1 ENERGY FLUX CONCEPT

The energy flux method can be employed to estimate longshore transport. The rate at which wave energy is transmitted across a plane perpendicular to the direction of wave advance is (U.S. Army Corps of Engineers, 1977).

$$\bar{P} = \frac{\rho g}{8} - H^2 C_g \tag{10.8}$$

where H = wave height and C_g = group velocity.

If the wave crests makes an angle α with the shoreline, the energy flux in the direction of wave advance per unit length of beach is

$$P_1 = \bar{P} \cos \alpha \sin \alpha = \frac{\rho g}{8} H^2 C_g \cos \alpha \sin \alpha$$

$$= \frac{\rho g}{16} H^2 C_g \sin 2\alpha \tag{10.9}$$

The surf zone approximation of P_1 is

$$P_{1s} = \frac{\rho g}{16} H_b^2 C_g \sin 2\alpha_b \tag{10.10}$$

where H_b = breaker height.

Several methods can be adopted to obtain useful formulae. These methods are based on evaluation of C_g and wave height H at the breaker position. It is the standard practice to approximate the group velocity C_g by the phase velocity C at breaking. The phase velocity may then be approximated by linear wave theory as (U.S. Army Corps of Engineers, 1977).

$$C = \frac{gT}{2} \tanh \left(\frac{2\pi d}{L} \right) \tag{10.11}$$

Figure 10.8 represents the longshore component of wave energy flux in a dimensionless from $P_1/(\rho_g^2 H_b^2 T)$ as a function of breaker steepness (H_0/gT^2) and the angle the wave crest makes with the shoreline in either deep water α_0

243

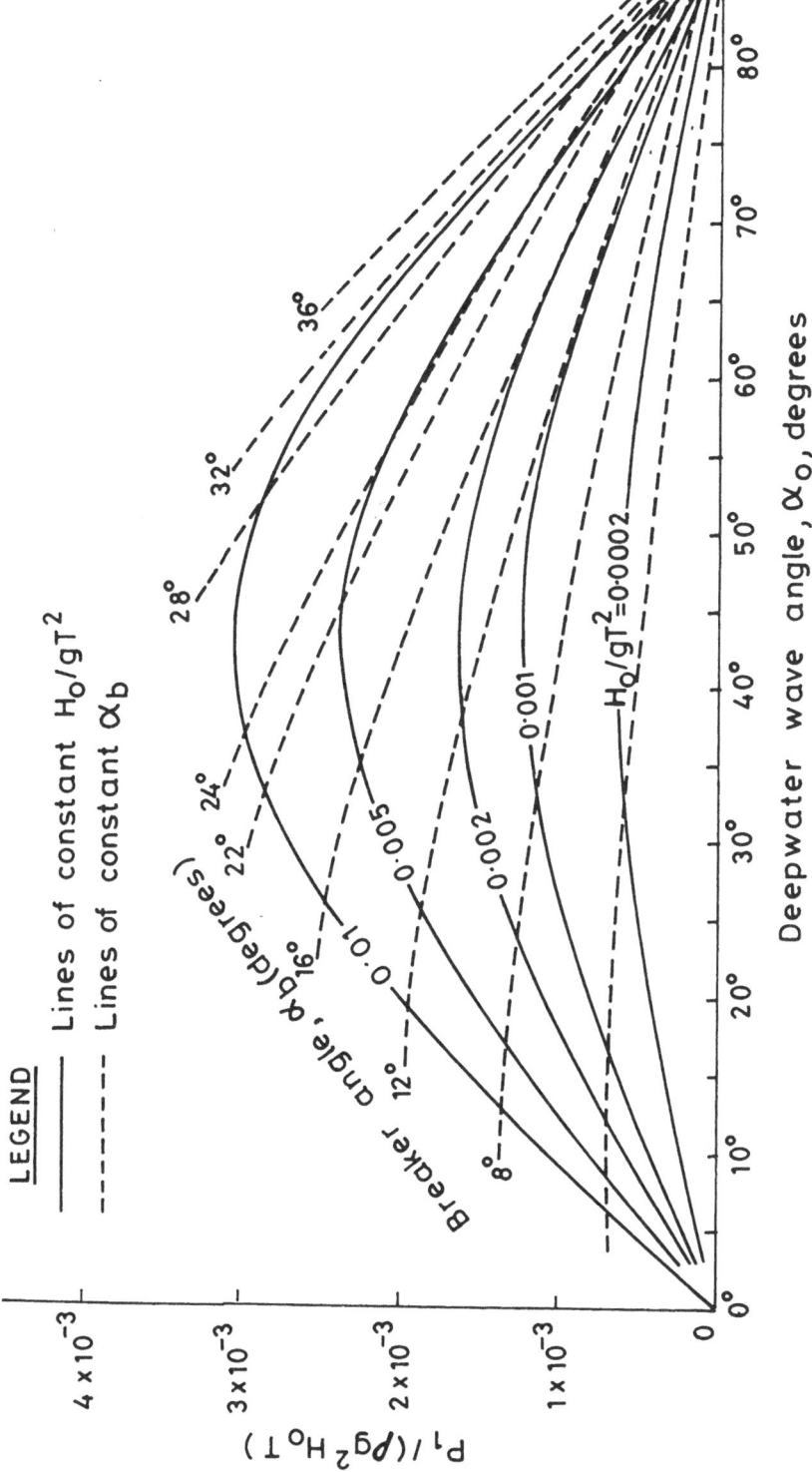

Fig. 10.8.

or at the breaker line α_b. For linear theory in shallow water, $C_g \approx C$ and

$$P_{1s} = \frac{\rho g}{16} H_b^2 C \sin 2\alpha_b \qquad (10.12)$$

where H_b and α_b are the wave height and direction at breaking and C = wave speed evaluated at a depth equal to 1.28 H_b from eqn 10.11.

For offshore conditions, the group velocity is equal to 1/2 the deep water wave speed C_0, and

$$P_{1s} - \frac{\rho g^2}{64} \cdot T(H_0 K_R)^2 \sin 2\alpha_0 \qquad (10.13)$$

where K_R = refraction coefficient.

Figure 10.9 also presents the longshore component of wave energy flux in a dimensionless from $[P_1/\rho g^2 H_0{}^2 T]$ as a function of deep water wave steepness (H_0/gT^2) and the angle that wave crest makes with the shoreline in either deep water, α_0 or at the breaker line designated as α_b. Here refraction by straight parallel bottom contours is again assumed. Once the P_{1s} is obtained, the sand transport rate is estimated from the relationship

$$I_e = KP_1 \qquad (10.14)$$

where, I_e = submerged weight sand transport,
P_1 = longshore component of wave energy flux, and
K = dimensionless constant with an average value of 0.35.

For conversion of I_e to Q_1, i.e., from weight to volume transport, the following relationship is used (U.S. Army Corps of Engineers, 1977).

$$Q_1 = \frac{I_e}{(1\frac{p}{100})(\rho_s - \rho_w)g} \qquad (10.15)$$

where, g = acceleration due to gravity,
p = porosity of sand in place expressed in percentage
ρ_s = sediment density, and ρ_w = water density.

Assume $p = 40$ per cent, and the specific gravity of sand $S_s = 2.65$, $\rho_s = 2.65 \, \rho_{fw}$ where ρ_{fw} = density of fresh water, $\rho_{fw} = 1000$ kg/m³

$\rho_s = 2650$ Kg/m³ (salt water)
$\rho_w = 1025$ kg/m³ (salt water)

$$Q_1\left(\frac{m^3}{yr}\right) = \frac{I_e\left(\frac{kg\,f}{yr}\right)}{(1 - \frac{40}{100})(2650 - 1025)9{\cdot}81}$$

$$Q_e\left(\frac{m^3}{yr}\right) = \frac{1}{(0{\cdot}6 \times 1625)}\left(\frac{m^3}{kg\,f}\right) I_e\left(\frac{kg\,f}{yr}\right)$$

$$I_e\left(\frac{kg\,f}{yr}\right) = (975)\frac{kg\,f}{m^3} \cdot Q_1\left(\frac{m^3}{yr}\right)$$

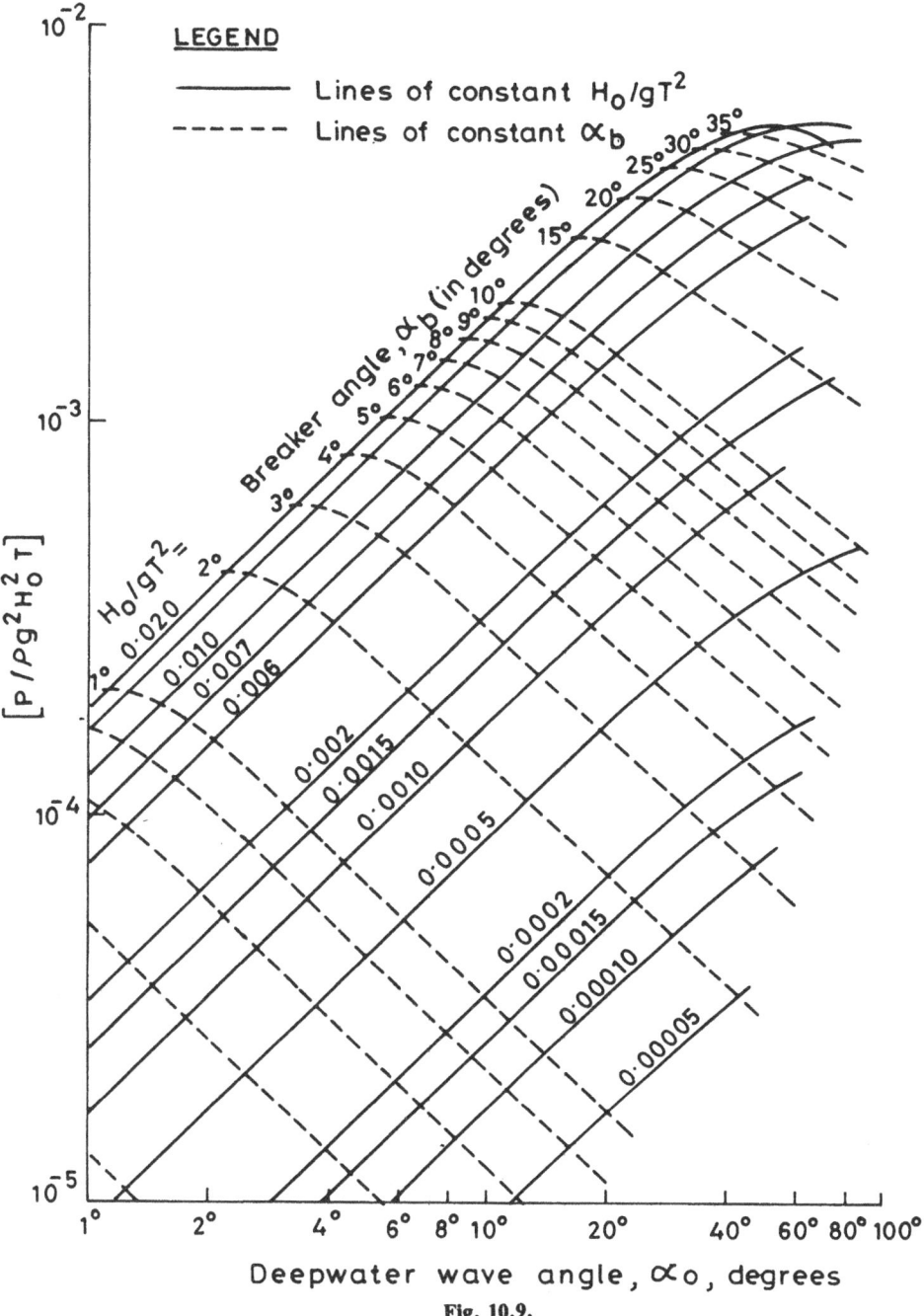

Fig. 10.9.

$$= KP_1 = \left(\frac{m\,kg}{m\,yr}\right)$$

$$Q_1 = \left(\frac{m^3}{yr}\right) = \frac{K}{975\left(\frac{kgf}{m^3}\right)} \cdot P_e\left(\frac{m\,kg}{m\,a}\right)$$

$$Q_1 = \frac{K}{975} \cdot P_1$$

With $K = 0.35$

$$q_e = 3.58 \times 10^{-4} P_1 \tag{10.16}$$

10.7 Design consideration of jetties

The construction of a jetty is an extremely costly venture for fishery, yet it is necessary in order to prevent or reduce shaoling of coastal inlet. Usually such a construction is also advantageously utilised for navigation. The jetties are usually of two types, i.e., rubble mound and sheet pile. The rubble mound structure is a mound of stones of different sizes and shapes which are randomly dumped or placed by hand. Its side slopes and stone sizes are designed to resist the expected wave action. Timber, steel, and concrete sheet piles have been used for construction of sheet piles in places where wave condition is not severe. These include a single row of piling with or without pile buttresses double walls of sheet piles held together by rods with space between the walls filled with sand or stones, or cellular steel sheet pile structures which are modifications of the double wall type.

In the analysis of forces exerted on structures by waves, a distinction is usually made between the action of breaking and nonbreaking waves. Structures located in an area or zone in which waves will break directly are designed to withstand greater forces and moments than those which will be subjected to breaking waves. Wave characteristics are determined first for deep water and then extended by computations shoreward to the structure. The stability and height of a wave at the structure will be dependent on the controlling water depths at the structure. Wave computations are generally made for the most significant wave height which has a statistical value equal to the average height of one third wave height of a deep water train.

The vertical pressure distribution due to a nonbreaking wave as shown in Fig. 10.10 is primarily hydrostatic which can be estimated with the help of the following relationship,

$$\text{i.e.,} \quad p_1 = \left(\frac{1+\lambda}{2}\right)\frac{\gamma_s H_1}{\cosh(2\pi d/L)} \tag{10.17}$$

where H_1 = height of original free wave in water with depth d,

λ = reflection coefficient $\simeq 1$ and

γ_s = specific weight of seawater.

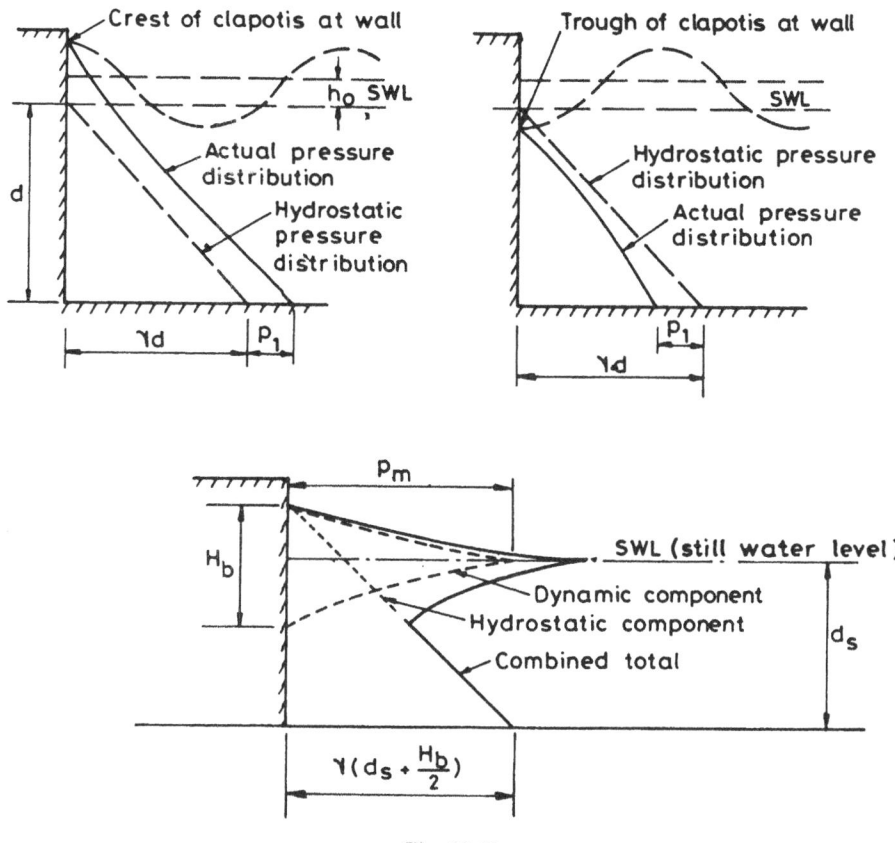

Fig. 10.10.

When the crest is at the wall, pressure in crests varies from zero at the free surface to $(\gamma d + p_1)$ at the bottom. When the trough is at the wall, pressure at crests varies from zero at the water surface to $(\gamma d - p_1)$ at the bottom. The nonbreaking wave is likely to be experienced by structures located at a depth greater than about 1.5 times the expected wave height.

Pressure caused by a breaking wave varies widely with the shape of wave as it breaks at the structure. The Minikan theory (1963) assumes breaking with an entrapment of air beneath the forward bending crest of the wave. Minikan's method can give wave forces which are 15 to 18 times of those calculated for non-breaking waves. Although there are limitations on its accuracy, the method is still generally used to determine breaking wave pressure. The maximum pressure according to Minikan is concentrated at still water level and is given by

$$P_m = 101 \frac{\gamma H_b}{L_D} \frac{d_s}{D} (D + d_s) \tag{10.18}$$

where P_m = maximum dynamic pressure
H_b = breaker height
d_s = depth at the toe of the wall
D = depth at one wave length in front of the wall, and
L_D = wave length of water with depth D.

The distribution of dynamic pressure is shown in Fig. 10.10 and the force represented by the area under the dynamic pressure distribution is

$$R_m = \left(\frac{P_m \, H_b}{3} \right)$$

In the case of rubble mound break waters, the weight or size of the armour units, side slopes, density of armour material and degree of interlocking between units are all interrelated and compose the essential factors in its design. Readers who are interested in its design are referred to the various literature, reports and books available on the subject such as those given by U.S. Army Corps of Engineers (1977) and Ippen (1966). To determine the stability of armour units the following formula developed by Hudscn (1961) is generally used

$$W = \frac{\gamma_r H^3}{K_D(S_r - 1)^3 \cot \alpha} \tag{10.19}$$

where W = weight of armour units in pounds
γ_r = unit weight of armour unit, lb/ft^3
H = design wave weight in feet,
S_r = specific gravity of armour unit relative to water,
α = angle of breakwater slope from horizontal,
K_D = stability coefficient which varies depending on slope, roughness, sharpness of edges and degree of water locking of the armour units.

REFERENCES

Boyd, C.E. (1982): *Water Quality Management for Pond Fish Culture*. Elsevier Scientific Publishing Company.
Bruun, P. and F. Gerritsen, (1958): Stability of coastal inlets, *ASCE. Proc.* 84 (WW3): May.
Buck, D.H. (1956): Effects of turbidity on fish and fishing. *Trans. N. Amer. Wildl. Conf.*, 21: 249–261.
Chaudhuri, A. (1984): Project Report of Mangrove Ecosystem of Sunderbans in Virgin and Reclaimed Areas with Special Reference to Productivity, Department of Science and Technology, Govt. of India.
Hudson, R.Y. (1961): Wave forces on Rubble Mound Breakwater. *Trans. American Society of Civil Engineers*, ASCF, Vol. 126 pt, 2.
Ippen, A.T. (1966): *Estuary and Coastline Hydrodynamics*, McGraw Hill, New Work.

Kadowaki, S., T. Kasedo, T. Nakazono and Hicata, H. (1978): Continuous records of DO contensts by cruising in the coastal culture farms. II. Diffusion of suspended particles by feeing. *Mem. Fac. Fish.* Kagoshima Univ. 27 (i): 281–288.

Krone, R.B. (1962): *Flume Study of the Transport of Sediment in Estuarial Process.* Rep. HEL Univ. of California, Berkeley.

Manickam, S. (1983): Sedimentation process in the Loire river and its estuary entrance, France. A Bio-Chemico-Sedimentological Approach. *Proc. 2nd International symposium on River Sedimentation,* held in Nanjing, China.

Minikan, R.R. (1963): *Winds Waves and Maritime Structures Studies in Harbour Making and in the Protection of Coasts*, Griffon, London.

O. Brien (1931): M.P. Estuary tidal prisms related to entrance areas, *ASCE, Civil Engineering*, May.

Shore Protection Manual (1977): Volume I & Volume II, U.S. Army Coastal Engineering Research Centre, Corps of Engineer.

Wallen, I.E. (1951): The direct effect of turbidity on fishery *Bull. Oklahoma A. and M. College*, 48, (2): 27.

Zhizhong, Z., Yunju, W. and Zhigang, Xu. (1983): Some flocculation characteristic of fine sediments of the Yangtze estuary. *Proc. 2nd International Symposium on River Sedimentation,* held in Nanjing, China.

CHAPTER 11

Basic Principles of Aquacultural Practices

11.1 Objectives of fish farming

The main objective of farming of fin fish, shell fish and other aquatic organisms is to have control over their growth and propagation or breeding by a judicious rearing of these organisms. Rearing is thus intended not only to increase the quantity but to improve the quality of the product as well, and the whole process should also be economical so that the farmer can make some profit.

11.2 Farming in the coastal areas

Farming of different fin fish, shell fish, etc. in the saline coastal belt is described under different names of which the use of the terms 'mariculture', 'marine cultivation', 'marine farming' and 'sea farming' have become very popular. 'Brackish water aquaculture' or culturing of aquatic organisms mainly near the estuaries is also an important system of farming in the coastal areas.

Though fresh and brackish water fish farming has been developed for a long time, the practice of culturing marine organisms has come into being only since the latter part of the nineteenth century when there was a renewed interest in oceanography (Schuster, 1949; Hickling, 1962, 1968, 1970). Even then only about 2 per cent of the marine species are at present being cultivated throughout their life-cycles under controlled conditions (Kinne, 1970), the rest are not cultured due to the lack of either suitable technique, equipment or financial support. Thus, ultimately, successful cultivation of a marine organism depends comprehensively on the basic and applied research on ecological, engineering as well as economic aspects (Milne, 1972).

Sea farms can be run in different ways depending upon the climate, topography of the coast line, water exchange, natural food sources and the species to be cultivated. Thus while crustacea and fin fishes would need some sort of fencing for their culture in the intertidal zones and the sea bed, the culture of sedentary organisms in the same zone would not require such enclosures. Among the different coastal zones six, namely the shore, intertidal, sublittoral, surface floating, mid-water and the sea bed, have been successfully adapted for sea farming. The first step to be taken in selecting a suitable site on the coast line for a marine farm is a critical assessment of the quality and quantity of sea water supply as well as tests for detection of pollutants and toxicants, and the exposure of the site to storms. In addition, any alterations in the environment that the construction of a farm might instill have also to be considered. In the case of intertidal ponds, an exchange of water through sluices, or provision of pumps are necessary. With netted enclosures in the sea, the mesh size in relation to the water current has to be correctly adjudged for the assessment of stocking density. In estimating the design loads and stocking densities, due weightage has to be given to the marine fouling growth on the coastal structures and the netting. Adequate measures, both physical and chemical, have to be taken to control predators as well so as to prevent a serious loss of stock. One of the common techniques to combat the predation of sedentary organisms normally cultivated on the sea bed, is to use off-bottom culture methods, which prevent attack by whelks, starfish and crabs. However, the most difficult problem faced by the marine farmer is perhaps fish diseases. Since some of these are also caused by dietary deficiencies and certain physical and chemical conditions of the environment, a careful management can prevent or control them to some extent.

In recent times there is a world-wide trend for a development of coastal areas for aquacultural practices, and the Food and Agricultural Organisation (FAO) is giving due importance and encouragement to such ventures. Essentially, appropriate utilisation of the coastal aquatic resources is a bio-engineering venture, requiring close cooperation between the biologist and the engineer (Bell, 1970). The biologist must study the ecology of the farmed species, that is, their environmental tolerances for spawning and growth. The engineer is required to produce economical design and construction methods to provide a suitable marine environment for the cultured species.

In its true sense, husbandry or farming of fish is possible in the ponds only where it is possible to regulate the different ecological parameters and consequently to supervise and manipulate the reproduction, feeding, quantitative growth and control of the size of the fish as well as the stocking and maintenance of the ponds for the maximum benefit, instead of leaving all this to nature. This chapter therefore mainly deals with the pond culture systems. To obtain higher production, proper attention should be paid to the following aspects right from the selection of the species to be cultured, to the pond preparation, improvement of the natural fertility of ponds, maintenance of optimum physico-chemical parameters of the water, application of fertilisers, manure and feed, manipulation of stock, and control of parasites and diseases.

11.3 Criteria of selection of culturable species

To obtain higher production in the culture ponds, the species of fish to be cultured must be selected very carefully, taking the following points into consideration.

i) natural ability of the species to grow fast,

ii) its efficiency to utilise organic production available in the particular aquatic system and its conversion into fish flesh,

iii) amiability to live with other species of fish cultured along with it,

iv) compatibility with the other species of fish, specially in food and feeding habits,

v) tolerance to changes in physical and chemical conditions of water,

vi) ability to adapt to crowded conditions,

vii) ability to adapt to the lentic conditions of the pond environment,

viii) they should be hardy and disease resistant,

ix) cost of production should be low,

x) easy seed availability,

xi) easy to handle and harvest,

xii) ready acceptability to the consumers and high market demand.

Fish intended for culture must reach full size quickly. Small sized species are not suitable for rearing, even if they reproduce well in ponds and accept the diet offered.

It is also desirable for the fish being raised to have a short feeding chain, in order to reduce, as far as possible, the loss of energy resulting from the passage of one link of production to the next. In the food chain, the conversion rate of food from one stage to the next is in the ratio of 10:1. Thus, to make 1000 kg weight of zooplankton organisms, 10,000 kg of phytoplankton are required. If this 1000 kg of zooplankton is utilised then only 100 kg of small animals would be produced. If all of these small animals are consumed by some small fish, their total weight would be 10 kg. So, a carnivorous fish eating the small fish of a total weight of 10 kg would increase in body weight by only 1 kg. This means that to produce 1 kg of carnivorous fish, 10,000 kg of phytoplankton are required. If it was a fish which could directly utilise phytoplankton, with this 10,000 kg of phytoplankton 1000 kg of fish could have been produced. Therefore, carnivorous fish are not generally chosen for culture. Herbivorous or detritivorous fish are thus considered best for culturing purposes and zooplanktonic feeders are considered the next best for culture.

Another important quality of a culturable fish is its amiability to live with other species of fish without creating trouble for them. Apart from carnivorous fish, there are some fish which affect the survival and growth of other fish. A typical example is *Oreochromis mossambicus* which by its prolific breeding and multiplication leaves no space and food for other fish in the pond.

In the fish farm, generally the dissolved oxygen, temperature, turbidity, carbon dioxide, pH, salinity etc. are likely to change frequently, so that fish which can tolerate changes in the physical and chemical conditions to a greater extent are considered more suitable for culture purposes.

Culturing a single species in a water body is also not profitable and economical. Each species are very specific in their feeding habits and eat only a particular type of food in the water and reject others. It is always necessary to select a group of fish each having different feeding habits from the other so as to make use of all the food available in the different zones of the water. In other words, in selecting species for culture it is important that the species selected should be compatible and should not compete with each other for food.

Also the fish which require lotic condition of flowing water like a river, cannot be cultured in ponds in which a lentic or stagnant condition prevail.

Some fish have burrowing habits, some may take shelter inside mud, and some others may have dangerous spines; fish having these types of characteristics make handling and harvestation very difficult and therefore cannot be cultured.

11.4 Growth of fish in ponds

The total food intake of a fish after assimilation is not utilised for growth only. A major share goes for the maintenance of metabolic activities and replacement of worn parts of the body, and it is only after fulfilling these requirements that the remainder is used for the growth of the body and the building of reserves. Under normal conditions the relation of growth requirements to those of maintenance lies between 1/1.5 and 1/3.2 (Huet, 1972).

The growth of fish is dependent on various factors, many of them inherent, for example, growth rate, ability to utilise the available food and resistance to diseases. However, certain external factors can also affect the growth of fish, some these being the environmental temperature, quality and quantity of food, water quality and availability of space.

The activity of a fish and its relative size determine the amount of energy utilised for the maintenance of the body. Thus under stress conditions that interfere with the normal bodily functions the increased activity shown by a fish uses up almost three times the energy required for its normal respiration.

Given a sufficient amount of food, the speed of growth is found to be directly related to its size; the smaller the size, the greater the growth rate, age being of little consequence in this regard. Thus, an older specimen that is smaller due to insufficient food will catch up with a younger fish by growing rapidly if finally given enough feed. Also, a dwarfed specimen through a lack of enough nourishment and space will attain sexual maturity at more or less the normal age. It is not always possible therefore to determine the age of a fish by its size only.

The fish are found to be quite resistant to hunger, probably due to their ability to directly assimilate organic substances dissolved in the surrounding water through their gills. Though the quantity taken up is minute and inadequate, it is however sufficient for their maintenance for quite a long period even in the complete absence of food. Such food deprivation does not adversely affect the health of the fish or its subsequent growth, and though a prolonged starvation may lead to weight loss, a fish will start growing again once the food supply is restored.

11.5 The biological cycle in ponds

A fish is the ultimate product of a complex biological cycle occurring in ponds. The cycle starts with the mineral nutrients in solution in the water. These nutrients are derived by the running or stagnant water from either the soluble substances present in the soil it is in contact with, or carried to it by exogenous detritus or rainfall. This inorganic material is transformed by the green vegetation into organic matter with the help of the sun's heat and light and carbonic acid in solution. This forms vegetable tissue, which is consumed, either in the living condition or even dead as fine detritus, by numerous small animals, which again serve as food for larger water animals. These, as well as certain larger plants, living or dead, are now eaten by the fish (see also section 2.1). This chain or cycle of natural fish production includes the following links: mineral nutrients, plant production, intermediate animal consumption and production, the final product being the fish. The last stage in the food chain is the return in solution of all the dead components of organic matter, both plant and animal, brought about by bacteria through the mechanism of mineralisation, and their subsequent re-integration into the biological cycle.

It is therefore evident that any increase in the production of the cultivated species of fish cannot be done directly. First, the vegetable production should be enhanced to get the desired effect, as well as to strengthen the weakest link in the biological cycle existing in the pond. For this purpose, the use of fertilisers have been found to be effective. A proper stocking and adequate upkeep of the ponds is also to be assured so as to enable the fish to live under the best possible conditions.

Ponds can be classified into the following categories: *oligotrophic*, *eutrophic* and *dystrophic*. An oligotrophic pond is a water body deficient in essential minerals like calcium, phosphorus and nitrogen, and in which the production of organic matter is also poor. The water is however clear, with the dissolved oxygen (D.O.) content high even at greater depths. A eutrophic pond is rich in nutrients and consequently there is abundant production of organic matter. The water is generally alkaline, thus promoting planktonic production giving it a green or brown-green colour. However, the D.O. content is reduced at depths, and during summer and winter, may even be nil at times. Dystrophic ponds contain humic matter mostly in the colloidal form and are not very productive with poorly developed vegetation. The water is acidic and coloured yellow to brown.

From the point of view of the origin of the nutritive substances present in the ponds, these can be classified as *autotrophic* (substances present in the pond itself) or *heterotrophic* (nutrients derived from an outside source).

11.6 Augmentation of production by biological process

For the augmentation of production different biological processes or techniques can be adopted as follow:

i) Selection of the appropriate species suitable for particular water body under cultivation;

ii) Maintenance of proper water quality;

iii) Proper stocking;

iv) Induced breeding and controlled reproduction;

v) Multiple cropping;

vi) Periodic partial harvestation; and

vii) Eradication/control of fish enemies, diseases, etc.

11.6.1 SELECTION OF THE APPROPRIATE SPECIES

Selection of an appropriate species for a particular pond is the most vital factor in getting the desired production from a water body. The criteria for selection of culturable species have been discussed in section 11.3. Among the species fulfilling these conditions, the ones that will most suit the specific conditions with regard to the available food and the environment present in the particular water body to be brought under cultivation have to be selected.

11.6.2. MAINTENANCE OF A PROPER WATER QUALITY

Survival, growth and reproduction of fish are all dependent on the physico-chemical characteristics of water (see section 11.9 for details). Proper water quality not only enhances the growth rate of fish but it also lowers the mortality rate by reducing the chance of attack by pathogens.

11.6.3 PROPER STOCKING

Stocking the pond with proper number of fish would ensure maximum quantity of harvest of desired quality. Stocking density essentially depends upon the productivity of the species under cultivation and its carrying capacity. While a lower stocking density would leave much of the organic production in the pond unutilised, a higher stocking density on the other hand would result in shortage of food and space reducing the quantity and deteriorating the quality of fish as well.

11.6.4 INDUCED BREEDING AND CONTROLLED REPRODUCTION

To obtain quality fish seed for stocking in required number and at the proper time, artificial breeding of the culturable variety of fish is very important. Seeds of brackish water varieties of fin fish and shell fish are mostly collected from tidal creeks in countries like India and Bangladesh where, consequently, a supply of adequate number of quality material cannot be ensured. In Japan or Taiwan however, where commercial hatcheries are being operated, a high stocking density can be maintained in ponds which are even used more than once a year for growing prawns.

If the culturable fish is allowed to breed profusely in an uncontrolled manner in the culture ponds, the quality and quantity of harvest are bound to suffer. It is therefore essential that the reproduction be controlled in such cases, as done through monosex culture in *Oreochromis mossambicus*.

11.6.5 MULTIPLE CROPPING

In order to get the maximum production from a water body, it has to be kept under cultivation for the maximum period possible. For this purpose, raising several fish crops a year is very profitable and it leaves the pond unutilised for a very limited period only. However, periodic drying of a pond is indispensable for cultivation as it favours the mineralisation of the mud and improves the sanitary condition of the pond, but this should be kept at a minimum through proper planning.

11.6.6 PERIODIC PARTIAL HARVESTATION

Periodic partial harvestation is necessary whenever the population density reaches a maximum, manifesting itself by a decline in growth. Such thinning out of the fish population ensures adequate space and food for the remaining fish so as to enable them to start growing again, till a saturation point is once again reached, when another partial harvestation becomes necessary. Thus, it is possible to maintain the ponds at their maximum capacity through periodic partial harvestation. However this cannot be practised indefinitely since periodic drying of the ponds is also necessary to control the population, to remove unwanted species that may have entered the water, and to restock the pond rationally.

11.6.7 ERADICATION/CONTROL OF FISH ENEMIES, DISEASES

Periodic drying of the pond and such measures as are generally taken to maintain a healthy water quality also help in the eradication and control of fish enemies and fish pathogens. Different types of carnivorous fish, unwanted low value fish, snakes, crabs, unwanted aquatic weeds etc. need removal from time to time. Similarly periodic disinfection of the pond helps to control fish pathogens.

11.7 Control of unwanted aquatic weeds to maintain high productivity

Presence of aquatic vegetation in ponds is essential for the production of fish. However, an excessive growth of such vegetation becomes harmful to fish life. Such growths hamper easy movement of fish, accelerate silting, keep the water from warming up, decrease D.O. content at night, provide shelter and nourishment of the competitors and enemies of fish, create a pond bottom rich in cellulose, and in general, decrease the productivity of the pond. Control of aquatic weeds is therefore a regular part of the maintenance regimen of a pond under cultivation, and should be done at regular intervals to avoid excessive choking up of the pond, and is thus more economical. Elimination of excessive aquatic vegetation increases the natural food available and improves the sanitary condition of the pond.

Weeds can be removed by both mechanical and chemical means. Sometimes biological methods can also be used.

11.8 Enhancement of the natural fertility of the pond

The natural fertility of a pond is enhanced by draining out the pond, drying the bed and generally working the soil, dredging it where necessary. Such procedures improve both the physico-chemical and biological conditions of the pond and thereby increase fish production. Thus, they give rise to fertile and finely colloidal mud that favour biological production; break down and reclaim the residual roots of plants which would otherwise have been lost as a source of nutrient from the biological cycle; help in the mineralisation of the excessive mud, particularly that rich in cellulose; destroy cysts as well as the intermediate hosts of the ogranisms causing disease in fish, and also destroy the larvae and adult forms of harmful insects and other enemies of fish.

However, the drying period should not be unduly prolonged and should only cover the period required for the mineralisation of mud as evident by the cracking of the surface. To make the drying effective, an effective network of primary and secondary drainage ditches should be present. Ponds that are never dried out gradually lose their productivity and their exploitation finally ceases to be profitable. Adequate advantage of the drying out period should also be taken by carrying out any repair and maintenance work required, for example, checking and repair of damaged dikes or other areas of water loss, if any; inspection of the inlet/outlet structures; any renovation of the pond bed; removal of unwanted plant parts, dry leaves or excess mud and indigenous or exogenous detritus; liming the muddy areas; and repairing and painting of screens and fences. Any mud removed from the drained ponds can be used for agriculture in the adjoining areas.

When the pond bed is sufficiently dry and firm, it is superficially ploughed so as to work the soil. However, if the bottom is not covered with vegetation, turning over the soil must be avoided so that the unproductive layers do not come up to the surface. When the silt accumulation is substantial so as to significantly raise the bottom of the pond diminishing the culturable area, as found in ponds that have not been dried for a number of years, merely drying out and working the soil is no longer effective, and dredging of the pond becomes necessary.

11.9 Maintenance of optimum physico-chemical conditions of water

A water body used for the cultivation of fish will not yield maximum production if the physico-chemical conditions are not optimal for the growth of fish and other aquatic organisms forming the food chain. The physico-chemical parameters that have important bearing on fish life have therefore to be identified and controlled so as to maintain a proper environment for the cultured species.

11.9.1 TEMPERATURE

Changes in temperature affect both the plant and animal life in many ways. The minimum, maximum and optimum temperatures for their survival vary in the different species of fish. For example, fish of temperate regions can survive even under ice in winter; tropical species however, cannot withstand such temperatures though some of them can survive temperatures up to 40°C. Normally there is a minimum point below which death will occur. Above this point, the growth increases with increasing temperature, but the relationship is rarely linear. This goes on till a maximum point is reached beyond which any further increase in temperature decreases production till the temperature in reached when death occurs. Generally, the temperature versus growth curve is described by a second or third degree equation and often displays a rather parabolic shape (Wheaton, 1977).

The effect of temperature on the production of any organism is basically its influence on the different chemical reactions catalysed by the enzymes that collectively mainifest into the life process. As is most often found, these chemical reactions in a living organism usually proceed at an increased rate with a rise in temperature due to the increased kinetic energy possessed by the atoms. However at some point, the kinetic energy becomes high enough to start the breaking down of some of the chemicals necessary for life, which results in a decreased growth rate, till the point is reached where a total breakdown occurs leading to death. Many enzymes start to lose their catalytic properties at about 35°C and have lost them completely by the time temperature reaches 60°C (Devlin, 1966). Similarly, at some low temperature also, certain critical chemical reactions necessary for life either will not occur or take place at too slow a rate to sustain life. However, besides the effect on the chemical reactions, there may be other causes of thermal death also.

The thermal effects are thus important enough to be considered when planning and designing aquacultural facilities. For this purpose, at least three characteristics have to be considered: thermal death point, acclimatisation temperature and temperature shock. Thermal effects on fish are very complicated, and there is no single temperature at which thermal death will occur. Fish are poikilotherms (cold-blooded animals) and their body temperatures change with the outside temperature. The chief place for the occurrence of thermal transfer is the gill surface, where the fish blood and the surrounding water come in close contact to allow the transfer of oxygen. This thermal transfer is very efficient due to the high convective ability of the water, and the energy requirement being too great, fish consequently cannot survive with a body temperature much different from that of the surrounding water. However, fish are adapted to change their living processes to various temperatures matching the outside temperature, though within limits, the process being known as *acclimatisation* or *acclimation*. The acclimatisation process should be slow enough so that the physiological processes of the fish are allowed enough time for adaptation to the changing temperature, or a thermal shock will result. Acclimatisation from a low temperature to a higher one is much more rapid than going from a high to lower-temperature. Thus, while an increase of the acclimatisation temperature by 5°C may take place within hours or at most a couple of days, reducing it by a similar amount may take several days. Also low oxygen concentrations have been shown to inhibit acclimatisation to higher temperatures in some fish (Brett, 1956). The temperature characteristics of fish are shown in Fig. 11.1 The zone of tolerance defines the combination of acclimatisation and environmental temperatures in which a fish can live indefinitely. The zone of resistance defines the temperature range the fish can tolerate for limited periods only. Outside the zone of resistance, death is essentially instantaneous, apparently through the effect on the central nervous system, though the specific cause is not quite clear (Brett, 1956). By changing the acclimatisation temperature, the lethal temperature can be changed: an increase in the acclimatisation temperature also increases both the upper and lower lethal temperatures, while a decrease has the opposite effect.

At temperatures somewhat below the upper zone of resistance or above the lower zone of resistance, fish are under stress and their normal efficiency of body functions is affected. So in the culture systems, where the aim is to obtain maximum economic return, care should be taken to minimise the stress as far as possible.

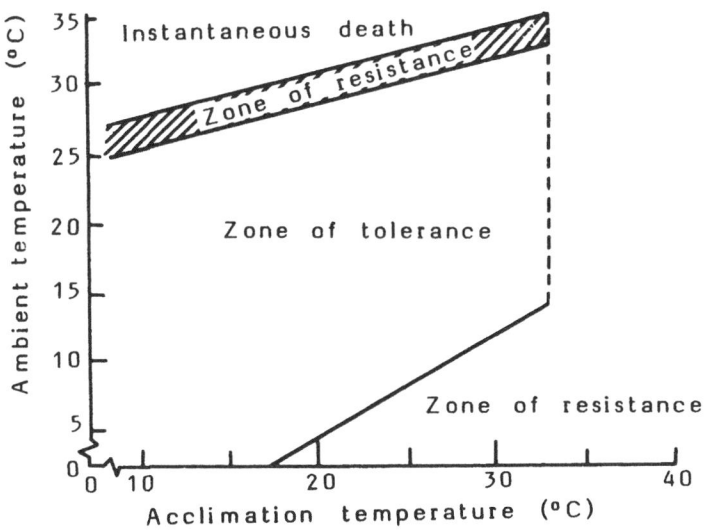

Fig. 11.1 Temperature tolerance diagram for the roach *(Rutilus rutilus)*. (After Wheaton, 1977)

11.9.1 CONTROL OF TEMPERATURE

It is far more difficult to influence the temperature of a pond than it is to improve other means of bettering production such as fertilisation and feeding. For decreasing the temperature, the depth of the pond can be increased or cool water from deeper zones of the pond can be sprinkled over the surface, which not only decreases the temperature but also increases oxygen. For increasing the temperature, manipulation of the pond depth, erection of wind breaks and introduction of warm water from the power generating station have been found to be effective.

11.9.2 OXYGEN

With the exception of some anaerobic bacteria, almost all aquatic organisms are dependent upon oxygen for respiration. The oxygen dissolved in the water is used for this purpose. The dissolved oxygen (D.O.) is therefore a critical factor determining the success rate of an aquaculture system, and the aim is always to keep it at a sufficiently high level. The D.O. content depends upon temperature, salinity, as well as the quantity of organic matter present and submerged aquatic vegetation.

11.9.2.1 Source of oxygen

The two major sources of oxygen for water are the aquatic plants which produce oxygen by photosynthesis and the atmosphere. Oxygen comprises 20.95 per cent of the total volume of air, but is only sparingly soluble in water. The oxygen content of air-saturated water is a function of temperature and salinity (or chlorinity). Dissolved oxygen concentrations of water when saturated in contact with atmospheric air at 760 mm pressure are greatest at 0°C (14.16 mg/litre) and decrease with increasing temperature, being 8.84 mg/litre at 20°C and 7.04 mg/litre at 35°C. The solubility of oxygen in water also decreases with increasing salinity: for each increase in salinity of about 9 ppm, the reduction in the solubility of oxygen is roughly 5 per cent of that in fresh water.

The efficiency of transfer of oxygen from the air to water depends on the water temperature, salinity, degree of saturation of the water, and turbulence of the air-water interface, of which, the first three factors directly influence the uptake of oxygen, while an increase in the last factor effectively increases the air-water surface, thereby increasing the transfer. Without the help of water circulation, current and turbulence, the diffusion rate of oxygen is too slow to be of any practical value. For example, to raise the oxygen content at a depth of 10 m from zero to 0.4 ppm by diffusion would require about 600 years (Reid, 1961).

The second major source of oxygen in aquatic systems is through photosynthesis by aquatic plants. Photosynthetic activity generates oxygen as a byproduct and is added to the water by the aquatic vegetation. Since photosynthesis ceases in the absence of light, the same plants that release oxygen in the day time utilise some of the supply during the periods of darkness for their respiration. During sunny days, oxygen production may be as high as to supersaturate the surface layers, when oxygen may even be given off from the surface to the atmosphere. The oxygen content of all natural and pond waters is at the minimum level at or slightly after day break since photosynthesis necessarily lags slightly behind the daily radiant energy schedule. The maximum oxygen concentration occurs in the middle to late afternoon.

11.9.2.2 Oxygen content of coastal water

There are large variations in the D.O. content of coastal waters. Thus, oxygen concentration may be very low in the deep anaerobic areas, but up to saturation or even supersaturation in areas near the coral reefs. As such, sea water has lower saturation concentrations for oxygen compared to fresh water due to its high salinity. Due to temperature differences, there is again variation in the saturation concentration between the tropical surface waters and the polar waters, the former being in the range of 6 to 7 ppm, while the latter more than 11 ppm. There is also vertical variation in the oxygen concentration, which is particularly marked in the subtropical waters. It is highest in the surface layers where there is a dual supply from the atmosphere and photosynthesis. Below the photosynthetic level, however, the D.O. concentration drops rapidly.

In the estuaries, the oxygen concentration fluctuates widely. The chief influencing factors are temperature and salinity which themselves vary in the estuaries and an oxygen gradient usually forms along the length of the estuary, with the highest oxygen levels occurring in the upstream areas. Any influx of fresh water increases the oxygen level while dry periods that promote evaporation of surface waters decrease D.O. content due to an increase in salinity.

The depth of the estuary, its stratification, vegetation, tidal effects, the season, as well as the time of the day, all influence the oxygen content of the estuaries. Of these, the seasonal and diurnal variations in the oxygen content are dependent on the morphology of the basin, tidal effects and the fresh water source. Seasonal variations in D.O. in the estuaries are also affected by the changes in the river discharge, the tides, length of the day and biological aspects.

11.9.2.3 Oxygen requirements of aquatic organisms

Aquatic organisms are generally well adapted for extracting oxygen from the water even if the concentration is low. The gaseous exchange in respiration of fish takes place by simple diffusion. In the gills the pressure of oxygen in the fish blood is lower than that in the water so that the haemoglobin present in the blood picks up the oxygen. In the tissues the oxygen is quickly used up and then it is replenished from the oxygen carried in the blood. In return, carbon dioxide is loaded into the blood and this is released to the water at the gill region since the pressure of CO_2 in the venous blood is more. The amount of oxygen required by a fish is dependent on its genetic make-up, its level of activity, stress experienced and the temperature of the surrounding water. However, for any fish a D.O. level below 3 ppm is fatal, though some shell fishes and other aquatic organisms have been known to tolerate lower D.O. levels for limited periods. Between 3 and 5 ppm, some species of fish can survive indefinitely, while some others can tolerate it for limited periods. Above 5 ppm, however, most aquatic organisms can live indefinitely, provided other environmental factors remain favourable. Usually, the young specimens of any species are less tolerant to a lower D.O. than are the adults, and among the latter those species that are lower in the evolutionary scale (for example, crabs or oysters rather than the fin fishes) are more tolerant towards a lower oxygen concentration.

Due to rapid growth and vigorous activity, oxygen consumption is generally quite high under optimum temperature conditions. With an increase in temperature, the oxygen consumption becomes higher still because of the stresses experienced by the organism, which triggers off its warning and defence mechanisms that require more oxygen. Scaring the organism, or otherwise applying stress conditions also induces similar reactions, regardless of the prevailing temperature.

Oxygen consumption of fish also determines its level of activity. As long as the oxygen concentration is not so low as to make the fish unable to extract sufficient oxygen to meet the basal metabolic needs, thus leading to death, the organism restricts its activity to meet the oxygen demand. However, at a low oxygen concentration, fish are under stress, leading to an increased oxygen demand and aggravating the condition. At such low concentration range, fish are more vulnerable to predators and diseases and a prolonged exposure to sub-lethal oxygen concentration leads to a reduced survivality and growth rates, also increasing the chance of bacterial infections (Snieszko, 1973; Plumb *et al.*, 1976). However, in ponds where oxygen concentration is low, there is usually high ammonia and carbon dioxide concentrations and according to Walters and Plumb (1980), all the three factors are collectively responsible for causing infections in fish. Whenever possible, fish move away from low oxygen concentration areas, but within such zones they typically move up to the surface to gulp air and show a general apathy. Fish have thus got to be kept at oxygen concentrations above this range when they exhibit unrestricted activity. However, if oxygen concentration far exceeds this optimum range, a supersaturation of the blood with oxygen may give rise to gas bubble disease that may affect both shell and fin fishes, the condition causing gas bubbles to form within the tissues and may in severe cases, even lead to death.

Taking into account the different factors influencing the oxygen demand in fish, oxygen budgets of fish ponds can be prepared in advance (Boyd, 1982).

11.9.2.4 *Maintenance of the desired D.O. level*

To ensure that there is sufficient oxygen in the pond water from the start, a shower system, a sprinkler system, cascades or water falls can be built into the supply channel or alternatively, paddle wheels can be installed. When the water reaches the pond it can be divided and spread out by means of passing over a perforated screen which will permit an increased contact with the air so as to enrich the water with oxygen. If the farm design and culture pattern followed so permit, good results can also be obtained by introducing submerged plants in the pond or semi-submerged plants in the water supply channel. There are indirect methods also by which D.O. level can be increased. Thus, the soil can be worked and limed to reduce the load of organic material. In a pond that has lower oxygen concentration, the quality of the water supplied should be improved by aeration. Projecting the water into the air with a pump, using an outboard motor on a boat circulating on the pond, or a compressor for injecting air into the water are common ways of aerating the water. The water can also be oxygenated with an aerator-mixer.

11.9.3 pH

The water used for fish cultivation is not chemically pure and contains, in solution, different substances which give it an acid, neutral or alkaline reaction which is measured by the pH value. The water best fitted for the cultivation is neutral to slightly alkaline (pH 7.0 to 8.0). Exposure to waters more acidic than pH 6.5 or more alkaline than pH 9 to 9.5 tend to diminish the reproduction and growth of fish (Swingle, 1961; Mount, 1973). The acid and alkaline death points of a fish are at about pH 4.0 and 11.0 respectively. However, since most aquatic organisms are adapted to an average pH value and are unable to tolerate sudden or wide fluctuations of pH, maintenance of a stable pH condition in the culture ponds is most important.

11.9.3.1 Control of pH

The pH is usually controlled by liming the pond, which increases the alkalinity and ensures sufficiently stable pH conditions.

11.9.4 ALKALINITY

Alkalinity of the water is determined by all the carbonates and bicarbonates of alkaline and alkaline earth metals present in solution. The most common basic ions so found which are also indispensable for fish cultivation are calcium and magnesium. Usually however, alkalinity of normal culturable waters is equivalent to the quantity of calcium carbonate and bicarbonate. Thus, total alkalinity is defined by the total concentration of bases in water, expressed as mg/litre of equivalent calcium carbonate, and natural waters containing more than 40 mg/litre of total alkalinity are found to be more productive (Moyle, 1945; Mairs, 1966). In the fertilised ponds, however, total alkalinity values in the range of 20 to 120 mg/litre have been shown to have little effect on fish production (Boyd and Walley, 1975), though such ponds containing 0 to 20 mg/litre of total alkalinity show increased production with increasing alkalinity. Thus, it seems desirable to keep total alkalinity values above 20 mg/litre in the fertilised ponds. Very high alkalinity (>200 mg/litre) is also not suitable for good production since it gives rise to calcareous furs in certain cases and hampers the development of the biological cover.

11.9.4.1 Control of alkalinity

It is controlled chiefly by liming the ponds which increase the alkalinity to bring it to the level considered best for the cultivation of fish.

11.9.5 CARBON DIOXIDE

Carbon dioxide (CO_2) is principally utilised in photosynthesis. In the natural waters, CO_2 may be derived from various sources such as respiration of plants and animals, decomposition of organic material, chemical reactions occurring within the water body between acids and different carbonate/bicarbonate compounds, and the atmosphere which contains 0.03 per cent CO_2. CO_2 may be directly absorbed from the atmosphere at the air- water interface, or it may be carried into the water by raindrops falling through the atmosphere. CO_2 has an acidic reaction in water:

$$H_2O + CO_2 = H_2CO_3$$
$$H_2CO_3 \quad = H^+ + HCO_3^-$$

About 1 per cent of the CO_2 in the water forms carbonic acid, which strongly dissociates as above. It is thus usual to include both the CO_2 as well as the carbonic acid when considering the total CO_2.

Pure water at 25°C has a total CO_2 concentration of 0.48 mg/litre, which gives it a pH of 5.66. With increasing CO_2 concentrations, pH decreases, being 4.8 at 30 mg/litre CO_2 level, though it generally never falls below 4.5 through high CO_2 values.

CO_2 is not appreciably toxic to fish; most species will survive for several days in waters containing up to 60 mg/litre, provided D.O. is plentiful (Hart, 1944). However, when the D.O. is low, the presence of a high level of CO_2 hinders oxygen uptake by fish. CO_2 concentration usually increases during the night when photosynthesis ceases but respiration goes on and decreases during the day with the onset of photosynthesis.

Particularly high concentrations of CO_2 are found after phytoplankton die-offs or loss of thermal stratification and during cloudy weather.

11.9.6 AMMONIA

Ammonia enters the water through fertilisers, fish excrement, and from microbial decay of nitrogenous compounds. Un-ionised ammonia is highly toxic to fish while the ammonium ion is harmless. In well-stocked ponds where supplemental feed is used, ammonia concentrations can go up to undesirably high levels. However, plants can rapidly absorb ammonia, and some bacteria can oxidise ammonia to nitrate. In water, un-ionised ammonia is found to exist in a pH and temperature dependent equilibrium with ammonium ion (Boyd, 1982):

$$NH_3 + H_2O = NH_4^+ + OH^-$$

The sum of un-ionised ammonia and ammonium is called total ammonia nitrogen. The proportion of the total ammonia nitrogen existing as un-ionised ammonia increases with increasing temperature and pH. Of the two factors, pH has greater effect on un-ionised ammonia concentration.

Colt and Armstrong (1979) have studied the toxic effects of ammonia on fish. With an increase in the ammonia level in the water, ammonia excretion by the fish decreases thereby increasing its level in the blood and tissues, resulting in the elevation of the pH of blood, and having adverse effects on membrane stability and enzyme catalysed reactions. The permeability of the fish is affected, and its internal ion concentration is reduced. It also increases the oxygen consumption by tissue, while damaging gills and reducing the oxygen transporting ability of the blood. Exposure to sublethal concentrations of ammonia induces histological changes in the kidneys, spleen, thyroid tissues and the blood of fish and probably also increases its susceptibility to disease. Ammonia is found to be more toxic when the D.O. concentration is low (Merkens and Downing, 1957). According to the European Inland Fisheries Advisory Commission (1973), for most species, a short-term exposure to 0.6 to 2 mg/litre of NH_3-N proves toxic.

11.9.7 NITRITE

Nitrites are harmful to fish and occur in concentrations of 0.5 to 5 mg/litre $NO_2^- - N$ occasionally in ponds. It is commonly believed that nitrite accumulates in ponds due to an imbalance in the nitrification reaction. However, Hollerman and Boyd (1980) have suggested that nitrite originates from the reduction of nitrate by bacteria in anaerobic mud or water.

When absorbed by fish, nitrite reacts with the haemoglobin of the blood to form methaemoglobin which is not effective as an oxygen carrier, so that a continued absorption of nitrite may lead to hypoxia and cyanosis. Since methaemoglobin is brown in colour, nitrite poisoning of fish is commonly referred to as "brown blood disease".

11.9.8 METHANE

Methane, or 'marsh gas' as it is popularly called, occurs in natural waters through incomplete bacterial decompositions of organic matter under anaerobic conditions. It is not produced in well-aerated waters, where a complete decomposition can occur forming carbondioxide and water. Methane can be formed in waters up to a temperature of 5°C or even less. While it is not toxic to fish by itself, since methane is produced only under anaerobic conditions, it is a sign of depleted oxygen in the pond which demands immediate attention to prevent large scale mortality of the cultured organisms. However, it is not unusual to find a few isolated pockets of anaerobic decomposition and methane formation in any pond system.

11.9.9 HYDROGEN SULPHIDE

Hydrogen sulfide is also a product of decomposition of organic materials, often occurring during the summer months. Unlike methane, however, it is directly toxic to many aquatic organisms even in low concentrations, and has thus to be avoided in culture ponds by maintaining a well-aerated system. However, being easily detected by its characteristic noxious smell, the existing gas and conditions leading to it can be eliminated by proper aeration of the water. There are certain bacteria which can combine oxygen and hydrogen sulphide to produce elemental sulphur, water and energy.

$$2H_2S + O_2 \rightarrow 2S + 2H_2O + energy$$

This sulphur accumulates inside the bacterial cell, and after the H_2S source is exhausted, the sulphur can be utilised by the bacteria to produce sulphuric acid and energy.

$$2S + 2H_2O + 3O_2 \rightarrow 2SO^{2-}_4 + 4H^+ + energy$$

However, another group of bacteria are able to utilise the sulphate and other sulphur compounds as well, to produce energy and hydrogen sulphide.

11.9.10 TURBIDITY

Various factors lead to turbidity in a culture pond. The turbidity caused by planktonic growth generally favours fish production by stimulating the growth of fish food organisms and also by suppressing the growth of undesirable underwater weeds. Plankton turbidity also improves fishing because the suspended particles limit the vision of fish, making them less wary (Swingle, 1945). However, turbidity resulting from high concentrations of humic substances is not desirable in ponds. While not directly harmful to fish, such waters are usually dystrophic because of acidity, low nutrient levels, and limited light penetration for photosynthesis, leading to poor growth. The most undesirable type of turbidity is probably that resulting from suspended particles of clay. Clay turbidity restricts light penetration thus adversely affecting the productivity, and particles that settle down to the bottom smother fish eggs and destroy the benthic community.

Maintenance of moderate to good fisheries is possible in waters containing 25 to 80 mg/litre suspended solid particles. However, other factors remaining the same, waters with turbidity less than 25 mg/litre show more productivity. Turbidity of 80 to 400 mg/litre does not support good fisheries, while waters containing more than 400 mg/litre of suspended solids will show very poor results (Alabaster and Lloyd, 1980).

11.10 Application of lime

Liming is routinely done during pond preparation and as a part of pond maintenace, since it has a favourable action on the biological factors of production and on the health of fish. The important functions of liming are:

1) It destroys the parasites present either directly in the water or at the pond bottom, or inside the infected fish, intermediate host or even in its encysted form. It also kills the adult insects as well as their larvae that are fish enemies. Algae and other aquatic plants with roots not too deep can also be destroyed by liming.

2) It maintains a desirable alkalinity by increasing the pH so that fish remain healthy and free from stress, and the whole biological cycle in the pond remains under optimum condition.

3) Liming ensures the presence of sufficient quantities of calcium in the medium to support the growth of vegetation and molluscan and crustacean shells. Present in adequate quantities, calcium can also neutralise the harmful action of magnesium, sodium and potassium salts.

4) Liming considerably improves the pond bottom through the liberation of bases, the neutral reaction, the increase in biological activity, the acceleration of decomposition of the mud and its cellulosic components. Concomitantly, liming minimises the risk of spreading different parasitic diseases and reduces the chance of oxygen deficiency, through the mineralisation of organic matters.

5) It precipitates excess organic material present in suspension in the water, and thereby again reduces the chance of spreading some diseases and preventing oxygen depletion.

6) Liming helps in the nitrification of ammonium compounds into nitrites and nitrates.

Thus, liming is done whenever the pH and alkalinity of water are too low, the pond bottom is too muddy, its organic matter content is too high and there is a threat of contagious diseases, or a danger of oxygen depletion.

In addition to quicklime (CaO), powdered limestone ($CaCO_3$), slaked lime (or caustic lime $Ca(OH)_2$) and calcium cyanamide ($CaCN_2$) are also applied in aquacultural ponds as and when necessary.

11.10.1 LIMING PROCEDURE
Liming is done in the following three different ways:

  i) liming the dried pond bottom;
 ii) liming the pond water; and
iii) liming the water flowing into the pond.

11.11 Use of fertilisers

Application of fertilisers in the culture ponds increases the production of fish without the accompanying risk of enhancement of dietary diseases and at the same time improves the hygienic condition of the pond.

The main function of the fertilisers is to increase the quantity of natural food, which in turn would take care of the productivity and stocking of ponds exploited extensively or semi-extensively for fisheries. The quantity of natural food ultimately depends upon heat and light, and certain essential inorganic nutrients that form the basis of the growth of vegetation. The essential elements that are present in minimum quantities in the ponds and need augmentation through the use of fertilisers are phosphorus, potassium and nitrogen.

Fertilisers do not act directly on either the water or the organisms in suspension, but on the pond bottom, which actually is the production laboratory of the pond. The fertilisers are absorbed by the mud at the bottom, and consequently being in a good state of decomposition, release the nutrients little by little in solution in the water above, whereby they are taken up by the phytoplankton and other aquatic vegetation and through them reaching the animal population indirectly. Fertilisers have a prolonged and continued action on the productivity of ponds.

Use of mineral fertilisers are recommended for all ponds in which extensive or semi-intensive culture of fish is practised. Sometimes fertilisers are also used in the intensive culture ponds to encourage the decomposition of excrement and alimentary residues. However, fertilisers can only be used in the ponds where strong water currents are not found.

11.11.1 Amount of fertilisers to be applied

The type and quantity of fertilisers to be applied depends upon the existing conditions of the pond and may vary considerably from region to region and between different farms. Fertilisers have to be used only to compensate for those substances that are found in too small a quantity, and here also there is a limit to the increase in productivity by the addition of a single mineral since this will ultimately lead to another substance falling to a minimum and a consequent disbalance of factors will result. For example, if phosphate fertiliser is used in excess, it leads to the formation of iron and aluminium phosphates which precipitate out. So the aim should always be to achieve an equilibrium condition with judicious application of different fertilisers. Physical factors like light and temperature may also limit the production, and finally economic factors must also be taken into account when using fertilisers.

11.11.2 Preconditions for using fertilisers

The following preconditions need to be fulfilled before the application of fertilisers to obtain the maximum benefit:

1) Since acidity in the soil reduces its absorption capacity and fertilisers have to be absorbed by the soil for their action, the soil and water of the pond need to be neutral or slightly alkaline. If the action is acidic the pond has to be limed before fertilisation.

2) The bottom mud should comprise fine detries of submerged plants and algae, rich in colloids and not too thick in consistency. The mud resulting

from decomposed emergent plants is too rich in cellulose, too thick and is not very productive.

3) Vertical vegetation should be cleaned out by repeated cutting and by the application of herbicides so that they do not compete indirectly with the fish for food by using up the fertilisers for their own growth, and also do not interfere with the penetration of light and heat. If parts of a pond cannot be cleared of the vegetation properly, these areas should not be fertilised.

11.11.3 DISTRIBUTION OF FERTILISERS

Fertilisers are applied either at the time when the pond is being prepared and the bed is dry, or just after filling it with water. In case of the latter, a boat should be used for the application to ensure uniform spreading. The texture of the fertiliser should be fine, and the amount to be used so fixed as not to harm the fish.

Mineral fertilisers can be used in one or many spreadings. Lime, and fertilisers containing substantial amounts of calcium, should not be used in combination with ammonium sulphate or organic manure rich in ammonium ions, such as dung or liquid manure. At least eight to fifteen days of interval should be kept in between the application of superphosphate and lime, since the presence of the latter hampers the dissolution of the former. Fertilisers like superphosphates which dissolve readily, can be applied as soon as the water begins to warm up after the winter months. When preparing the nursery ponds, fertilisers are to be applied two to three weeks prior to stocking to help the development of the natural food.

11.11.4 PHOSPHATE FERTILISERS

Most pond waters generally lack phosphorus, and consequently the use of phosphate fertilisers in fish cultivation is very effective. In ponds rich in lime, with the mud in a good state of decomposition, the effect of the application of phosphates can be readily observed with the naked eye, as the water turns green with algal growths often reaching bloom conditions. Phosphates are held in the soil and released slowly. Generally, renewal of water is stopped for about five days after spreading phosphates, particularly superphosphates, to prevent their being washed away.

11.11.5 POTASSIUM FERTILISERS

Sufficient quantities of potassium are usually present in the soil and waters of most ponds, and the application of potassium fertilisers is only needed under the following conditions:

1) ponds that are particularly poor in potassium;
2) ponds with low alkalinity; and
3) ponds with hard bottom and poor growth of aquatic plants.

Potassium fertilisers seems to favour the growth of submerged plants that are beneficial to production while the harmful vertical vegetation are eliminated.

11.11.6 NITROGEN FERTILISERS

The relative proportions of nitrogen and phosphorus is important for good fish production, the ideal ratio of P/N being 1/4. In ponds with an alkaline bed and well-mineralised water, the P/N ratio may be 1/8. However, it is usually a deficit of phosphorus that prevents the full utilisation of nitrogen that is already present in the water. Ponds with a good colloidal mud bottom produce enough nitrogen by themselves, and application of further nitrogenous fertilisers is unnecessary in these cases. But in the new ponds either without or with little mud, the use of nitrogenous fertilisers is desirable.

11.11.7 ORGANIC MANURE

Application of solid or liquid organic manure brings about an increase in production in the following manner:

1) It supplies nearly all of the nutritive substances indispensable for the biological cycle;

2) Organic manure usually has a favourable action on the structure of the soil;

3) Presence of organic manure promotes the growth of bacteria in pond water, which, when not excessive, in turn favours the development of zooplankton; and

4) Presence of organic manure is necessary for the proper action of phosphate and potassium fertilisers.

The application of organic manure, however, poses some problems as well. Thus, there is a risk of oxygen deficiency occurring particularly in the early hours of the morning, and a threat of certain diseases.

Organic manure is periodically distributed in the water in small quantities at a time, and if possible, spread uniformly over the surface of the pond. It can also be distributed regularly on specified spots along the banks. However dung should not be evenly spread, as this would hamper the biological activity of the soil, but distributed in heaps or rows. It can only be spread evenly on sterile virgin soil which requires a coat of rich colloidal mud.

11.12 Artificial feeding

Use of artificial feed depends upon the intensity of culture which may be extensive, semi-intensive or intensive. For intensive cultivation artificial feeding is usually exclusively used. Artificial feeding is one of the principal methods of increasing fish production, and allows a higher stocking density and a consequent better utilisation of the natural food which also increases through the utilisation of unconsumed artificial feed and the excrements of a denser population.

The economics of artificial feeding depend upon the costs of food used and its rate of conversion. The rate of food conversion is the amount (in kg) of the food necessary to produce 1 kg of fish. Two terms, namely, the *absolute conversion rate* and the *relative conversion rate* are generally used for this purpose. The absolute conversion rate or simply the 'conversion rate' is obtained by dividing the quantity of food distributed by the extra growth obtained, and implies that the extra growth in totality depends only upon the addition of the feed. Thus, 'absolute conversion rate' is not exact since it does not take into account any error, like the loss through death and the increase in production due to higher stocking density which is invariably done with artificial feeding and leads to a better utilisation of the natural food. The 'relative food conversion rate' is however obtained by dividing the quantity of food distributed by the total production, that is, production due to natural food, manuring, and artificial feeding.

Apart from the amount and quality of the food distributed, the rate of conversion depends upon other factors like the individual weight and age class of the fish, their state of health and the stocking density, temperature of water, etc. It also depends upon the frequency of distribution and the manner of application (that is, spreading) of the food.

Feeds are formulated to the specific requirement of the particular fish under culture and are usually made from materials of plant as well as animal origin. On the basis of the development of the mouth parts of the fish, the sizes of the pelleted feeds are also changed. Recently micro-encapsulated feeds are being prepared. For the proper utilisation of the food given, a strict feeding schedule has to be maintained. Also, to minimise unconsumed food, different devices like automatic feeders, or demand feeders, have also been designed.

11.13 Proper stocking

By proper stocking it is meant that an optimum number of fish of each species and age class is to be released in a culture so as to obtain the highest possible production, both in quality and quantity, in the most economical way. Stocking therefore has to be very precise, and it depends upon the size and productivity of the pond. In an intensive culture system where the fish depend entirely upon artificial feed, stocking is independent of natural productivity and is determined entirely by the existing physico-chemical conditions and the quantity of the artificial feed.

11.14 Control of fish enemies and diseases

In the culture ponds, fish may have many enemies belonging to different classes of animals like insects, other fish, amphibians, reptiles, birds and mammals. Different groups have different control measures, but the common aim is not to disturb the culturable varieties of fish while destroying the fish enemies, so that production is not hampered.

Fish may also die either due to pollution of the culture water, when all types and sizes of fish are simultaneously affected, or through diseases which usually attack a single variety or age class. In any case, mortality can be high and rapid. The problem of disease is particularly high in intensive culture systems. The diseases usually occur and spread under poor culture conditions, high stocking density, and the small number of species cultured that may give rise to a specific type of disease. In the wild state, where fish are not aggregated together in a confined area, diseases are not noticed very often, and the risks of contamination or loss of life are also less.

The main disease causing organisms include protozoans, fungi, bacteria and viruses. There may be nutritional diseases too. The pathogens are controlled by appropriate chemotherapy, antibiotics, and recently by vaccines also. However, the most important preventive measure is by maintaining a healthy environment in culture ponds.

11.15 Harvestation

Harvestation of cultured fish can be carried out by either drying the pond or by using nets and traps without drying. For total harvestation at the end of a culture period, the pond is usually dried out, which also helps in its preparation for the next culture operation. For partial harvestation from time to time while the culture process is still under progress, various suitable net traps are used. To avoid or reduce loss while capturing, grading and handling, the following precautions should be taken:

1) Feeding should be stopped two or three days prior to drying;
2) Harvestation should be done only in cool weather, preferably in the morning, unless it is cloudy or raining; and
3) Fish should not be heaped up.

11.16 Transport of fish

Transport of fish is of particular importance in fish culture, specially in case of live fish. Unless proper care is taken, mortality of live fish, particularly at their younger stages, occurs due to the following reasons:

1) depletion of D.O. in the water medium,
2) accumulation of toxic gases such as ammonia,
3) physical injury caused by careless handling of fish,
4) hyperactivity and strain experienced while in confinement, and
5) sudden fluctuation of water temperature,
6) diseases.

To avoid the transport mortality of fish seed, the following steps should be taken:

1) Prior to packing and transport, the fish seed are generally caught and kept in high concentrations in a limited water space overnight for the purpose of:
   a) conditioning before despatch,
   b) emptying their guts which otherwise spoils the transport water due to pollution as a result of their excretory products,

c) helping in eliminating weaker or injured fry, as they may die during transport and consequently pollute the water, and

d) making the fish more resistant to temperature fluctuations and the stress of handling.

2) Size-wise and species-wise segregation of seeds prior to transport to avoid cannibalism.

3) Transportation should be carried out either early in the morning, or at night, when the temperature remains comparatively low.

4) Cool and clear water with high D.O. content is to be used.

5) In hot weather, the carriers are to be covered with sacks or cloth soaked with water.

6) If necessary, water is to be changed on the way while transporting.

There are 'open' and 'closed' systems of transport. The open system is the transportation in open containers where the water gets oxygen from the atmosphere. In the closed system, fish seed are transported under oxygen packing. To decrease the oxygen demand anaesthetics are now used.

The transportation of freshly caught and killed fish is possible, but there are difficulties because of the speed at which their flesh decomposes. One way of dealing with the danger of decomposition is to transport them at low temperatures.

This will stop the growth of bacteria. The temperature should ideally be between 0 and 4°C and for this purpose, fish are packed in crushed ice.

REFERENCES

Alabaster, J.S. and Lloyd, R. (1980): *Water Quality Criteria for Freshwater Fish*. FAO, U.N. Butterworth, London.

Bell, M.C. (1970): Fisheries engineering. In: *Marine Aquaculture*, W.J. McNeil, ed., Oregon State University Press.

Boyd, C.E. (1970): *Water Quality Management for Pond Fish Culture*, Elsevier Scientific Publishing Company.

Boyd, C.E. and Walley, W.W. (1975): Total alkalinity and hardness of surface water in Alabama and Mississippi. Auburn University Agricultural Experiment Station, Auburn, Alabama, Bull. 465, p. 16.

Brett, J.R. (1956): Some principles in the thermal requirements of fishes. *Quarterly Review of Biology* 31(2): 75–87.

Colt, J. and Armstrong, D. (1979): *Nitrogen Toxicity to Fish*, Crustaceans and Molluscs, Department of Civil Engineering, University of California, Davis, California, p. 30.

Devlin, R.M. (1966): *Plant Physiology*. Reinhold, New York.

European Inland Fisheries Advisory Commission (1973): Water quality criteria for European freshwater fish. Report on ammonia and inland fisheries. *Water Res.* 7: 1011–1022.

Hart, J.S. (1944): The circulation and respiratory tolerance of some Florida freshwater fishes. *Proc. Fla. Acad. Sci.* 7: 221–246.

Hickling, C.F. (1962): *Fish Culture*, Faber and Faber, London.

Hickling, C.F. (1968): *The Farming of Fish*, Biology in Action Series, Pergamon Press.

Hickling, C.F. (1970): Estuarine fish farming. *Adv. Har. Biol.* 8: 119–213.

Hollerman, W.D. and Boyd, C.E. (1980): Nightly aeration to increase production of channel catfish *Trans. Amer. Fish. Soc.* 109: 446–452.

Huet, M. (1972): *Text Book of Fish Culture, Breeding and cultivation of fish*. Fishing News Books Ltd., Surrey, England.

Kinne, O. (1970): Cultivation of marine organisms and its importance for marine biology. *Hel. Wiss. Meeresunters*. 20: 1–5, 707–710.

Mairs, D.F. (1966): A total alkalinity atlas for Maine lake waters. Limnol. *Oceanogr*. 11: 68–72.

Merkens, J.C. and Down, K.M. (1957): The effect of tension of dissolved oxygen on the toxicity of un-ionized ammonia to several species of fish. *Ann. Appl. Biol*. 45: 521–527.

Milne, P.H. (1972): Fish and shellfish farming in coastal waters. Fishing News Books Ltd., Surrey, England.

Mount, D.I. (1973): Chronic effect of low pH on fathead minnow survival, growth and reproduction. *Water Res*. 7: 987–993.

Moyle, J.B. (1945): Some chemical factors influencing the distribution of aquatic plants in Minnesota. *Amer. Midl. Natur*. 34: 402–420.

Plumb, J.A., Grizzle, J.M. and DeFigueriredo J. (1976): Necrosis and bacterial infection in channel catfish (*Ictalurus punctatus*) following hypoxia. *J. Wildl. Disease*. 12: 247–253.

Reid, G.K. (1961): *Ecology of Inland Waters and Estuaries*. Reinhold, New York.

Schuster, W.H. (1949): Fish culture in brackishwater ponds of Java. Indo-Pacific Fisheries Council, Special Publication, No. 1, India.

Snieszko, S.F. (1973): Recent advances of scientific knowledge and developments pertaining to disease of fishes. *Adv. Vet. Sci. Comp. Med*. 17: 291–314.

Swingle, H.S. (1945): Improvement of fishing in old ponds.*Trans. North Amer. Wildl. Conf*. 10: 299–308.

Swingle, H.S. (1961): Relationships of pH of pond waters to their suitability for fish culture. *Proc. Pacific Sci. Congress* 9(1957): 10, 72–75.

Walters, G.R. and Plumb, J.A. (1980): Environmental stress and bacterial infection in channel catfish, *Ictalurus punctatus* Rafinesque. *J. Fish Biol*. 17: 177–185.

Wheaton, F.W. (1977): *Aquacultural engineering*. John Wiley and Sons.

CHAPTER 12

Different Culture Techniques

As discussed under section 11.2 there are six zones available for coastal farming, namely the shore, intertidal, sublittoral, surface water, mid-water, and the sea bed, which are being utilised for fisheries through different types of culture techniques. If left undisturbed, the natural cyclical system is found to work efficiently, though a trifle slowly, by which production of materials sought after by man occurs. In the natural system, waste disposal and feeding are no problem since wastes from one organism provide food for another. However, production occurs slowly and the aim of the aquaculturist would be to augment this production as far as possible by applying different techniques evolved and without disturbing the natural rhythm.

To increase the productivity of aquatic systems, many different culture techniques can be adopted. These techniques can be broadly categorised into the following three types: open system, semiclosed system and closed system. An open system culture technique is defined as the production in a natural body of water with or without any modification. In a semiclosed system culture technique, the water from a natural body is passed through the system and utilised for production with or without its modification and again returned to its natural source after one such passage. A closed system culture technique is one in which the water once taken inside the system is either never replaced or is replaced only at widely separated intervals.

12.1 Open water culture systems

The most commonly practised aquaculture technique, and also the oldest one, the open water culture system owes its popularity to the low cost that is involved and the minimum requirements for its management. However, at present, the more advanced types of open system require considerable expertise and managerial skill and the capital costs may also thus be quite high.

The simplest form of the open system is the natural system, and if the harvesting pressure is not too high, nature can be left to take its own care, any outside management being redundant. One of the major problems of the open system is the danger of predators to the cultured organisms. Of the different methods that are followed to keep out the predators, the most common one is to put up mechanical barriers like bubble curtains, nettings, screens, dikes, and thus fence off an area or a bay mouth. The procedure then is to eliminate the predators within this zone by chemical or any other means before starting culture operations. Any further attack would be successfully fended off by the fencing. However a major constraint of the open water culture system is diseases which are very difficult or sometimes even impossible to control.

12.2 Semiclosed systems

Compared to the open system, a major advantage of the semiclosed system is that better management is possible in the latter, particularly in relation to the physical parameters of the water used, such as temperature, volume, or velocity, though a complete control over predators, diseases and pollution is still not possible here, the water being obtained from a natural source. However, one advantage of using natural water is the availability of plankton as a free source of food in the semiclosed systems.

Most semi-enclosed systems have structures specially constructed to meet the specific demand for that particular system: these however increase the capital costs. The final product which is uniform and of higher quality as well as not dependent upon the weather and other natural factors usually amply compensates the higher capital cost. The advantage, however, is offset to some extent by the fact that the higher density of organisms that are present in semi-enclosed systems tends to increase the danger of infections and diseases, and the aquaculturist has to be alert always to a possible stress condition arising out of such situations. Also the higher investments in comparison to the natural systems calls for good management practices for a better return which may not always be forthcoming due to a dearth of adequate expertise.

Many semi-enclosed systems need pretreatment of the water used for culture, which is effected usually through filtration and purification procedures. For the control of diseases and predators, the incoming water may be treated with ultraviolet light and chemicals like chlorine. Before returning the water to its natural source, it might require anti-pollution treatments, particularly if the culture involves a high density of organisms.

12.3 Closed systems

In closed systems where the water once taken in is seldom or never exchanged with its outside source, it is circulated within the system with the help of pumps and other devices. Thus, closed systems permit a thorough monitoring of the environmental conditions which accounts for its higher efficiency with respect to the economy, as well as the production and control of stress conditions. Energy requirements are also less in closed systems than in the semiclosed systems, despite the additional running of pumps. This is because the large amount of heat (or cold) that is discarded with the outgoing water in semiclosed systems is not lost here, thus making the control of water temperature also more efficient as well as economical.

Control measures for diseases are more feasible in the closed systems since constant observation makes it possible to detect diseases early enough, and treatment procedures more easily adapted. It is also more economical than in either semiclosed or open systems, since outside contamination is mininised. For the same reason, a pretreatment of the water for preventive measures is also possible. Though higher population typical of closed systems augments the disease problem, successful control of environmental factors reduces animal stress thereby increasing their resistance to diseases.

Though the advantages of closed systems over either the open or semiclosed systems are many, the main constraints seem the high capital and running cost and the requirement of good management practices. The high capital costs come from the need for filtration and purification equipments and growing tanks or ponds. Filtration and purification apparatus are a must for the removal of metabolities from the water before recycling. Purification equipments also include ultraviolet light units for the control of diseases. These are, however, as mentioned earlier, also required in the semiclosed systems. The increased operational costs that are typical of the closed systems are due to running of the pumps, which though may be minimised in well designed systems, cannot be entirely dispensed with. Also in the high density cultures, power may be needed for aerating the water. However, the major expense item in the closed systems is the artificial feed, which is essentially free in the open systems due to the utilisation of the natural food, and used only as supplement to the natural foods in the semiclosed systems. It is mainly the economic constraints that have kept the closed systems in restricted use, there being only a few commercially operated systems existing at present. Good management practice is also a key factor to successful operation of a closed system. Present knowledge places good management of closed systems more in the category of an art than a science (Wheaton, 1977).

At present, fin fish, shell fish and aquatic plants are cultured with profit in the open systems. For some species and several types of operations, the semiclosed systems are economically viable and are being practised. The closed systems are now profitable for only a few species and that only under very special circumstances. However, increasing fishing pressure on the natural water bodies, increasing fuel costs for running fishing fleets, rising labour costs and labour shortages, together with the rapid advances made by researchers in this field to minimise operational costs while increasing mechanisation, will ultimately enhance the economic feasibility of the highly controlled closed systems for aquaculture.

12.4 Water supply and its pretreatment
Whatever be the type of culture technique that is being followed, and each

has its own merits and demerits, the ultimate success for each depends upon the quality of water that is being used, and also its adequacy for the purpose.

The quantity of water should be enough to compensate for losses through evaporation and seepage, to provide a sufficient amount of oxygen to the organisms cultured, and also provide enough flushing for the removal of waste products. Loss of water due to evaporation depends upon the temperatures of water and air, the vapour pressure, total surface area of the water and the wind velocity. Regardless of the type of soil, any earthen impoundments lose water through seepage. Though depleted oxygen levels, if detected in time, can be rectified with a supply of oxygen-rich water, the volume of such water added should be enough or else the uptake of oxygen will not be rapid enough to prevent mass mortality. Similarly, the noxious waste products that are produced by the cultured organisms if not removed by enough flushing with clean water in sufficient quantities, will lead to mortality of these organisms. The amount of waste products produced depends upon population density, type and amount of feed given, temperature of the water and other management factors. The amount of waste increases with increasing stocking density, the feeding rate as well as the water temperature till the temperature detrimental to the growth of fish is reached. Any increase in temperature beyond this point means metabolic activities and the production of waste matter decrease. Overfeeding increases BOD, thus decreasing the oxygen level. It also augments the production of noxious wastes by bacterial decomposition of the excess feed which then needs higher volumes of water for the supply of oxygen and flushing. The quantity of water needed also varies according to the species cultured and their densities and management practices,.

The water should be free from pollutants and able to support the growth of the cultured organisms while maintaining high sanitary standards. Pollutants are materials, either dissolved or suspended, the presence of which in water is detrimental to the organism under cultivation or even to its consumer. However, a substance that is a pollutant for one species may not be for another. Thus, the solid material of raw sewage is not considered a pollutant in algal cultures, while it is in prawn hatcheries. Similarly, oil present in the culture waters of aquatic organisms that are able to utilise it, is a nutrient rather than a pollutant. Further, the addition of heat energy to water may either be considered a pollutant or beneficial energy as the case may be.

In coastal aquacultures, the use of brackish and sea water has several drawbacks. Since salt water is highly corrosive, the pipes, tanks, pumps and other equipment coming in contact with the water must be either lined or constructed with materials that do not corrode easily. While most plastics, epoxy resins and rubber compounds meet these standards, many of them cannot be used in aquacultural practices due to the presence of toxic compounds affecting either the species cultured or its human consumers. Most metals corrode rapidly when in contact with salt water. While aluminium is better than most steels, both types corrode eventually. Stainless steel may be used in most systems, but here also care must be taken so that the chromium present in the stainless steel is not dissolved out and concentrated in the bodies of the growing organisms, since many organisms, shell fish in particular, concentrate heavy metals in their bodies.

Another drawback of coastal aquacultural farms is the problem of salt water fouling. Many aquatic organisms during some part of their life-cycles attach themselves to any hard surface available, be it a pipe, tank or pump. Such growths, or fouling, require removal and though several methods are available for this, some problems are associated with each of such methods.

The chief constraint of the chemical removal of fouling is that, even if the chemical added to the culture water is non-toxic to the crop, some of it may be absorbed inside their bodies and if it is toxic to man (for example, mercury) the crop has to be rejected. Other anti-fouling methods used are change of temperature of water, sterilisation of the water by UV, increasing water velocity to prevent the attachment of the fouling organisms, mechanical or hand scraping and the use of duplicate plumbing systems.

The intensive and closed-water culture systems that are being increasingly used calls for a flow-through system of water that is used to capacity while maintaining a high quality, together with tight control of effluent discharge. This, therefore requires efficient filtration of the water used. The most widely used filters at present, that primarily convert ammonia to nitrate, are the biological filters. Charcoal filters are also used often to remove dissolved organic materials. In the salt water systems, for the same purpose, foam fractionation units are becoming increasingly popular. The most practical approach that is being followed presently in coastal waters is the employment of the following arrangements in a series: a settling basin to remove the large particulate materials, a foam fractionation unit that can remove the dissolved organic materials, and finally a biological filter that is utilised for nitrification. Monitoring is needed for the removal of inorganic ions by foam fractionation. Control of nitrate levels can be achieved by the addition of water. For the aquariums and other systems, where very clear water is required for better visualisation diatomaceous earth filters are often used (Wheaton, 1977).

12.5 Important coastal aquaculture techniques

The culture technique to be adopted is determined on the basis of the particular zone of the coastal area which is to be utilised and the species to be cultivated. According to suitability, the system to be adopted may be of the open, semiclosed or the closed type.

Some of the important coastal aquacultural techniques and suitable representative species thus cultivated have been described as follows. Though many countries now utilise coastal areas for aquaculture, the techniques followed in the countries where the engineering aspects have received due importance and have been developed to the level of established technologies augmenting production, have mainly been discussed here.

12.5.1 CULTURE OF FIN FISH

12.5.1.1 Mullet culture

The most common species of mullet, the grey mullet (*Mugil cephalus*) is cultured in brackish water ponds along with other brackish water fish. It is also often cultured in fresh water ponds, where other fresh water varieties of fish are cultured simultaneously. Culture of mullets is popular in the Indo-Pacific regions, Italy and other Mediterranean countries and Israel. Culture practices of grey mullet, including its artificial propagation developed in Taiwan (Chen, 1976) is described as follows.

Artificial propagation collection of breeders

The breeders, usually four years of age, weighing 1.0 to 2.1 kg with a length of 32 to 50 cm are caught during the winter from among the mullets that annually migrate at that period for breeding purpose. These specimens are collected in dark coloured plastic bags filled with sea water and are transported under oxygen packing to the stock tanks. A stock tank is an indoor concrete structure measuring $5 \times 7 \times 1.5$ m, into which fresh sea water is continuously being introduced and there is a provision for adequate aeration. The males and females are kept separated in these tanks by means of nylon nettings.

Induced breeding

Spawning is induced by the injection of either of the following: (i) Extract of pituitary gland, preserved in acetone at 5°C; the glands are obtained from mature mullets either male or female, and (ii) Synahorin, a mixture of chorionic gonadotropin and mammalian hypophyseal extract. The female is given the first injection within an hour being released in the stock tank and a second injection follows within the next 24 hours for best results. If there is no response, a third or even a fourth injection follows. The males caught for breeding are usually fully ripe and do not need and hormone injection for milting. However, those collected towards the end of the spawing season may require hormone treatment.

A ripe and healthy female readily responds to the hormone injections which is evident by its distended abdomen from which eggs freely come out through the genital aperture by the slightest pressure applied on it or even without pressing. A little sample of eggs is pipetted out for examination under a microscope. If found to be transparent and completely round in shape with one oil globule, the eggs are ready for fertilisation. The female is then manually stripped and the eggs, numbering about 1 to 1.5 million per female of 1.5 kg body weight, are collected in a plastic basin. Simultaneously the male is stripped by a second person and the milt is allowed to fall upon the egg mass, a third person mixing the eggs and milt gently by means of a feather. After fertilisation, the eggs are washed repeatedly in fresh changes of sea water to remove traces of blood and any other foreign matter. The fertilised eggs are then introduced into hatching tanks with water provided with aeration for hatching.

A fertilised mullet egg is round, 0.93 to 0.95 mm in diameter, transparent, non-adhesive, and with an yellowish oil globule of about 0.38 mm diameter. These eggs float about near the surface when the water is being aerated, while some may sink slowly in stagnant water. Dead eggs sink to the bottom.

Hatching

The indoor hatching tanks may be of plastic with a capacity of 0.5 or 1.0 t or made of concrete with the dimensions $5 \times 7 \times 1.5$ m. The water quality is kept good by continual change and aeration. The temperature is maintained at 20 to 24°C. After an incubation of 16 to 30 hours, the eggs contain well-developed embryos with black pigments. Hatchlings come out in 34 to 38 hours with the water temperature at 23 to 24.5°C, and in 49 to 54 hours when the water temperature is kept between 22.5 and 23.7°C. The salinity is between 30.1 to 33.8 ppt. The rate of success depends upon the maintenance of a high dissolved oxygen content of water, minimum variations in temperature, with a gentle water movement and overall cleanliness.

Larval rearing

The new hatchlings are small (2.5 to 3.5 mm), transparent, with dark chromatophores scattered throughout the body, complete fin fold, colourless eyes, mouth and digestive tracts not yet well developed. They may swim weakly in belly up head down position, sometimes even with sharp up and down darting movements. These larvae generally keep out of strong light, congregating in darker corners. Older larvae move freely in schools. After the fortieth day of hatching, the larvae that by now have grown to fingerlings of lengths of about 1.5 to 2.0 cm, are ready for release in the outdoor ponds.

The feed given to hatchlings varies according to their progressive growth and development. The feed includes fertilised oyster eggs, trochophore larvae, rotifers, copepods, *Artemia* nauplii as well as adults, cooked egg yolk, rice bran, wheat flour etc.

Pond culture

Mullets are stocked simultaneously with grass carp, silver carp, big head carp, common carp, tilapia, etc. in polyculture ponds, in Taiwan. The stocking density of mullets varies according to the pond conditions and its fish community. Mullets are benthic feeders with a stomach (resembling the gizzard of a chicken) capable of grinding the bottom detritus the fish feed on. Thus stocking density of mullets can be higher in ponds with bottoms rich in organic material. Generally, however, 1,000 to 2,000 fingerlings are stocked per hectare in polyculture ponds, and 4,000 to 10,000 fingerlings/ha in mono-culture ponds (Chen, 1976). After one year of growth in ponds, *M. cephalus* attains the weight of 0.3 kg, after two years 1.2 kg, and if left for three years, it may attain a weight of 2.0 kg at the end of this period. A yield as high as

2,508.8 kg/ha has been reported from a pond in Hong Kong, and a yield up to even 3,500 kg/ha has been reported to have been achieved in case of intensively managed ponds (Bardach *et al.*, 1972). In India, mullets are cultured in brackish water polyculture ponds.

12.5.1.2 Red sea bream culture in Japan

Culture of the red sea bream (*Pagrus major*) is largely dependent upon the seed collected from the sea, for which juvenile specimens 30 to 100 mm in length are sought. Seeds are, however, also produced artificially.

These fish are cultured in floating net cages (Fig. 12.1). These cages are fitted in places in the sea where the water current is between 5 and 15 cm/s, water temperatures fluctuate between 10° and 29°C in a year, without rapid changes, and a depth of sea more than double the depth of the net cage. The optimum water temperature for the growth of red sea bream is between 20 and 28°C. The nets should also be protected from rough seas, red tides or pollution etc.

Immediately on placing the fish inside the floating net cages, the feeding is started, and the fish become accustomed to feeding in their new surrounding within 12 hours. The food given is inexpensive either fresh or frozen minced fish such as anchovy (*Engraulis japonicus*), mackerel (*Scomber japonicus*) etc. Alternatively, formula feed may also be given, either exclusively or as a supplement mixed with the minced fish. The netting needs changing every 10 to 45 days, more frequent changing being needed in summer when the amount of organic substances increases. The physiological inactivity of the fish during winter does not necessitate such frequent changes.

12.5.1.3 Culture of grouper in Taiwan

A special method is used for the culture of grouper (*Epinephelus* sp.) in the Pescadores of Taiwan. This is through underground ponds, excavated in the intertidal zone where the flood tide can change the sea water and the ebb tide

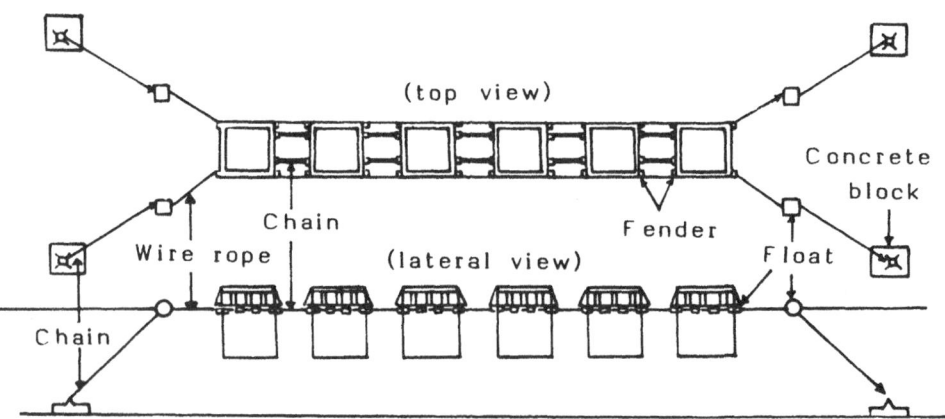

Fig. 12.1. An example of floating net cages for red sea bream culture

is utilised for feeding the fish (Fig. 12.2). The site selected is such that the ponds may be flooded with water every day as early as possible. Usually an underground pond is 10 m in length, 5 m in width and 2 m in depth. A 30 cm wide concrete surface is built along the top of the pond about 20 cm above the ground level, to which timber battens are fixed so that a nylon netting can be attached to cover the pond surface preventing the fish from escaping. The pond size is kept small since harvesting is done by pumping the water out at low tide. There is also no water gate or drainage pipe, and the water inlet is solely through the pond surface. The structural characteristics of underground ponds depend upon the texture of the soil which determine the bearing capacity of the foundation and the stability of the walls, the cost of construction being also dependent on it. Soft porous materials are avoided as far as possible since the wind and wave action can otherwise erode the walls. However, if the substrate is not found suitable, rocks may be used to re-inforce the walls from the top to the bottom of the ponds. Compared to the ponds above ground, underground ponds have certain advantages: cost of construction and maintenance is low, there is little temperature fluctuation, damages from wind and wave action is less, problem of marine fouling so extensive in the net cage cultures is minimal here, and the nets on the surface of the ponds can be cleaned easily.

The main disadvantage of underground ponds is the depletion of oxygen at low tides, when the oxygen remaining in the water is quickly used up. It has been reported by Tsai and Hu (1982) that in an underground pond ($10 \times 5 \times 2$ m) stocked with 800 groupers, the oxygen content decreased from 7.5 ppm to 2.5 ppm in 6 hours. However, with the advent of the flood tide, the oxygen content again went up to the normal level. Other problems encountered in underground pond cultures is the introduction of sediments into the pond water through wave action, and the limited period available for harvesting the fish. The species to be cultured, and its stocking density seems very important in determining the success of underground pond cultures. Groupers being comparatively inactive in habit and having a high economic value are a suitable species. The optimum stocking density is between five and 15

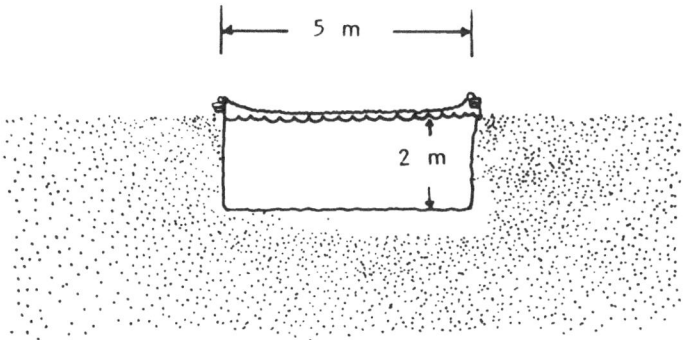

Fig. 12.2. Profile of intertidal under ground pond

individuals/m^3, depending upon the pond site. Not all intertidal underground pond cultures have met with success, many having failed due to the site being either too muddy or sandy, or due to a poor water exchange.

12.5.2 CULTURE OF SHELL FISH

12.5.2.1 Oyster culture

Edible oyster (*Crassostrea gigas*) culture is very popular and developed in Taiwan, Japan etc. The culture of *C. gigas* is discussed here.

Seed production

Seeds for oyster culture are collected from the sea where both natural and cultured oysters spawn. Specialised structures called oyster seed collectors, made up of oyster or scallop shells, known as cultches, strung on a metal wire (1.6 to 1.8 m in length) through holes drilled at their centres, are immersed into the sea water onto which the fully grown larvae or the spats attach themselves and grow up to the size of seed oysters. Individual strings of seed collectors are hung on racks constructed from bamboo and pine logs in shallow areas of the sea. Seed collectors have to be sunk at the appropriate period during the spawning season, when the emergence of spats is at its peak concentration, otherwise the collectors would become contaminated with barnacles, sessile diatoms or silt, that would hamper the attachment of spats. Oyster seed collection is largely dependent upon the existing environmental conditions. If the water temperature does not rise to about 23°C during the spawning period, spawning of oysters will be repressed despite full maturation of the gonads, and consequently the seed collection will also fail. For a cultch of shell length 10 cm, an average of 25 oyster seeds are usually attached.

Some of the seed oysters thus collected are transferred to the culture grounds where they are grown into marketable size in a year and are marketed during the winter and spring as one year old oysters. A major portion of the seeds are however kept for 'hardening' during the autumn and winter, and are cultured the following year, to be marketed as two-year-old oysters. The hardening process involves the fixing of the collectors by means of a wooden rack at the mean tidal level whereby the oyster seeds are exposed to wind and wave action. Hardening induces the ability to withstand changes in the environment and transportation over long distances though stunting their growth in the process. These hardened oysters, which are approximately 1 cm in shell height, however grow much faster the following summer than do the ones not hardened and raised as year old oysters.

Growing to marketable size

The oyster culture techniques for growing to marketable size as practised in Taiwan are described here.

The culture season is from February to September. This is partly because the spat collecting period is the winter and also because harvestation has to be completed before the onset of monsoon in October. Oysters cultured for the Taiwan market reach the preferred size, which is about 5 g of meat weight, in four to five months (Lin and Tang, 1980; Hu and Sheu, 1982).

Long-line culture

The long-line system of oyster culture involves the use of parallel horizontal lines made up of polyester ropes of 3 cm diameter, each set being 50 to 60 m long, attached to a series of floats arranged in rows, 10 floats per set at a distance of 20 cm from each other, each line anchored by 60 kg anchors at its ends, and fixed near the surface of the sea. From these parallel ropes a total of about 500 strings of oysters are hung 20 cm apart, each string being about 4 to 5 m in length and bearing 20 cultches. At first oil drums coated with pitch were being used as floats. However, since these require continuous maintenance and replacement every one or two years, styrofoam boxes and balls, and drums made of fibreglass which are strong and longer lasting are being used increasingly as floats. Styrofoam blocks covered with canvas or used-tyre rubber have been found to be more economical as floats, being 50 per cent cheaper than any other type (Hu, 1986). Also, initially four main ropes were used for hanging the oyster strings. It was later found that by increasing the distance between the floats, the use of two ropes instead of four would reduce loss when the sea is rough, while there is no requirement for a further increase of floatation equipment.

A further small modification effected by hanging the strings from ropes other than directly from the main ones greatly reduces direct stress due to wind and wave action on the string and prevents their breakage from the portion near the floats (Fig. 12.3) (Hu, 1986).

Hu and Sheu (1982) have reported that the best utilisation of the space and floating equipment is possible through the rotation of oyster culture methods by a combination of long-line (or raft) process and hardening of oyster seeds. Three rotations are usually recommended per season.

In the horizontal hanging culture method practised on the west coast of Taiwan, the oyster drill *Purpura clavigers* caused heavy losses (Kuo, 1964). In many places predation by oyster leeches caused even 25 per cent mortality in the long-line oyster cultures (Lin and Tang, 1980). However, by shifting the culture period from January to July, the abundant season for leeches could be avoided. Another serious problem of oyster farming is bio-fouling. Thus many species of algae, barnacles, mussels, and other organisms frequently get attached to the shells. As a control measure for both fouling and leeches, farmers expose the oysters to direct sunlight on open rafts for eight hours during the second or third months of culture (Hu, 1986).

Raft culture

In Taiwan, floating rafts are also used for oyster culture. These rafts measure about 12 m length-wise, being made up of 15 individual bamboo

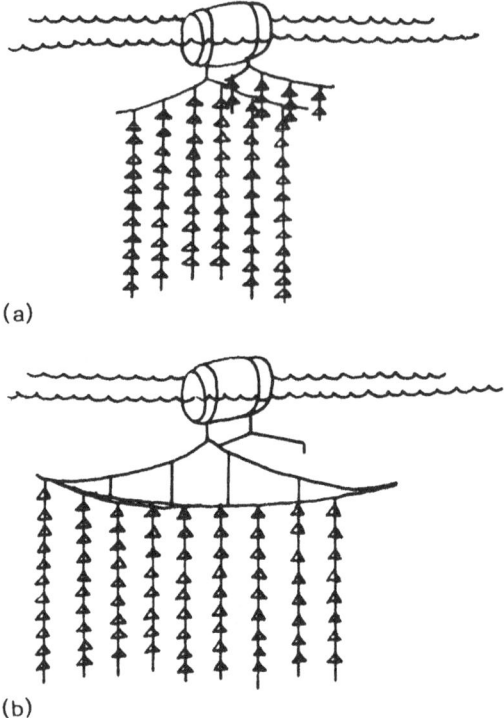

Fig. 12.3. Long-line oyster culture (a) and modification (b)

poles 10 to 15 cm in diameter, fixed about 60 cm apart, and 10 m width-wise comprising 60 such bamboo poles placed around 15 cm apart. The rafts are kept afloat due to the buoyant nature of the bamboo poles and the placement of floats under it. Strings (800 to 1000) of oysters are hung from each raft which is fixed into place by 60 kg anchors at either side. Sometimes extra floats have to be added to the rafts towards the end of a culture season, when the oysters get bigger and heavier. The rafts are brought inshore in the winter for maintenance. Bamboo poles usually last for two years. Sometimes instead of bamboo poles, plastic tubes with styrofoam inside are also used. The culture procedure and growth pattern of oysters in raft culture method is similar to that of the long-line culture system.

Submersible culture
    Oyster culture by the submersible method is practised in many parts of the coastal waters of Taiwan where the depth is more than 10 m so that a safe distance can be kept from the sea bottom. Introduction of submersible culture techniques have enabled the culture of oysters to be extended further into the sea where traditional farming could not be undertaken thus far. For submersible culture, both long-line and raft culture techniques can be followed.

In the long-line system, the main parallel ropes are kept 2 m under the water surface by fastening ropes with surface floats, and the oysters are hung at this depth. The floats are individual plastic balls filled with styrofoam, with a buoyancy of 45 kg, each being able to support eight strings of oysters. The oyster strings are lifted up to the surface for hanging the seeds and also for harvestation (Fig. 12.4). Each string of oyster weighs around 20 kg on an average at the time of harvestation.

Submersible oyster culture by the raft method is essentially similar to that for the submersible long-line system. Each raft requires 20 floats to keep it at the appropriate position on the water surface but preventing it from sinking further.

The growth rates of oysters in both the floating and submersible types of culture are comparable, but in the latter the loss through strong wave action or typhoons is much less than that in the former. However, in the submersible culture, constant checks have to be made of the floats and ropes so that the buoyancy is maintained at the desired level to keep the oysters in position. Extra floats have to be added, as required, whenever the float balls are found to be sinking (Tsao, 1984).

12.5.2.2 Culture of abalone in the intertidal zone in Taiwan

The small abalone *Haliotis diversicolor supertexta* which fetches a high price, is cultured in the shallow rocky areas. The prerequisites for selecting a

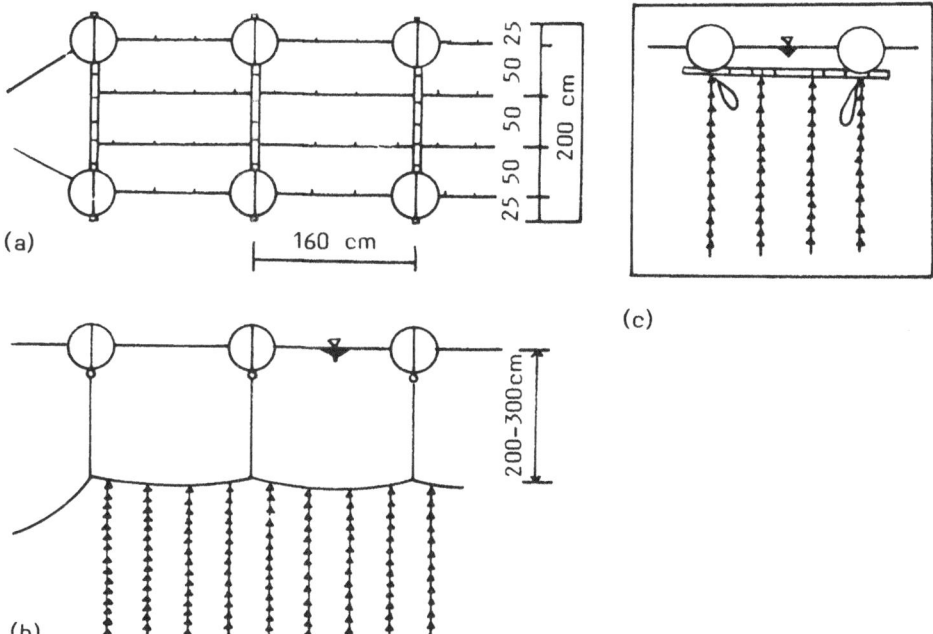

Fig. 12.4. Submersible long-time oyster culture; top view (a), profile (b) and at harvesting (c)

suitable site for a small abalone farm are: (i) clear sea water free from pollution and without dilution with fresh water, (ii) rocky substrate without drifting sand or mud, and (iii) a good water circulation to supply sufficient oxygen (Tzeng, 1977; Liao, 1979). The introduction of new techniques and equipment has now led to the commercial production of abalone seeds in Taiwan (Chen and Yang, 1979). According to the different topographical features of the site, the layout of a small abalone farm varies from place to place, but the general scheme is usually the same everywhere. The general layout of a small abalone farm in Taiwan is shown in Fig. 12.5. The ponds are usually constructed in the intertidal zones with little tidal range and are made of concrete for easy management and long durability. Ideally, a big rectangular pond of an area of 150 to 250 m^2 is constructed, which is divided into six or more ponds by means of 30 cm high brick walls. PVC pipes 10 cm in diameter are fitted at the bottom of these smaller ponds for the exchange of water. A series of PVC pipes passing through the walls of these ponds are connected at their other ends to the water inlet canal to admit water into them. To facilitate drainage of water, these 1.2 to 1.5 m wide inlet canals which are constructed along three sides of the pond are made 30 cm deeper than the rearing pond. The concrete walls of the ponds are 3 m high and the PVC pipes, which are 8 cm in diameter, are fitted at 1 m and 1.7 m above the bottom with a distance of about 1.5 m between them. The water exchange is affected through these and the inlet canal, and is controlled by a sluice gate at either side of the open mouth of the canal. The sluice gate is fitted with a screen inside to prevent the entry of predatory and other unwanted animals. A depth of water of about 1 m is generally always maintained in the ponds to minimise undue rise in temperature during summer. At one corner of the inlet canal, a small tank deeper by 20 to 30 cm than the canal is built to facilitate pumping out of the water. The bottom of the rearing ponds are laid out with 10 cm^3 stone pieces, fixed into place by reinforced iron to avoid disturbance during the typhoon season and to which the abalones can adhere.

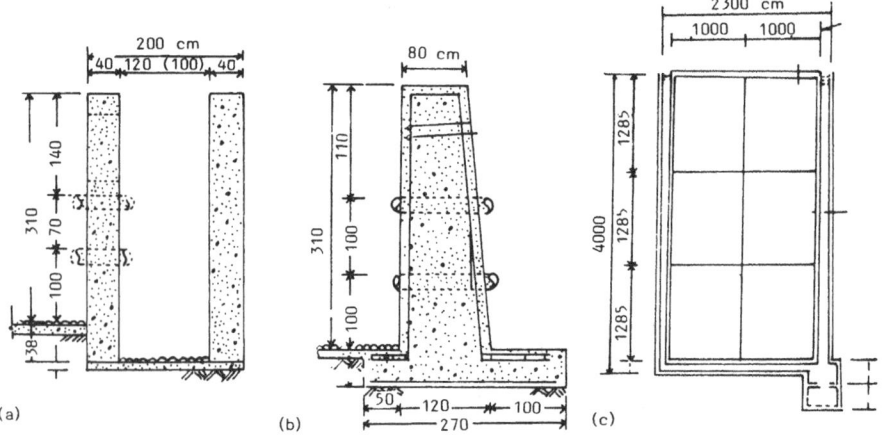

Fig. 12.5. General layout (c), pond wall (b) and canal (a) of a small abalone pond in Taiwan

The stocking density of small abalones is about 20 to 30 seeds/m² area. The seeds used are of the size of about 1.5 cm. The food given is *Gracilaria*, which has a conversion rate of 15:1. The small abalones reach marketable size after about six months of culture, when they are harvested.

12.5.2.3 *Shrimp culture*

The shrimp or prawn culture industry has developed immensely in Taiwan, Japan and the Philippines in recent years. For instance, the annual production of grass prawn exceeded 18,000 (Chiang and Liao, 1985) and its production rate is still going up. The important and popular penaeid prawns include *Penaeus monodon* (grass prawn), *P. japonicus* (Kuruma prawn), *Metapenaeus ensis* (sand shrimp), *P. penicillatus* (red-tail prawn) etc.

Technologies for cultivation of penaeid prawns have been developed in different countries. The culture process can be divided into several steps: maturation of the adult prawns, spawning, hatching, larval rearing and growing to marketable size. The culture technique of *P. monodon* has been described here. The different stages of the life history of this prawn species is illustrated in Fig. 12.6.

Maturation

Sexual maturity is rarely attained by *P. monodon* in the ponds. However, the males can spontaneously mature under captive conditions. Since the

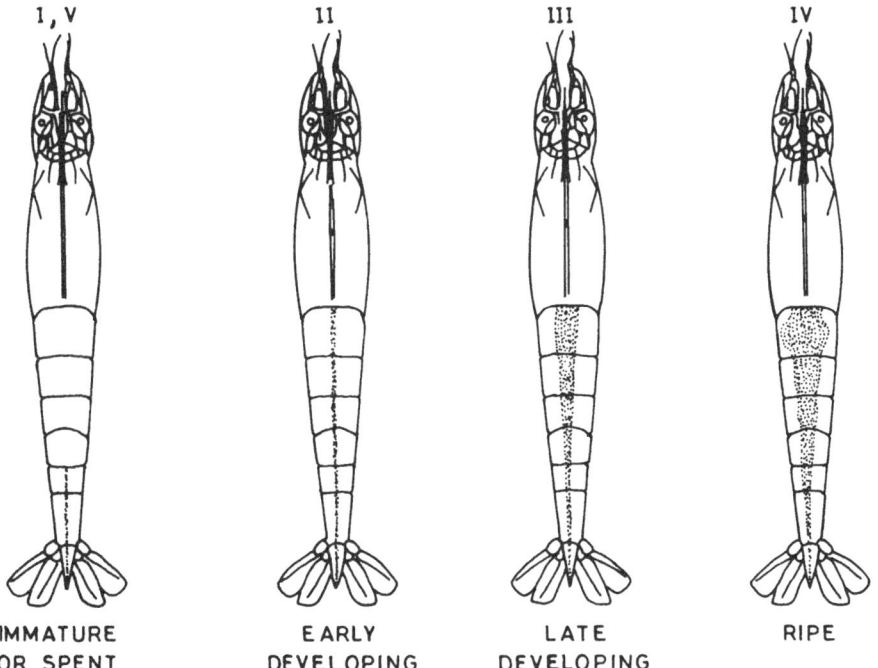

| I , V | II | III | IV |

IMMATURE
OR SPENT

EARLY
DEVELOPING

LATE
DEVELOPING

RIPE

Fig. 12.6.

eyestalks of decapod crustaceans are known to contain an X-organ that secretes moulting-inhibiting hormone (MIH) and gonad-inhibiting hormone (GIH) (Waterman, 1960; Adiyodi and Adiyodi, 1970), eyestalk ablation of females was found to be an appropriate procedure undertaken to obliterate the hormonal influence and enhance vitellogenesis. The technique followed at Tungkang Marine Laboratory (TML), Taiwan, which also seems to be the best method since it minimises the trickling out of the haemolymph from the wound (Liao and Chen, 1983), is to unilaterally sever one eyestalk at its base by heated surgical forceps. Though bilaterally ablated prawns show a more rapid development of ovaries, it is usually avoided due to a higher mortality rate of such prawns. Following eyestalk ablation, the females are kept either in indoor tanks or in outdoor concrete ponds without bottom substrate since non-burrowing, non-grooved forms of penaeid prawns, to which *P. monodon* belongs, do not require soft bottom substrates (Liao and Chen, 1983).

At the TML, the water used in the tanks is directly pumped from an underground well 7 to 10 m in depth, situated near the sea shore. During the winter months, the water of indoor tanks is heated by means of automatically controlled electric heaters. The pH of the water varies between 7.5 and 8.2, and the salinity range is between 17 and 31 ppt, the fluctuation being largely dependent on the local rainfall. The ponds are provided with air stones, and the dissolved oxygen concentration is always maintained above the 5 ppm level. Lighting of about 10 per cent of natural daylight is provided at the indoor ponds (Liao and Chen, 1983).

The diet provided to the maturing prawns is selected from the following: trash whole shrimp, live marine worms, different species of mussels or hard clams on half shells, fresh oysters, mashed squid mantles, or commercially available artificial feeds. The prawns are fed twice a day—at 8 a.m. and 5 p.m., the total amount given per day being about 10 to 15 per cent of the total biomass of the prawns.

The prawns gain maturity around two to three weeks after the eyestalk ablation, the average interval between the eyestalk ablation and the first spawning in *P. monodon* being 18 days (Chen, 1979). Others have reported that maturation of spawning may occur within five to eleven days after ablation and a rematuration may take place three to five days after the first spawning (Beard and Wickins, 1980). The development of the ovaries of *P. monodon* may be observed at night by means of a flashlight after scooping out the specimens from the tanks with a fine-meshed scoop net. This allows the inspection of the colour, texture and shape of the ovaries through the transparent exoskeleton. One practical method of measuring the stage of ovarian development is through the noting of its colour, and four clear stages (I to IV) can be recognised (Fig. 12.6). The colour changes thus observed are from a translucent whitish hue to a dark green colour.

Spawning

Maturation can either be induced, as described before, or mature specimens can be collected from the sea and used for spawning. For the latter, the conventional trawl nets provided with tickler chains are used. These are hauled for 15 to 30 minutes so that not too much strain is imposed on the spawners which are then collected in plastic buckets filled with sea water and subjected to disinfection (five spawners in a 20 litre volume of water, provided with aeration during the procedure). The disinfection is achieved through the addition of either formalin (50 ppm, treatment time 15 to 20 minutes) or furnace (3 ppm, treatment time one hour) after which the spawners are thoroughly washed with clean sea water and placed in the spawning tanks (Platon, 1978).

In case of induced maturation of spawners, when the ovarian development is found to reach stage III or IV on examination, these females are put into circular tanks of 500 litres capacity, made up of fibreglass, or any other suitable container, for spawning. One air stone is always provided per spawning tank to ensure adequate aeration for the spawn, and the temperature of the water is maintained at 28° to 30°C.

The spawning always occurs between 9 p.m. and 12 midnight (Chen, 1979). The spawners start their spawning behaviour by circular swimming movements on the sides of the tank wall just beneath the surface of the water, after which the actual spawning lasting about a minute takes place in deeper waters (middle depth) through vigorous movement of their pleopods (Liao and Chen, 1983). Extrusion of a substance by the spawners, evident as a yellow-orange scum on the surface of water or along the side walls, indicates successful spawning. However, some spawners do not show this behaviour when the occurrence of spawning can be ascertained by examining a sample of the water of the spawning tank taken into a transparent container to look for the presence of eggs. The spawner and spawning debris are removed by a coarse, net, leaving the eggs in the tank the following morning. A continuous flow of running water is then kept up in the tank, for washing the eggs, the water so introduced also being simultaneously siphoned off, a strainer being provided at the exit point to prevent the eggs from being washed out. The aeration is kept at a moderate level. Egg count is taken from a sample of water, a spawner weighing 90 to 200 g releasing about 500,000 eggs per spawning on an average, with a maximum recorded number of more than 1,000,000 eggs (Platon, 1978).

After 12 to 15 hours, the eggs hatch out to nauplius larvae. All the eggs do not hatch simultaneously. The hatching rate is calculated by counting the nauplii from a sample of water the following morning, it being 60 to 98 per cent in case of good spawners.

Larval rearing

Sea water supply system and pretreatment of water: In well-equipped systems, the water is obtained directly through offshore subsand wells that provide adequately prefiltered sea water to be introduced into the system. A 3 H.P. submersible pump is used for transporting the water from the beach to the hatchery site, usually situated at a distance from it. This partially clean sea water is pumped into two high capacity (about 120,000 litres) reservoirs for further settling of sediments. The reservoirs are fitted with standpipes that enable the upper water column to flow out by gravity, to the hatchery tanks. The rate of water flow into the tanks is enhanced by the presence of an in-line centrifugal pump. Before entry into hatchery tanks the water is passed through a series of pressurised filters for adequate purification. The first filter in line contains 1 to 4 mm of activated charcoal that filters out the large particulate matter, larger phytoplankton, and certain hydrocarbon toxins. The next pressurised filter in use is a 5.0 μm cartridge one which effectively removes most of the phytoplankton and fine particulate matter. After this, an in-line UV germicidal lamp encased in waterproof quartz filament housing, capable of treating 8.0 gpm, is placed for the treatment of water. The water is then passed through a second 5.0 μm cartridge filter before use (McVey and Fox, 1983).

Rearing in larval culture tank: While still at the N_4 substage, the larvae are generally transferred to the larval culture tanks, that is, conical bottomed tanks of 2 t capacity (Figs. 12.7 and 12.8), as this is found to be more convenient and less stressful for the larvae. The nauplii are collected by stopping the aeration in the tanks which allow them to congregate at the surface, when they are scooped out with pails and transferred to the larval culture tanks filled with 1 t of clean sea water prior to stocking. The stocking density ranges from 50,000 to 100,000 nauplii/t. Higher stocking densities may be permissible, but this requires a more rigorous water management technique and more intensive feeding. The volume of water in the tanks is increased by the daily addition of about 200 litres of clean sea water. The water management of larval tanks is dependent on factors like the density of diatoms, amount of particulate matter, metabolite concentrations, presence of infections, etc. A change of water as and when required is always done by using a strainer to prevent the escape of the cultured larvae.

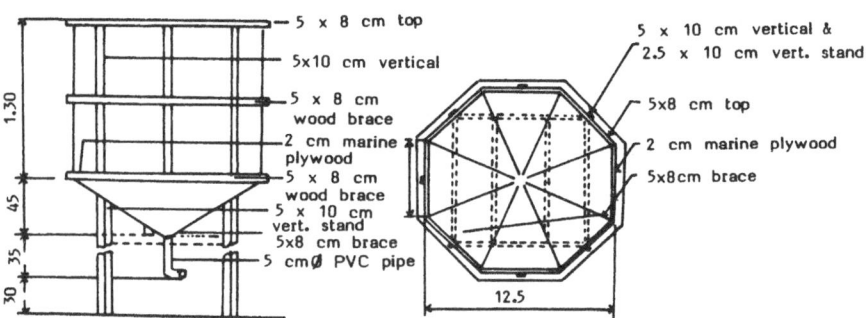

Fig. 12.7. Specification of a 2 t conical-bottom wooden larval rearing tank

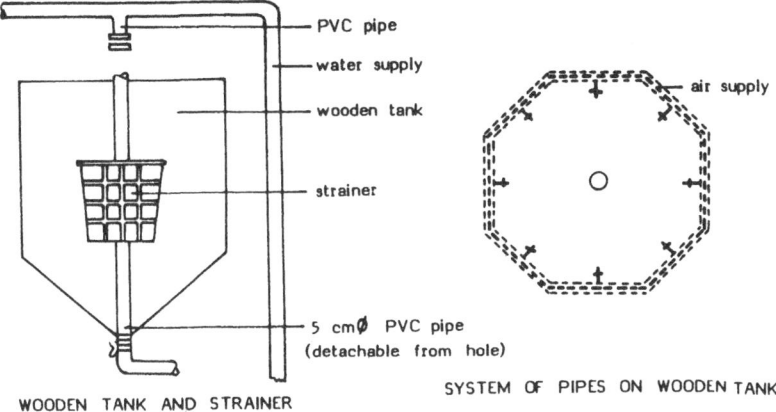

WOODEN TANK AND STRAINER

SYSTEM OF PIPES ON WOODEN TANK

Fig. 12.8. Details of the larval rearing tank showing the strainer and the supply lines for water and air

The nauplii metamorphose into zoea larvae about two days after hatching. From this point, the larvae start feeding. The feeding schedule should be monitored such that suitable food is available just prior to the metamorphosis of the nauplii into zoeae. Suggested feeding levels are shown in Table 12.1. The optimum water temperature to be maintained throughout the culture period is about 28°C. Lower temperatures delay moulting (Platon, 1978).

The zoeae metamorphose into the mysis stage in about five or six days. From this period onward half to two-thirds the volume of water in the tanks requires changing daily. The larvae pass through three mysis substages in four

Table 12.1
Feeding schedule for *P. monodon* larvae

Culture Period (days)	0	1	2	3	4	5	6	7	8	9	10	11	12	13	14	15	16	17	18

Larval Nauplius stage	Zoea			Mysis			Post larva		
	Z_1	Z_2	Z_3	M_1	M_2	M_3	PL_1	to	PL_5
Chaetocerous cells/mL	30,000	50,000	80,000	80,000	80,000	80,000			
Tetraselmis, cells/mL		5,000	10,000	20,000	20,000	20,000	5,000 to		10,000
Rotifer (*Brachionus*)/mL				5	8	10			
Brine shrimp/mL								5	

to five days, after which they become post-larvae. The main food consumed by the *P. monodon* post larvae of early stages (PL$_1$–PL$_5$, one day old post-larvae being designated as PL$_1$, two days old as PL$_2$ and so on) are newly hatched brine shrimps or finely ground fish meat. Algae are also given to improve the quality of water. Post-larvae of the stage PL$_5$ can be harvested for stocking into previously prepared nursery ponds (Platon, 1978).

Harvesting technique of post-larvae

At first about three-fourths the volume of water in the larval rearing tanks is drained out, a strainer being used at the outlet to prevent the fry from escaping. The total tank content is then slowly emptied into a 150 litres harvesting box by opening the drain valve. About one-fourth of the walls of the upper portion of this box is made up of a plastic screen that permits the overflow of water while retaining the fry inside it.

Pretreatment and transport of post-larvae

Prior to their transport for stocking into grow-out ponds, the post-larvae need certain pretreatment to minimise the stress involved in the procedure. For this, the fry are kept for 30 to 60 minutes at a water temperature 5°C less than the prevailing temperature in the larval rearing tanks. This is effected by introducing previously frozen sea water into the tank. After this, the temperature is brought down further, to about 20°C, when the fry are packed in 50 × 90 cm plastic bags containing 16 litres of chilled sea water, about 20,000 fry packed per bag. The bags are then pumped with oxygen, their mouths secured with rubber bands and placed in styrofoam boxes for transportation to the ponds for stocking (Platon, 1978).

The necessary data for designing a prawn hatchery system are given in Tables 12.2 to 12.5. The layout plan of a small-scale hatchery system is shown in Fig. 12.9.

Growing to marketable size

Since the production per hectare in Taiwan is very high, the development of the grow-out pond system, where the shrimp fry are stocked and grow to marketable size, in Taiwan is described here.

Initially irregularly shaped mud dike ponds (Fig. 12.10 a, i) were used for the polyculture of prawns and fish. These ponds often had uneven bottoms containing tree roots, rocks and other obstacles, and the production of prawns was dependent on environmental factors that could not be controlled.

Figure 12.10 (a, ii) shows slightly improved polyculture grow-out ponds for prawns and milk-fish that were in use before the 1960s. These ponds had surface areas of 1 to 3 ha with usually earthen dikes, though often with their inner walls lined partially with bricks or cement. The same water-gate was utilised for both the inlet and outlet of water, the supply and drainage of which was entirely dependent on the tides. Actually during this period, the prawns were considered to be only a byproduct of the milk fish ponds, giving

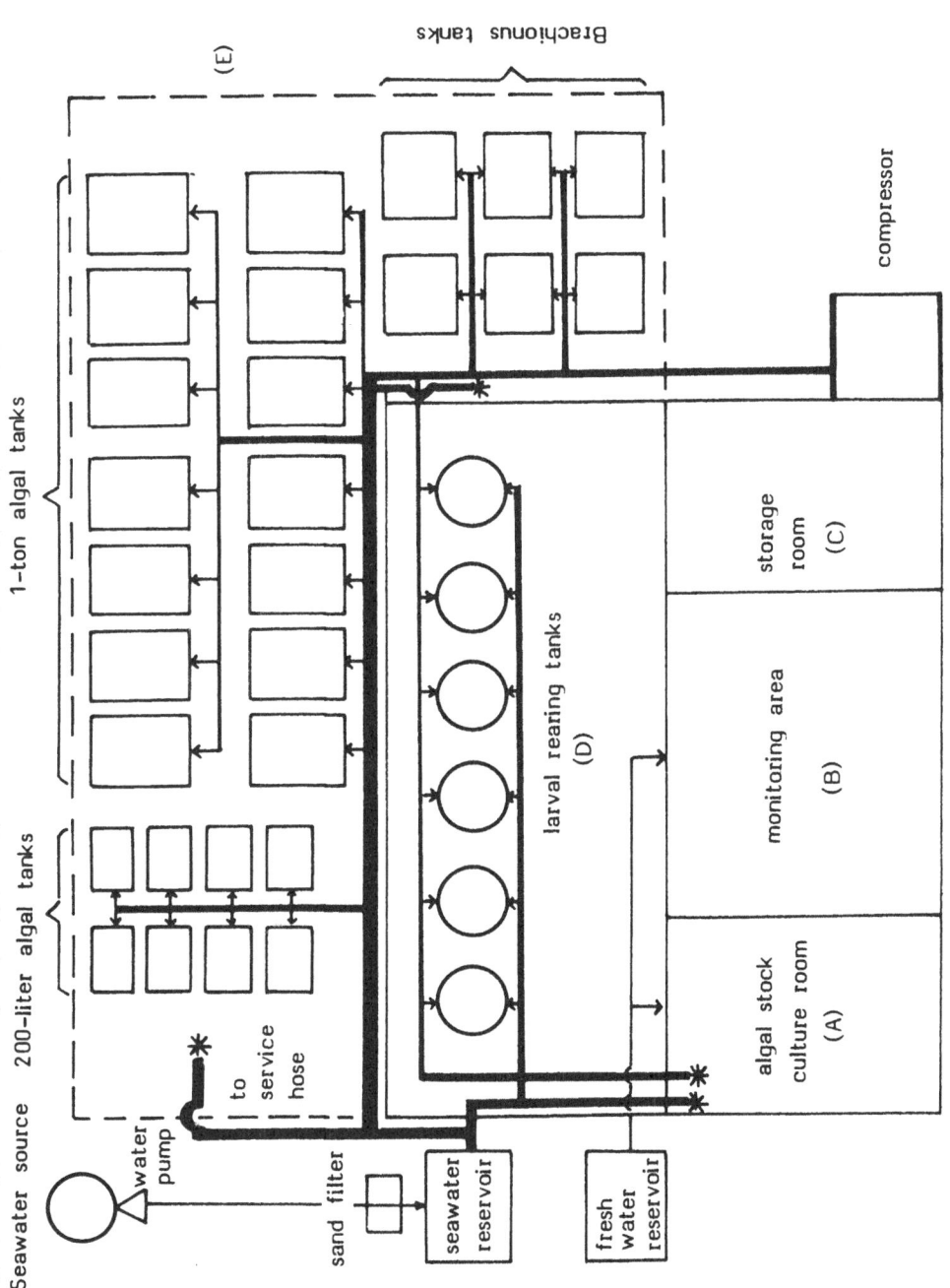

Fig. 12.9. Layout of a small-scale hatchery system

Table 12.2

Sea water and pump requirements for different hatchery capacities

Hatchery capacity	Maximum total consumption (L/day)	Recommended reservoir volume (L)	Total water consumption per run (including 15% of actual for washing), (L)	Theoretical pump hp requirement with 6 m head (kW)	Recommended pump hp (kW)
1 tank	3,000	1,500	36,900	0.025 (0.018)	0.1 (0.0746)
5 tanks	12,000	6,000	149,000	0.075 (0.056)	0.3 (0.224)
10 tanks	25,000	12,500	296,000	0.150 (0.112)	0.6 (0.448)
15 tanks	37,000	18,500	433,000	0.200 (0.149)	0.8 (0.597)

Table 12.3

Compressed air requirements for different hatchery capacities

Hatchery capacity	Theoretical hp requirement (kW)	Recommended actual hp (kW) at 0.2 kg/cm^2	Total air volumetric requirement, (L/min)
1 tank	0.16 (0.12)	0.64 (0.48)	430
5 tanks	0.305 (0.228)	1.22 (0.91)	820
10 tanks	0.533 (0.398)	2.13 (1.59)	1430
15 tanks	0.71 (0.53)	2.84 (2.12)	1910

Table 12.4

Algal and *Brachionus* tank requirements for different hatchery capacities

Hatchery capacity	*Chaetoceros* culture tanks (1-ton)	*Tetraseimis* culture tanks (1-ton)	*Brachionus* culture tanks (1-ton)	*Chlorella* culture tanks (1-ton)	200-L algal tanks
1 tank	3	—	1	2	5
5 tanks	6	3	3	2	5
10 tanks	9	6	6	2	7
15 tanks	12	6	8	8	7

Table 12.5
Area requirements of a hatchery installation for different capacities, in square metres

Hatchery capacity	A	B	C	D	E	F	Total
1 tank	6	8	4	10	45	7	80
5 tanks	6	12	6	30	60	16	180
10 tanks	9	12	9	50	90	23	183
15 tanks	9	12	12	70	120	25	248

A = Algai culture room, inside building
B = Monitoring area, inside building
C = Storage compartment, inside building
D = Larval rearing area, inside building
E = Outdoor algai and Brachionus production area
F = Combined area for compressor, reservoir, and sand filter

a production of only a few hundred kilograms per hectare (Huang, 1969; Liao and Huang, 1973; Liao, 1981). Prawn culture on a large scale only became possible with the development of larval rearing techniques around 1968 (Liao, 1981). In the ensuing years, the depth of water in the milk fish ponds was increased from 30–40 cm to 50–70 cm to allow extensive prawn culture and supplemental feed of the locally available trash fish, snails and soyabean cakes were given to augment the production. The stocking density employed was about 3 to 3.5 post-larvae/m^2 which gave a satisfactory harvest of 0.5 to 1.0 t year (Huang, 1969; Liao and Huang, 1973). Though the pond systems described above are no longer used in Taiwan, similar systems are still in existence in different southeast Asian countries.

Since the early 1970s, grow-out ponds specially designed for prawns began to be constructed. The ponds of the late 1970s were of the area of 0.1 to 0.5 ha with aeration equipment and well-developed water distribution and central drainage systems (Fig. 12.10b). Stocking densities as high as 30 to 40 post-larvae/m^2 with survival rates of 80 per cent, and harvests of 8.4 to 11.2 t/ha/crop have been achieved from this type of ponds (Liao and Chao, 1983).

Another popular type of grow-out pond (Fig. 12.10c) has been designed in 1981. Designed for segmental culturing, each unit consists of a small pond for raising fry of PL_{10-20} to the juveniles of 2 g size, a medium sized pond for growing juveniles of 2 g to about 5 g, and a large pond for growing prawns from 5 g to the marketable size. The water current is utilised for conveniently transferring the prawns from one pond to the subsequent one as they grow. At the same time, the pond bottoms can be maintained in better condition since the culture period in each pond is considerably shortened (Liao, 1986).

Pumping, water distribution and drainage systems
The water pumping and distribution/drainage systems of the prawn ponds

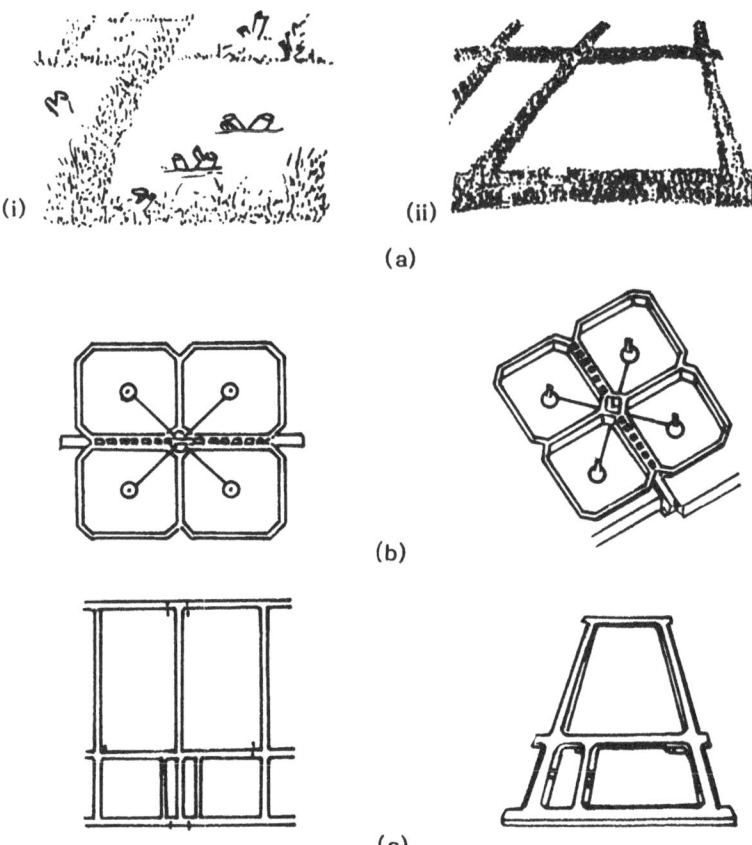

Fig. 12.10. Prawn grow-out ponds with mud dikes, central drainage and segmental culture system. (a) Two kinds of mud dike pond: (i) Simple and rough traditional pond, (ii) improved pond with water-gate; (b) central drainage pond, (c) segmental culture pond

are vital accessories to the culture system. Four major methods are used in Taiwan for the supply of water in the prawn farms: (i) for ponds constructed near rivers or estuaries, the water supply is effected by the high and low tides, supplemented with pumping, (ii) water supply through pipes extended into the coastal water and directly pumping the sea water without any filtration device to prevent the ingress of unwanted organisms (Fig. 12.11a), (iii) pipes buried into the sand of the beach supply the sea water, which is thus naturally filtered, into the ponds (Fig. 12.11, b). Sometimes a sea water well may also be dug on the beach for the same purpose, and the naturally filtered sea water so collected pumped to the ponds, and (iv) before being used for prawn culture, the underground water which is pumped up is oxygenated through the use of a water spilling ladder to expose the water with depleted oxygen level to the air (Fig. 12.11 c).

When the tidal difference in water levels is utilised for the filling and drainage of ponds, only one of the water gates is generally used (Fig. 12.12). To augment the flow of water, the design of the filter nets have been improved to increase the total area of the filter surface. The inlet and outlet water gates are also constructed at vantage points whereby maximum supply and drainage can be possible. However, a more effective and at the same time more flexible water system can be designed when pumps are used for the supply of water. For example, in the central drainage systems, a side gate is provided in one corner in addition to the central water-gate, by which complete drainage is possible at the time of final harvestation and sludge can also be removed.

Aeration device

There are several aeration devices in use, some of which are illustrated in Fig. 12.13. Among these, the most economical as well as popular is the paddle wheel. The use of such aeration devices can considerably increase the production by raising the dissolved oxygen level of ponds. Thus, while in the

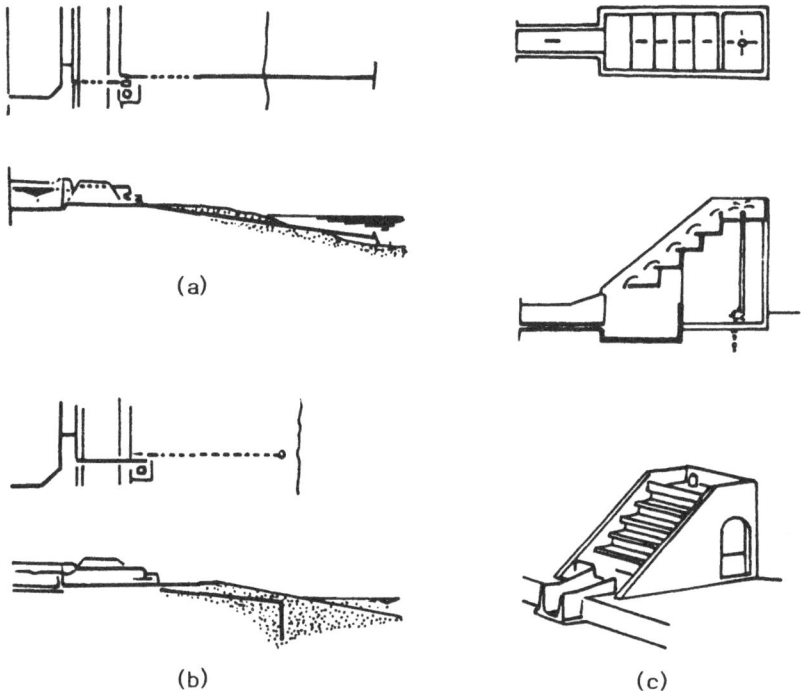

(a)

(b)   (c)

Fig. 12.11. Three systems for pumping sea water. (a) direct pumping from coastal water; (b) pumping from the coastal sand layers; (c) pumping from underground and exposing the water to air by a ladder raceway

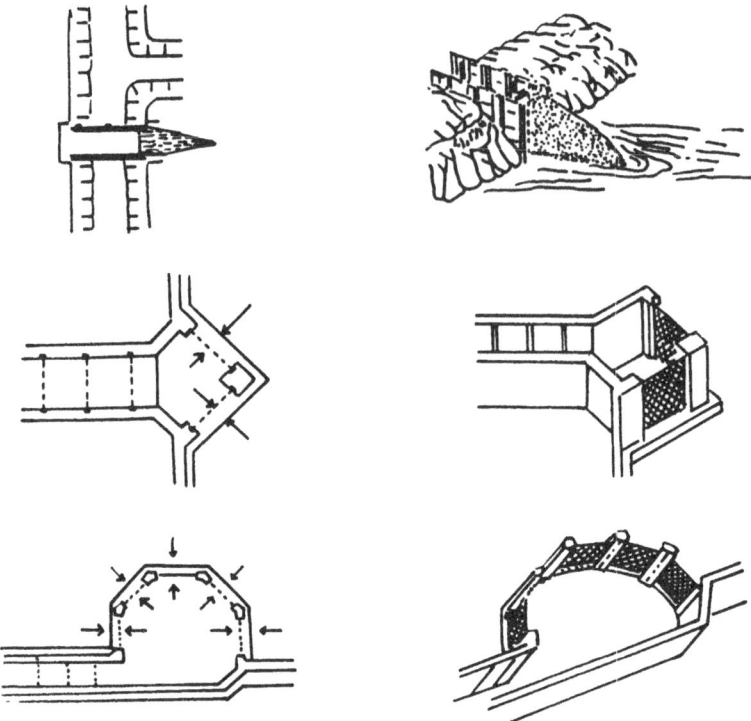

Fig. 12.12. Illustration of various water-gates for inlet and outlet

non-aerated ponds, the stocking density of prawns is between 30,000 and 50,000 per hectare, a modern well-designed and well-aerated pond can be stocked with 300,000 to 700,000 prawns per hectare.

12.5.3 CULTURE OF SEA WEED

12.5.3.7 Nori culture
Nori, as cultured in Japan is described here.

Preparation of nursery nets
Germlings of *nori* are first germinated from the spores released by the *Conchocelis* filaments. These germlings are planted on a twine of nets. When covered with 2 to 3 cm long germlings, these nets, called nursery nets, are set horizontally on the sea surface for the growth of fronds from which dried *nori* is prepared.

Growing of nori fronds
The culture period of *nori* fronds in Japan extends from October to April. Seas with a high nutrient salt content and strong water movement provide the most suitable cultivation grounds for *nori*. As long as the culture equipments

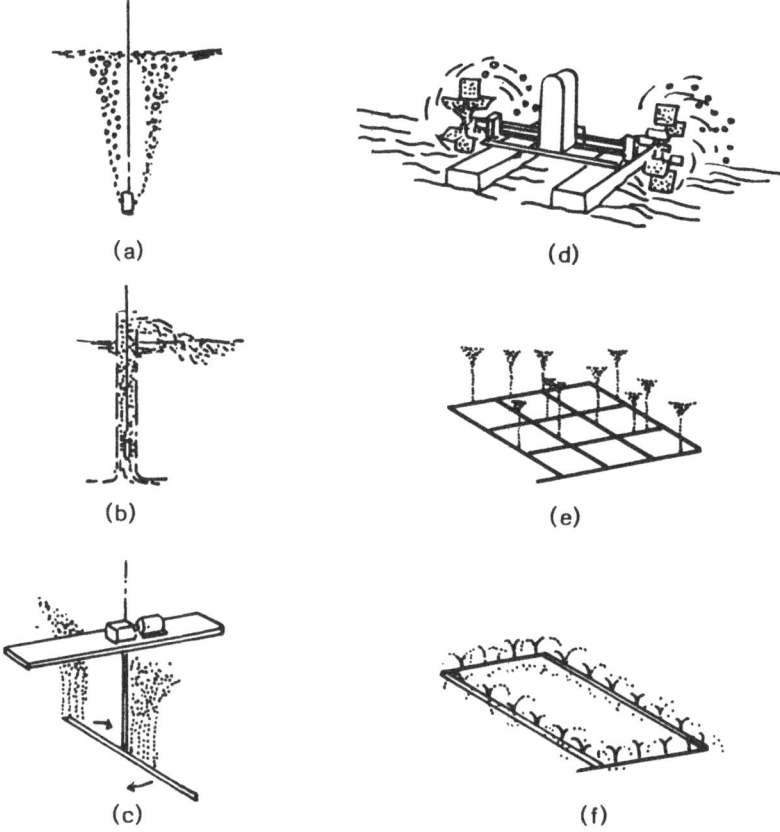

Fig. 12.13. Various aeration devices. (a) Aeration with air stone; (b) air lift; (c) rotating aeration with reduction motor; (d) paddle wheel; (c) pond bottom pipe supplied with air; (f) water surface pipe with tiny fountains.

are not destroyed, strong wind and wave actions are most desirable since they assure rapid water exchange around the growing fronds.

As mentioned before, the nursery nets are spread horizontally on the surface of the sea for the growth of fronds. Three methods described as follows, are used for keeping the nets in position (Fig. 12.14).

A) Fixed type with poles

In this method, the nursery net is secured tightly to bamboo poles planted in two rows in the shallow seas to a depth of about 10 m. The level of the net is kept within the tidal range, and is exposed to the air once or twice daily according to the prevailing tide conditions. This level is adjusted from time to time so that the growth of the fronds is as rapid as possible.

B) Movable type with poles

This method is preferred to the fixed type method in the cultivation

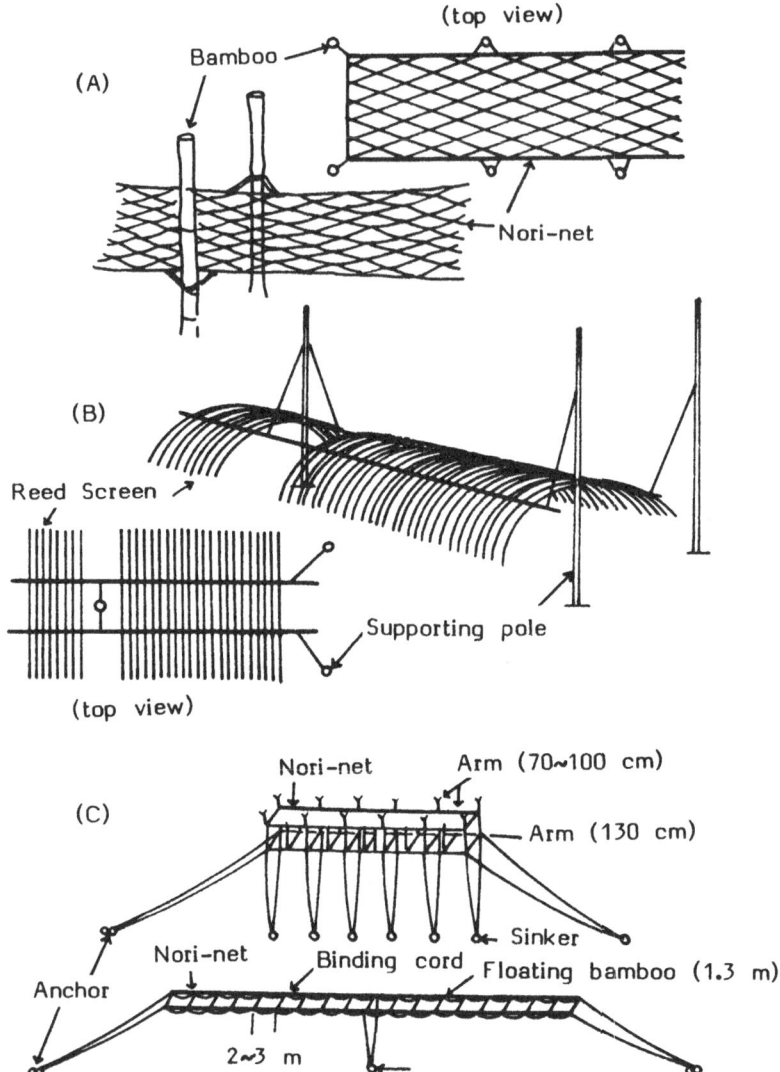

Fig. 12.14. Culture methods of *nori*. A, Fixed type with poles; B, Movable type with poles; C, Floating net culture

grounds which have a large tidal range. Here the nursery net is provided at the upper surface with cylindrical floats made up of bamboo poles or synthetic resin pipes that are fastened with binding cords to the poles. The cords are kept at such a length as to allow an up and down movement of the net according to the tide. The net is also exposed to the air for a fixed period once or twice daily, by adjusting the lengths of the binding cords.

C) Floating net culture

This method is employed in the seas where the depth is more than 10 m. The nursery net is tied to a floating rectangular frame fixed by anchors. Synthetic resin pipes in the form of cylindrical floats are tied at intervals to the net to provide the necessary buoyancy to it. In this method of *nori* frond culture, the nursery net is never exposed to the air.

Harvesting

Harvestation starts in November, 50 to 60 days after the onset of culture, when the fronds grow up to 15 to 60 cm in length. Harvestation is done from a small boat either by hand picking, or by a mechanical device resembling a vacuum cleaner. After thinning the dense growth, the smaller fronds left on the nets again start growing, and thus from the same net *nori* is harvested about five times, after which the net is replaced by a new one kept ready in low temperature storage (Kafuku and Ikenoue, 1983).

REFERENCES

Adiyodi, K.G. and Adiyodi, R.G. (1970): Endocrine control of reproduction in decapod crustacea, *Biol. Rev.* 45: 121.

Bardach, J.E., Ryther, J.H. and McLarney, W.O. (1972): Aquaculture. The farming and husbandry of freshwater and marine organisms. John Wiley and Sons.

Beard, T.W. and Wickins, J.F. (1980): Breeding of *Penaeus monodon* Fabricius in laboratory recirculation systems. *Aquaculture* 20: 79.

Chen, C.A. (1979): Preliminary report on the gonadal development and induced breeding of *Penaeus monodon* Fabricius, M.S. thesis, Institute of Oceanography, National Taiwan University, Taipei, Taiwan, R.O.C., 43.

Chen, J.P. (1976): *Aquaculture practices in Taiwan*. Fishing News Books Ltd., Surrey, England.

Chen, H.C. and Yang H.H. (1979): Artificial propagation of the abalone *Haliotis diversicolor supertext. China Fisheries Monthly* (in Chinese), 314: 3–9.

Chiang, P. and Liao, I.C. (1985): The practice of grass prawn (*Penaeus monodon*) culture in Taiwan. Presented at the 16th Annual Meeting of the World Mariculture Society, Orlando, Florida, USA, 13–17 January.

Hu, S.H. and H.H. Sheu, H.H. (1982): Studies on the increasing production of cultured oysters *C. gigas* in Penghu-II. Growth of oysters combined two seeding and of rotating culture. *Ann Coll. Rep. Penghu Fish. Lab.* TFRI, (in Chinese), 2: 27–38.

Hu, S.H. (1986): Introduction to the design of coastal aquaculture systems in Taiwan. *Aquacultural Engineering* 5: 235–252.

Huang, R.L. (1969): Prawn culture. *Bull. Fish. Assoc. Taiwan* 1: 54–60.

Kafuku, T. and Ikenoue, H. (1983): *Modern Methods of Aquaculture in Japan*, Developments in Aquaculture and Fisheries Science, 11. Elsevier Scientific Publishing Company.

Kuo, H. (1964): Economic molluscs of Taiwan, JCRR Special Publication no. 38 (in Chinese).

Liao, I.C. (1986): General introduction to the prawn pond system in Taiwan. *Aquacultural Engineering* 5: 219–233.

Liao, I.C. (1981): Status and problems of grass prawn culture in Taiwan. Presented at ROC—Japan Symposium on Mariculture, Taipei, Taiwan, ROC, 13–24 Dec.

Liao, I.C. and Huang, T.L. (1973): Experiments on the propagation and culture of prawns in Taiwan. In: *Coastal Aquaculture in the Indo-Pacific Region*, T.V.R. Pillay, ed., Fishing News Books Ltd., London, pp. 238–354.

Liao, I.C. and Chao, N.H. (1983): Development of prawn culture and its related studies in Taiwan. Presented at First International Biennial Conference on Warm Water Aquaculture—Crustacea, Brighan Young University, Hawaii Campus, Laie, Hawaii, USA, 9–11 Feb.

Liao, I.C. and Chen, Y.P. (1983): Maturation and spawning of penaeic prawns in Tung Kang Marine Laboratory, Taiwan. In: *Handbook of Mariculture* Vol. I, Crustacean Aquaculture, J.P. McVey, ed., CRC Press Inc., Boca Raton, Florida.

Liao, W.T. (1979): Culture of small abalone. *Harvest Farm Magazine* (in Chinese) 29(4): 30–1.

Lin, Y.S. and Tang, H.C. (1980): Biological studies on cultured oyster in Penghu. *Bull. Inst. Zool. Academic Sinica*, 10(2): 15–22.

McVey, J.P. and Fox, J.M. (1983): Hatchery techniques for penaeid shrimp utilized by Texas A and M—NMFS Galveston Laboratory, Program. In: *Handbook of Mariculture*, Vol. I, Crustacean Aquaculture, J.P. McVey, ed., CRC Press, Boca Raton, Florida.

Platon, R.R. (1978): Design, operation and economics of a small-scale hatchery for the larval rearing of sugpo, *Penaeus monodon* Feb. SEAFDEC, Aquaculture Extension Manual No. 1, Philippines.

Tsai, W.S. and Hu, S.H. (1982): Cultivation of groupers of 'Intertidal underground pond'—Introduction of a new culture method. *Ann. Coll. Rep. Penghu Fish. Lab*. TFRI, 2: (in Chinese) 2: 103–7.

Tsao, H.C. (1984): Oyster culture of submersible raft method. *Harvest Farm Magazine* (in Chinese) 34(4): 16–17.

Tzeng, W.N. (1977): Studies on the growth and rearing environment of the abalone *Haliotis diversicolor supertext*. *China Fisheries Monthly*, (in Chinese) 292: 2–7.

Waterman, T.H. (1960): *The Physiology of Crustacea*, Vol. I. Academic Press, New York, p. 437.

Wheaton, F.W. (1977): *Aquacultural Engineering*. John Wiley and Sons.

CHAPTER 13.

Post Harvest Technology

13.1 Microbiology of fish

Spoilage of fish is primarily due to the action of bacteria on dead fish although enzymes present in fish tissue may play an important role in certain types of spoilage as in the development of a rancid flavour in oily fish. Flesh and body fluids of newly caught healthy fish are generally found to be bacteriologically sterile (Shewan, 1949). However many of the surveys carried out on the bacterial flora on fish indicated that the bacteria responsible for spoilage of fish are normally found on fresh fish (Reay and Shewan, 1949). Such observations led to intensive study of the microflora of fish with a view to controlling the activity of microbes which are mainly responsible for the spoilage of fish. The commensal bacterial flora on fish are found on the skin, the gills, integuments and within the alimentary tract of healthy fish. These bacteria then spread to different parts of the body, aided by damages caused to the fish through improper handling and favourable temperature of the environment. Fish from warm waters carry a larger number of bacteria than those from cold water and the bacterial flora of the former has a larger population of mesophilic bacteria (Shewan, 1977). This fact is also evident from the bacterial count on shrimp as shown on Table 13.1 which is caught from both warm and cold water, although the counts may be affected by the contamination of mud in the case of bottomdwelling organism like shrimp.

Table 13.1
Bacterial content of shrimp from different locations

	Number per gramme
Shrimp (cold water)	$10-10^3$
Shrimp (tropical)	1000-above 10^6
Shrimp (Pond culture)	$1500-16^6$

Source: Liston J., 1980.

The bacterial flora on freshly caught fish depend primarily on the environment from which it is captured. Shewan (1977) observed that micro-flora on the skin of all the nine cases of cold water fish comprised over 80 per cent gram-negative rod bacteria and had a preponderance of bacteria belonging to generic groups of *Moraxella, Acinetobacter, Pseudomonas, Flavobacterium* and *Vibrio* (Liston 1980). Fish from warm water marine fish carry gram-positive bacteria mostly micrococci, coryneforms and bacillus (Table 13.2).

It has been observed that fish living in unpolluted water do not carry the kind of bacteria commonly found in human intestines. However, the fish caught from tropical waters have shown the presence of salmonella and coliform. *Vibrio parahaemolyticus* was detected on fish from the marine waters of tropical regions.

The method of capture of fish influences the bacterial counts of fish harvested from the same water. It has been demonstrated that trawled fish carried 10 to 1,000 times more bacteria than fish caught with a baited line (Horseley, 1973). The high bacterial count for trawled fish may be due to contamination from sediment materials when the catch is dragged over the sea floor or due to the excretion of intestinal fluids as a result of pressure caused by crowding in the cod end of the trawl net. Surface counts ranged from 10^2 to 10^5 organisms cm^2 on skin with adhering slime, 10^3 to 10^7 on gills and 10 to 10^8 per gramme of intestinal fluid. A higher count was found in the intestinal fluid of feeding fish. The bacterial flora of elasmobranch differs from that of

Table 13.2
Bacteria of various generic groups in fish and shell fish

Fish type	Percentage of total flora				
	Pseudomonas	Vibrio	Achromobacter	Coryneform	Others
Shrimp (North Pacific)	10	—	47	3	40
Shrimp (Gulf ocean)	22	2	15	40	21
Shrimp (Pond culture)	—	5	2	83	13
Scampi (U.K.)	3	—	11	81	5
Haddock (North Atlantic)	26	2	45	4	23
Flat fish (Japan)	20	13	30	17	20
Pescada (Brazil)	32	—	35	4	29
Mullet (Autralia)	18	—	9	12	61

Source: Liston J., 1980.

other fish from the same waters. Venkataraman and Sreenivasan (1952) noted that *Pseudomonas* spp, was absent on them though these bacteria were numerous in the samples of water of the region where from they were captured.

13.1.1 THE MECHANISM OF BACTERIAL SPOILAGE

It has been conclusively demonstrated that fish muscle which has been aseptically obtained and is held in a sterile atmosphere does not develop either a flavour or an odour associated with spoilage (Shewan, 1971). Autolytic enzymes as such do not have any significant role in spoilage except perhaps in case of a belly-bursting phenomenon observed with fish that are heavy eaters such as oil sardines and herring, in certain seasons.

The growth curve of bacteria on fish as indicated on Fig. 13.1 shows the classical pattern for bacteria. The rate of growth of bacteria is found to be a function of storage temperature as well as of the species of fish involved. The lag period of the growth curve coincides with the development of *rigor mortis* and may be due to the lowering of pH which inhibits bacterial growth (Shewan, 1977). Growth of bacteria follows the resolution of *rigor mortis*. It appears that only a few strains of bacteria found on fish are capable of growth below pH 6. Optimum growth occurred between pH 6.5 and 7.5.

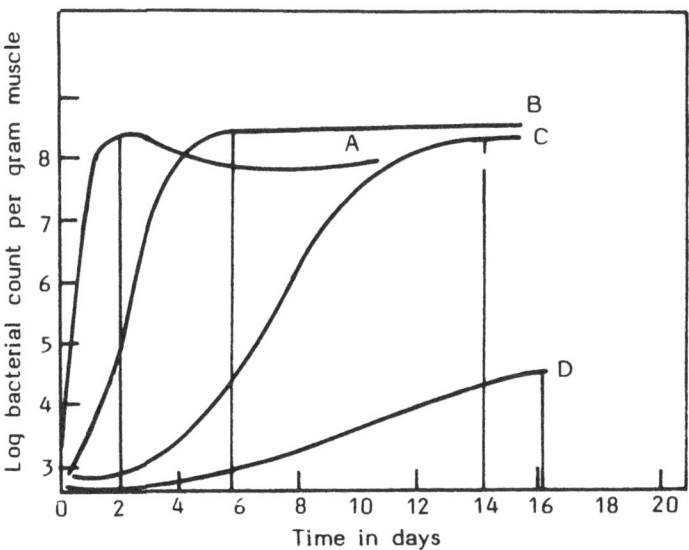

Fig. 13.1. The effect of temperature on bacterial growth: $A = +25°C$, $B = +7°C$, $C = 0°C$, $D = -4°C$. (Shewan, J.M. Fish as Food, Vol. 1, pp. 487–560, ed. Georg Borgstrom)

Temperature is primarily responsible in determining the shelf-life of fresh fish. The minimum temperature of growth of the species of micro-organism present on fresh fish is an important factor for bacterial spoilage of fish.

Laycock and Regier (1970) found that there is a shift in bacterial types during the holding period at 3°C. *Pseudomonas* rapidly assumes a dominant position in micro-flora. *Achromobacteria* and *Flavobactrium* also persist but on a decreasingly relative level. A number of types of bacteria that produce spoilage odour have been identified by growing a pure culture of the specified bacteria on sterile fish muscle or press juice and their mode of action being studied. Liston (1980) has summarised the general bacteriology of spoilage of fish in most cases in sequence as follows:

1) Spoilage bacteria are naturally present on fish.

2) Amino acids and other non-protein nitrogen substrate pools are also present.

3) Selective growth of the organism (mostly *Pseudomonas*) which actively and oxidatively deaminates amino acids.

4) Repression of proteinase production depressed by selective use of amino acids by *Pseudomonas* bacteria.

5) Amino acid recruitment to the substrate pool by bacterial hydrolysis of protein.

6) Ammonia and volatile fatty acid production sharply increases due to (5)

7) Specific 'spoiler' types of bacteria produce sulphur containing and other odorous compounds.

13.1.2 MICROBIOLOGICAL CONSIDERATION IN PRESERVATION OF FISH

Microbiological activity being the most important cause of spoilage of fish, microbiology of fish processed for preservation and the control microbes for prevention of spoilage have been the subject of intensive study in fishery science and technology.

13.1.2.1 Wet fish handling before processing

Handling of wet fish before it reaches the processing plants and butchering and filleting of fish affect the micro-flora on fish both quantitatively and qualitatively. Although bacteria on fish are confined to skin, gill and intestinal tract, removal of them should result in a significant lowering of bacterial count on fish. But in actual practice the fish flesh gets contaminated not only with bacteria present on fresh fish but also with those from the extraneous environment. Fillets prepared under good factory conditions and washed with chlorinated water have been found to have a count of 10^3 to 10^5 per gramme of fillet flesh. The major change from primary processing is an increase in relative proportion of gram-positive bacteria and the appearance of bacteria associated with human beings including *Staphylococci* and entric bacteria (Liston, 1980).

The beneficial effect of beheading shrimp, on the microbial quality, has been reported by many workers. However, Alvarez and Koburger (1978)

noted that microbial flora of shrimp stored whole or headless showed no significant difference. Generally microbial count increased after peeling and sorting and decreased after cooking, washing or brining. The proportion of gram-positive bacteria increased following each step.

Due to the rising cost of fish, recovery of more meat from fish in the form of minced meat will be increasingly resorted to. Although expected to be more contaminated with microbes and of poorer quality than the raw fish from which the mince is obtained, the Japanese fish sausage industry has demonstrated that with proper sanitary precautions very good quality products both microbiologically and organoleptically may be obtained. However, Raccach and Baker (1978) study shows that while bacterial contamination of minced fish may increase by one log scale, the products spoiled rapidly when the frozen material was thawed and held at 3°C.

13.1.2.2 Freezing

Freezing does not sterilise in the bacteriological sense, but the processing by freezing and subsequent holding in cold storage at below freezing temperature causes significant reduction in the bacterial population. However, the behaviour of different types of bacteria during freezing is very erratic and not easily predictable. Gram-negative bacteria are more sensitive to freezing than gram-positive bacteria and bacterial spores. Raj and Liston (1961) found that *Salmonella* and other members of entrobacteriocea are amongst the more sensitive ones. Cold sensitive bacteria such as *V. Parahaemolyticus* on depurated oysters may survive after freezing and also during subsequent storage at −30°C (Liston, 1980). Quick freezing and prevention of fluctuation of cold storage temperatures which is necessary for protecting the quality of frozen food helps in survival of bacteria. Hence it will be correct to state that freezing only maintains the status quo as regards the bacterial quality of fish on freezing.

Studies on tropical prawns iced immediately after harvesting showed that there was steady increase in bacterial count in frozen storage. With uniced prawn there was an initial decrease followed by an increase in count (Lekshmy *et al.*, 1962 and Bose, 1969).

The microbiology of frozen prepared seafoods has been given great importance due to increasing consumer demand for such products. As many of these products require incorporation of starch from different sources, flavouring agents and breading, the presence of extraneous bacteria particularly those of public health significance have received particular attention. Possible introduction of *Salmonella*, through egg, often used as a breading component is a matter particular concern. Although Raj (1968) detected the presence of *Staphylococci* in most samples, though not of *Salmonella*, the general bacteriological quality of frozen prepared seafoods has shown considerable improvement over the years.

13.1.2.3 Canned products

As canned food products depend, for their quality (from health point of view) on the prevention of growth of pathogenic and food-spoilage bacteria, they have to be sterile with respect to such micro-organisms. The fact that food poisoning traced to canned fish has been rare, speaks of a better understanding of the factors which control the heat sensitivity of the target bacteria. Liston (1980) has reviewed the microbiology of canned sea food. The problem of bacterial spoilage of canned sea foods arises from the presence of an excessive number of bacterial spores in unprocessed material which cannot be destroyed by heating process based on the assumption of a lower number. This calls for strict vigilence on raw fish handling and sanitation. The other important cause of spoilage of canned products is the leakage of cooling water into the cans due to improper seaming or contaminated cooling water. Swollen or 'blown' cans often contain *Clostridia*, such as *C. sporogens*, which survive due to inadequate processing or gain entry through leakage. Thermophiles such as *Bacillus stearothermophyllus* which escape destruction and grow during slow cooling or in storage at suitable higher temperatures may cause flat-sour spoilage of canned seafoods.

The most feared of faulty processing of canned fish is botulism caused by *Clostridium botulinum*. Fortunately there have only been a very few instances of such incidences traced to commercially packed canned fish although botulinum poisoning has been reported to originate from some domestically canned fish and fish products.

13.1.2.4 Salted (cured) and dried fish

Preservation by salting or drying is still widely practised throughout the world particularly in maritime states in Asia, Africa and Latin America and provides a relatively cheap protein food to poorer sections of populations. The preservative effect of this process depends primarily on the reduction of water activity to a level, below 0.90, where growth of microbes is arrested. Higher concentrations of salt (NaCl) may however be lethal to many microbes. The micro-flora on dried fish is dominated by micrococci and gram-positive rods (Liston and Shewan, 1958). Presence of *Salmonella* has been detected on samples of dried fish particularly on the fish and shell dried on sea beaches in tropical countries.

Salt-tolerant bacteria and moulds may grow on salted fish. They grow when the water activity is still above 0.75 and infect locally dampened portions of otherwise dried fish. They cause pink or red discolouration and decay of fish.

13.2 Handling and transportation of fresh fish and fishery products

One of the most easily spoiled foods, fresh fish requires careful handling and transport to reach the consumer in a sound state. Temperature is by far the most important factor in determining the rate of spoilage of fresh fish. The spoilage at 5°C will be twice as fast as at 0°C. The essential requirements for maintaining unfrozen fish in a sound condition are (i) temperature as close to but above the freezing point of fish flesh to reduce to a minimum the bacterial and enzymic changes. For practical purposes this temperature is 0°C though the limit is slightly lower, (ii) an atmosphere saturated with water vapour immediately surrounding the products to prevent desiccation, and (iii) an environment free from agents that may aid the development of an off-flavour in fish by oxidation (formation of rancidity in fish lipids) or by absorption of extraneous smell.

13.2.1 USE OF ICE

The most satisfactory and effective method of keeping fish cool and thereby reducing the chances of spoilage is to use melting ice. Other methods of cooling, for example, by cold air or solid carbon dioxide, have not been found to be effective and efficient. Cold air blown across the surface of fish will cause evaporation of water in fish resulting in the loss of moisture and the attractive appearance of fish. Further cold air does not penetrate below the top layer and hence it may take a very long time for fish in the interior layers to be cooled down to the desired temperature.

Solid carbon dioxide (dry ice) does not have the disadvantage of melting ice resulting in slush and an unsanitary appearance. But dry ice cannot be mixed directly with fish as its low temperature ($-80°C$) will freeze the fish that are in contact with it. Indirect cooling through the medium of air is necessary with a resultant drop in efficiency in the utilisation of refrigeration.

Besides efficient cooling, water ice has the following advantages:

1) The water covering as a result of melting prevents surface drying of fish and helps in the retention of the characteristic glaze of fresh fish.

2) It helps to remove slime and dirt from the surface of fish.

3) The layer of water on fish prevents direct contact with air and thus reduces the chances of oxidation of the lipids in fish.

13.2.2 USE OF CHILLED OR REFRIGERATED SEA WATER

Use of chilled sea water, particularly on board the fishing vessels offers some distinct advantages over storage in ice. The harvested fish is cooled very rapidly and much less effort is required in stowing and unloading fish. There is no likelihood of fish being crushed under the pressure of overlying fish and ice. Sea water having a lower freezing point, it can safely be lowered to $-1°C$ without freezing the fish contained in it. The circulation of water helps in washing and bleeding the fish. Salt in water also tends to firm the flesh which may aid further processing. The lay out of a refrigerated sea water system is shown in Fig. 13.2. The storage life of most species of fish in refrigerated or chilled sea water is comparable with storage life in ice. Salt uptake can be a problem but will not be serious if storage is not prolonged or a proper fish to water ratio (4: 1) is maintained (Londahl, 1981).

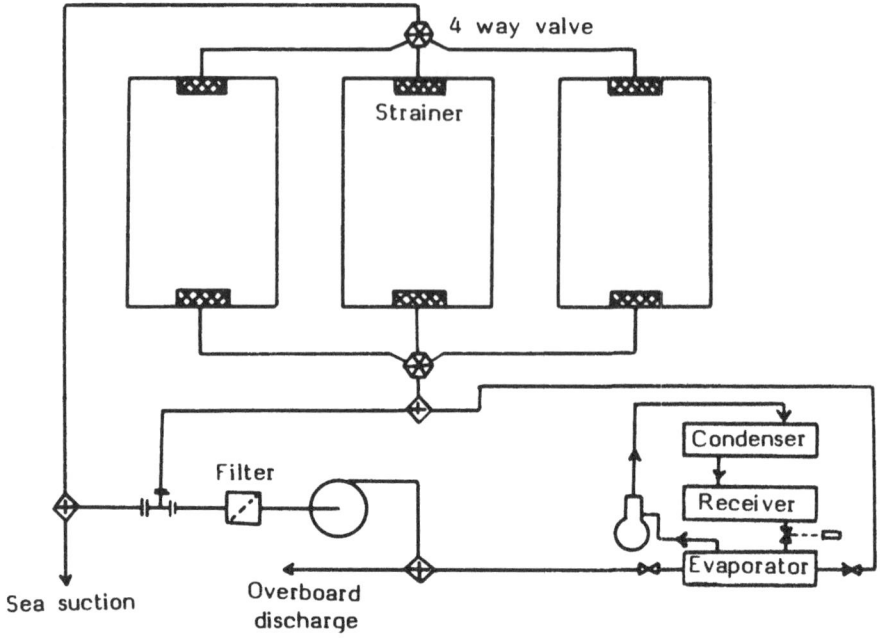

Fig. 13.2. Refrigerated seawater system

13.2.3 SUPER CHILLING

Super chilling means reducing the temperature of fish uniformly to a point slightly below that obtained with melting ice. Salt, blood, and other substances in a mixture of ice and fish depress the freezing point of ice to about − 0.55°C. When the temperature of fish, for example, of white fish containing 80 per cent water, is brought down to 0.55°C, all the water in the fish does not freeze. If the mass of fish and ice is further cooled some of the water in fish separates as ice. In the super chilling process, temperature is brought down to − 2.2°C, when about 50 per cent of the water in the fish is frozen. To this extent of ice formation, the number of ice crystals of larger size is not critical but at a temperature below − 2.77°C., 78 per cent of water is slowly frozen with a resultant increase in the number of ice crystals of larger size as well as excessive damage to the fish. Hence, a very close check on the super chilling temperature is essential if the adverse effects of slow freezing are to be avoided. At − 1.1°C, the shelf-life of cod fish is extended to about 20 days from 15 days in melting ice. The shelf-life goes up to 26 days at − 2.2°C and 35 days at − 2.7°C. But at the latter temperature, damages due to ice formation makes it unsuitable for filleting or smoking.

13.3 Cold chain concept

The preserving action of low temperature is effective as long as the commodity is kept at the desired low temperature from the initial handling or processing stage to the final stage with the consumer. The cold chain comprises the systems of units through which low temperature is applied and maintained with the purpose of preservation of fish in order to reach the consumer at a level of acceptable quality.

The precondition for successful operation of a cold chain may be summarised as (i) the commodity must be in sound condition as low temperature cannot improve the quality of a food material. It only stops or slows down further deterioration, (ii) the commodity should be cooled to the desired temperature as quickly as practicable and economically justifiable, (iii) the product should be kept under appropriate refrigeration throughout its travel, and (iv) the temperature of holding at every stage in the cold chain should be appropriate for the commodity and the product held in refrigeration. The construction and operation of different units of the cold chain system should conform to the above mentioned requirements.

The cold chain comprises two types of units: fixed and mobile. The fixed units are essentially refrigerated storages maintained at different temperatures depending on the specific function of the unit while mobile units are insulated or refrigerated railway wagons, road vehicles, ships and containers.

Cold stores play a crucial part in the effective operation of the cold chain system. The total space required by a cold warehouse, its capacity and arrangement of individual compartments will depend on the type of facilities which the cold store is intended to provide. The units may be intended for providing only refrigerated space for holding of fish or fishery products at the desired temperature for a sole processor or a party or may provide also for chilling, freezing, packaging, etc. prior to storage and may in addition extend the facilities to a number of parties. The cold storage in the latter case is much bigger in size. Substantial economy may be realised in such a set up. It has been observed that investment and operational costs for a unit of storage space of $50,000 \, m^3$ is half that of a $5,000 \, m^3$ unit (Londahl, 1981). The capacity of cold storage should be arrived at on the basis of maximum storage needs for a particular commodity taking into account seasonal variations in market arrivals and market demands. With the knowledge of the required capacity, the refrigerated space is determined on the basis of a loading factor (storage density) for the commodity. In the case of storage for different commodities the seasonal variation in production and market demands are computed and the desired capacity of storage is the maximum space required at any particular period taking the arrivals of all the commodities together. The annual utilisation index of a cold storage is determined as the portion of the space utilised out of the space available and will be less than unity. For cold storages accommodating a number of commodities the index is generally about 0.6 to 0.7 and is larger than that of storages for a limited number of products such as the one only for fish.

Total space for a cold storage unit cannot, however, be restricted to actual cold store space only. Requirement of space for reception and sorting of materials, prestorage treatment and processing, if any, and transport within the plant has also to be taken into consideration. Commodities such as those with strong penetrating odours which affect other food materials stored with them have to be segregrated into separate rooms. In addition space for workers' facilities, machine rooms, workshop, office etc. are to be provided. Generally single-storey structures are prefered for a cold storages plant unless the availability of space is limited or the cost of space is relatively high.

Two types of handling and stacking systems are used in cold sotrage practice (i) manual system, and (ii) palletised system. The latter may again be carried out in two ways (i) pallets may be piled one above the other, or (ii) pallets may be laid in shelved racks.

The manual system is suitable for small low rooms of height 3 to 4.5 m.

Pallets consist of loads representing a transport handling and storage unit. International (ISO) Standards of pallets prescribe a base of 800×1200 mm and 1000×1200 mm. The usual height of a loaded pallet is 1750 mm. For specific materials, pallets of other sizes are sometimes used.

The refrigerated space is determined by the storage density of each commodity. The space actually useable will, however, be less as necessary provision has to be made for openings between stacks, for floor clearance, for ceiling clearance and clearance below air ducts under ceiling coils, and for passages for movement of forklift trucks. The net effective capacity of cold room V_u may be obtained from the formula

$$V_u = KV_B \tag{13.1}$$

Where V_B is the refrigerated space equivalent to the geometric volume of the room and K is the possible utilisation factor to be taken from the Table 13.3.

Table 13.3
Utilisation factor K of storage rooms

Room floor area (m^2)	Room height (m)	Storage system	K
	< 4	Manual	0.60
100–250		Palletised	0.56
	> 4	Manual	0.68
		Palletised	0.63

13.3.1 REFRIGERATED TRANSPORT

The system of transport for refrigerated fish necessitates provision of thermally insulated spaces with or without refrigeration depending on the temperature at which products have to be maintained, the length of haul and the ambient temperature.

The refrigerated space holding the fish should be clean and free from debris to prevent microbial contamination. The vehicles should not impart any taint to the stored fish. If the same vehicle is used for other food commodities, removal of fishy odour from the vehicles becomes important.

The arrangement of refrigerated load in the vehicle should be such as to ensure proper and uniform circulation of air to prevent any local heating and to ensure that the desired temperature uniformity is maintained throughout the load. For chilled material, air should flow through and around the load, while for frozen fish, circulation around load is generally sufficient.

Transport vehicles for chilled or frozen fish are generally of two types (i) insulated, and (ii) refrigerated. For short hauls, chilled fish packed in ice, wooden or wicker baskets are extensively transported in uninsulated vehicles in many developing countries.

13.3.2 INSULATED VEHICLES

Insulated vehicles are classified as normally or heavily insulated on the basis of the overall heat transfer coefficient according to standards adopted by the U.N. European Economic Commission at Geneva in 1971.

As per standards, normally the insulated vehicle should have the overall heat transfer coefficient (k) less than or equal to 0.6 Kcal/m^2/hour/°C and k for heavily insulated ones should be equal to or less than 0.35 Kcal/m^2/hour/°C.

The transport vehicle should ensure the maintenance of the proper temperature for the chilled fish being carried in it. The temperatures of the products recommended for transport of such materials are shown on Table 13.4. The material should be at the recommended temperature at the time it is put in the refrigerated space on the transport vehicle (Table 13.4). Temperatures (°C) recommended for the transport of fresh fish.

Table 13.4

	Transport duration (days)	
Nature of product	1–3	4–6
Iced fish	0–2	0–2
Smoked fish	Not over 10	Not over 6
Prawns	2–4	Not recommended

13.3.3 REFRIGERATED VEHICLES FOR FROZEN FISH

The refrigerated vehicles may be two types (i) insulated ones using a source of cold other than a mechanical or absorption type of refrigeration unit, and (ii) insulated ones with compression or absorption type refrigeration unit. While the overall heat transport coefficient (k) of refrigerated units should be equal to or less than 0.35 Kcal/m^2/hour/°C the unit must be capable of maintaining an inside temperature of -18°C at an outside temperature of 38°C. With insulating materials such as polyurethane it is not difficult to get K values of 0.15 to 0.30.

i) Insulated types without inbuilt refrigeration source:
Frozen food including fish which is to be stored at a particular temperature

may be brought down to a temperature lower than that specified for storage and carried in properly insulated transport vehicles, provided that during transport, loading and unloading the temperature of the product does not go above the specified one. It has been demonstrated that quick frozen fish subcooled to − 30°C can be transported during the cooler months of the year for up to 24 hours without temperature of the cargo exceeding − 18°C.

ii) Refrigerated type:

These vehicles may be of two types depending on the method of supply of refrigeration, that is, (a) total loss of refrigeration system, and (b) mechanical refrigeration system.

a) Total loss refrigerant system is one in which refrigeration is supplied generally by melting of water ice but dry ice has also be used for frozen cargo.

Water ice usually in 10 kg blocks, are put into bunkers fixed at the ends of a railroad car or at the nose of road trucks. Forced air circulation has to be maintained to assure uniformity in temperature distribution in the lot as well as to maintain the cargo at the desired temperature level. Simplified diagrams of a railway wagon and a road trailer are shown in Figs. 13.3. and 13.4 respectively.

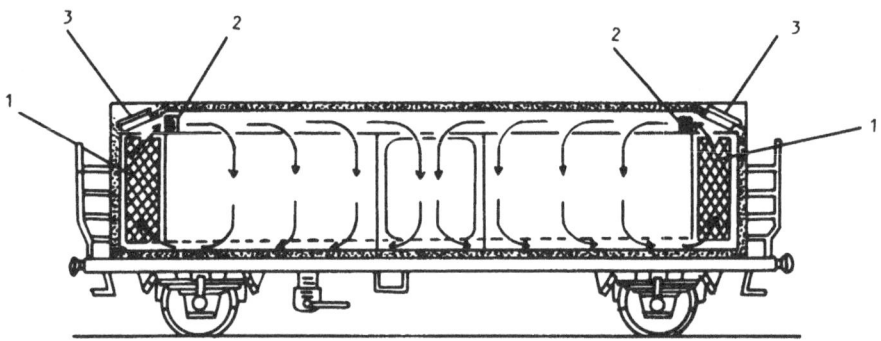

Fig. 13.3. Ice refrigerated rail car 1. Bunker. 2. Electric fan for air circulation. 3. Ice hatch

b) Mechanical refrigeration systems:

Mechanical refrigeration systems are much more widely employed than other systems because of the convenience of opration and comparable total cost with other systems. Although earlier such a system was used mainly for long hauls, it is now being increasingly used for shorter hauls also.

13.3.4 CONTAINER TRANSPORT

Containers are gaining popularity in transport of frozen fish particularly if the journey involves changes in the method of transport on the way such as from road truck to rail wagon or to mercantile ships. The packs remain intact and chances of mechanical damage are minimised. Containers for frozen fish are thermally insulated bodies which are generally provided with individual

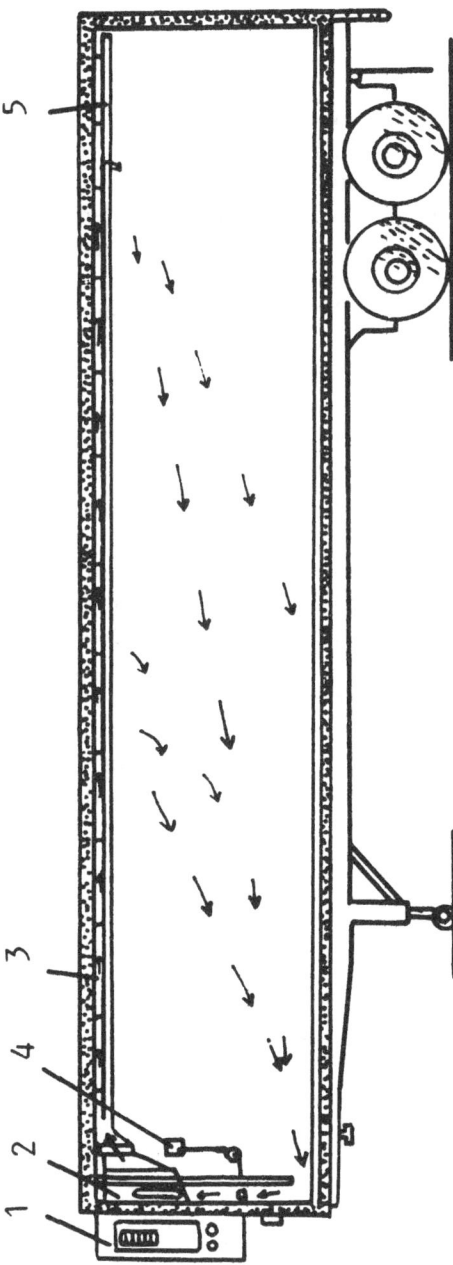

Fig. 13.4. Mechanically refrigerated semi-trailer 1. Air cooled condensing unit. 2. Forced draught air cooler. 3. Air duct. 4. Thermostat with temperature sensor placed inside and adjusting head outside the vehicle. 5. Overhead hanging rails

refrigerating equipment. They are designed with standard fittings and structures so that they can be temporarily fixed on and carried by different types of transport vehicles—road trailers and semi-trailers, railway flat cars, the holds of ships and cargo aircrafts.

13.4 Freezing of fish

Freezing is one of the widely used methods of preservation of fish. In 1983, 22.9 per cent of the world's catch of fish was frozen and this share of freezing has risen steadily during the last three decades. Frozen fish ranks highest in the international trade of fish and fishery products for human consumption in value as well as in weight. The popularity of frozen fish is derived from the fact that the freezing process helps to retain the quality and nutritive value of the raw fish better than any other commercially applied preservation method.

Lowering of temperature reduces the rate of putrefactive and deteriorative changes in fish and thus prolongs its shelf-life. The beneficial effect is commensurate with the lowered temperature at which fish is kept. Chilling fish to a temperature of 0°C or slightly lower without freezing, can extend the shelf-life of fish from a few hours to more than two weeks. However, for longer storage life of fish, the temperature is to be further lowered. Freezing and frozen storage can give a shelf-life of over one year if properly carried out. Bacterial action which is the primary cause of spoilage of fish is practically stopped at −10°C but chemical, biochemical and physical processes leading to irreversible adverse quality changes in fish still occur even though at a progressively lower rate at lower temperatures. Hence for longer storage life, fish has to be frozen and held at a much lower temperature, generally at or below −30°C. The storage life of a sample of lean fish at different temperatures is shown in Fig. 13.5 Table 13.5 gives the storage life of a few varieties of frozen fish stored at different temperatures.

Table 13.5
Practical storage life for fish. From IIR

	Storage life months		
	−18°C	−25°C	−30°C
Fatty fish, sardines, salmon ocean perch	4	8	12
Lean fish, cod, haddock	8	18	24
Flat fish, flounder, plaice, sole	9	18	24
Lobster, crabs	6	12	15
Shrimp	6	12	12

Source: Londahl, G., 1981: *FAO Fisheries Technical Paper No. 214,* Food and Agricultural Organisation of the United Nations, Rome.

Fish flesh does not begin to freeze before its temperature reaches about −1°C due to the depression of the freezing point of water caused by the

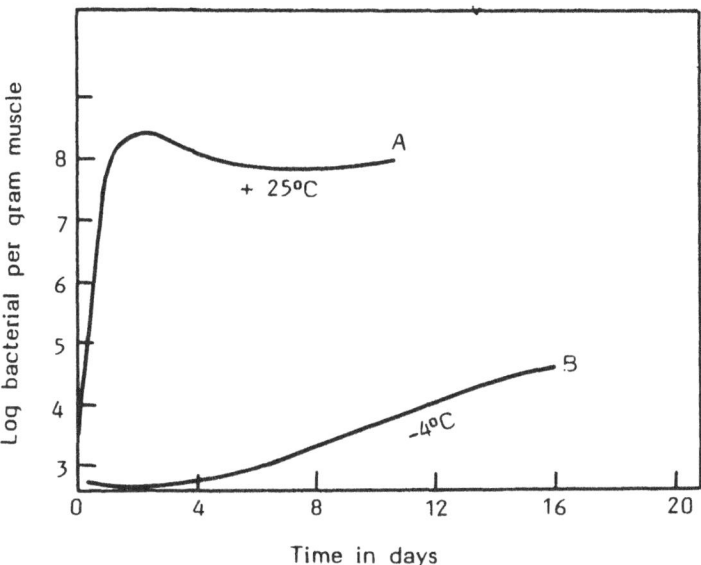

Fig. 13.5. The effect of temperature on bacterial growth

solution of the salts present in it. As the temperature continues to fall more ice separates from the remaining solution which thereby becomes increasingly concentrated in salts with a resultant depression in freezing point. Eighty per cent of the water is frozen between -0.8 and $-5°C$, 90 per cent at $-10°C$ and about 96 per cent at $-20°C$. However, after 90 per cent of the water is frozen, the condition of the rest is indefinite because much of this water is held by hygroscopic materials in fish tissue and some are bound as a chemical component of the molecules.

The rate at which ice formation takes place determines the size of the ice crystals that are formed during the freezing process. Slow freezing allows the first formed crystals to grow to a larger size. The faster the freezing, the more are the number of crystals-nuclei formed with a smaller average size of crystals. Ice crystals formed inside the cells of tissues will exert pressure and may puncture the cell wall. The greater the size of the ice crystals, the more is the chance of damage to the cell wall, as pressure will be exerted at a few locations instead of on the entire cell wall. Fluids oozing out of the damaged cells of the thawed frozen products are called 'drip', which besides causing fluid loss is also largely responsible for the poor appearance of thawed fish and the loss of flavour and firmness.

Quick lowering of temperature during the freezing process is necessary to control the size of the ice crystals formed. Most of the ice is formed between the temperatures of 0 and $-4°C$ which is termed the zone of maximum crystal formation. When the temperature of the substance being frozen passes through the zone in 30 minutes or less the process employed is called 'quick freezing'. The rate of freezing is defined as the velocity of the progress of ice front from the surface to the interior, expressed in millimetres per hour. In the quick freezing processes in use at present, the rate of freezing varies between 1 and 2 cm per hour and is found to be adequate to prevent undesirable changes during freezing process (Heen, 1981).

The important types of freezers employed in commercial practice may be put into four categories according to the mode of application of refrigeration:

1) Air blast freezer—stationary and moving belt,
2) Plate freezer—horizontal and vertical,
3) Immersion freezer, and
4) Cryogenic freezer.

1. Air blast freezer

The air blast freezer is essentially a device to freeze fish in a current of rapidly moving refrigerated air. The freezing is carried out in an insulated tunnel. The product itself may move in the freezer or be stationary. The temperature, velocity and turbulence of air should be reasonably constant in the freezing area for the sake of unformity in quality of the frozen product. An air temperature of $-20°C$ to $-30°C$ is favoured for blast freezing. An air velocity of about 6 m/s is generally used. Although some investigators claim that a speed of over 3 m/s does not provide any additional advantage, the rise in temperature of air as it passes over the product calls for a higher speed to reduce the time required to lower the temperature. Higher velocity will give higher freezing rate at the cost of larger energy consumption for blower operation and will also require dissipation of large amounts of heat generated by the blower fan. Good air distribution may be achieved with a number of smaller fans where if the length of the air path is shorter, higher volumes/velocities of air will not be necessary. An air speed of 10 to 15 m/s is recommended for a continuous freezer to reduce time or to increase freezer capacity. A sketch of an air blast freezer is given in Fig. 13.6.

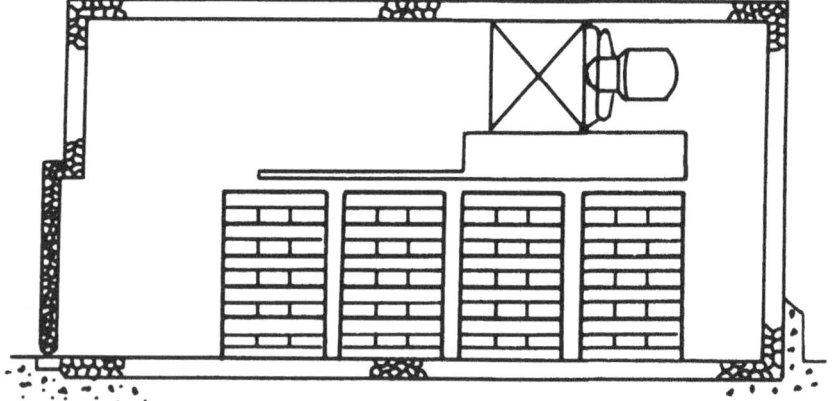

Fig. 13.6. Air blast freezer

Belt freezers are examples of air blast freezers in which products move on a belt in a refrigerated blast room. Since for large capacities, the single belt freezer will require ample space, the building requirement is reduced by having several tiers of belts one above the other. Better floor saving is achieved by a belt which can bend laterally around a rotating drum. The flow of refrigerated air is guided vertically through the freezing chamber. A diagram of a spiral belt freezer is given in Fig. 13.7. Belt freezers have been found to be particularly suitable for individually quick-frozen products. The relative advantage of a blast freezer lies in its versatility in accommodating products of all sizes and shapes. It is particularly suitable for freezing large single fish and also packs of irregular shape.

The two major drawbacks of an air blast freezer are (i) dehydration loss of the product during freezing, which may go up to 3 to 5 per cent (Astrom, 1971), and (ii) the need for frequent defrosting of the equipment due to freezing of evaporated moisture on coils. Tunnel freezers are also more expensive to operate than plate contact freezers.

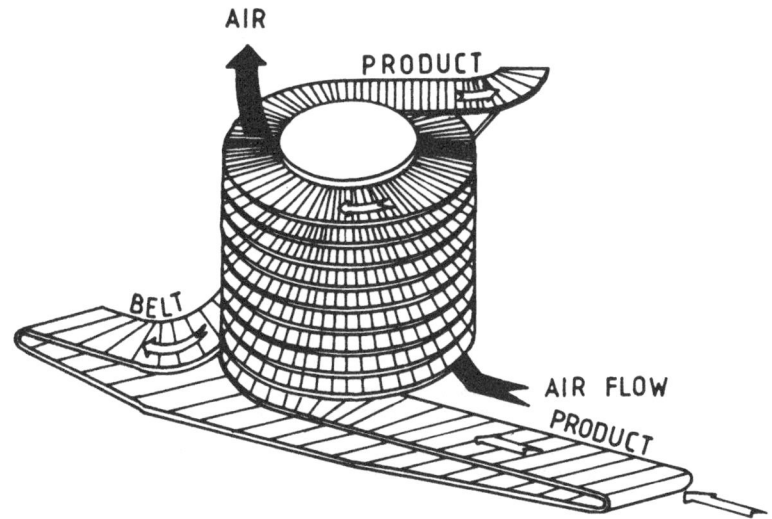

Fig. 13.7. Spiral belt freezer

2. *Plate freezer*

Plate contact freezers are a result of the development in direct freezing which is freezing by bringing the product in contact with a metal surface cooled by refrigerated brine or any other suitable refrigerant. The fish is frozen between two metal plates through which the refrigerant circulates. Modern plate freezers have their plates constructed from expanded sections of aluminium alloy arranged in such a manner as to allow the refrigerant to flow through the plates and thus provide heat transfer surfaces on both sides. Earlier models of plate freezers used vacuum plates constructed with an intregal pipe grid to contain the refrigerant and many of these are still in use.

All plate freezers are fitted with hydraulic systems which move the plates to compact the produce and give higher density blocks. They also improve the fish-to-plate contact area for quicker freezing and assist in the release of the block after freezing. Plate freezers are of horizontal or vertical types.

i) Horizontal plate freezer

The two main uses for this type of freezer are the freezing of prepacked cartons of fish and fish products and the formation of homogenous rectangular blocks of fish fillets called laminated blocks for the preparation of fish portions. The thickness of the package or frozen fish block is 32 to 100 mm. Horizontal plate freezers need not be defrosted after each use. Only a light brushing is sometimes necessary after each freezing cycle.

Horizontal plate freezers will operate satisfactorily if good contact is made on both the top and bottom surfaces of the pack or tray with the materials to be frozen. Initially a pressure of about 0.5 kg/m^2 is applied to the plates by means of a hydraulically operated piston for a better contact with the upper surface of the product but is increased to about 1 kg/m^2 as the fish expands during freezing Figure 13.8 shows a horizontal plate freezer.

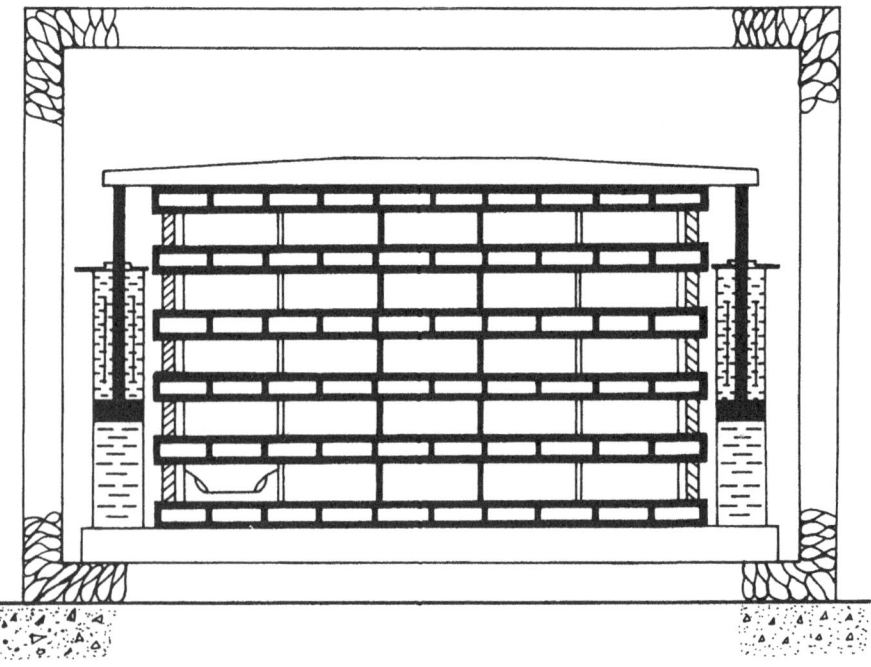

Fig. 13.8. Plate freezer

ii) Vertical plate freezers

The main advantage of using a vertical plate freezer is that fish can be frozen in bulk without package or tray. The plates form a bin with an open top. In this type of freezer, fish is loaded directly into the space. Such a freezer is therefore particularly suitable for bulk freezing and is used extensively for freezing whole fish at sea. The block thickness may vary from 25 to 130 mm. Maximum block dimensions and weight are limited by the physical effort required by the operator to life the block without causing any damage to the fish. Compact blocks with a density of 800 kg/m^3 can be expected from such freezers. It is advantageous to use wrappers and some water to fill the voids during freezing of fatty fish. Consequently the freezing time is increased due to the insulating properties of the wrapping material. Vertical plate freezers are defrosted to release the blocks of fish after each freezing operation. Otherwise there is a possibility of damage to the plates, if brushes are used to remove frozen blood, slime etc.

Vertical plate freezers can be made with arrangement for top, side or bottom unloading of the blocks. Generally, top unloading models are preferred since the block is lifted clear of the plates and presented at a suitable height for handling by the operator.

3. Direct contact freezers

Freezing of fish by immersion in sodium chloride brine is an old practice. Because of the chances of salt uptake by fish, the use of immersion method is now practically restricted to freezing of tuna on board the fishing vessels for subsequent production of canned tuna in brine. A eutectic solution of 22 per cent NaCl at −21°C is generally used.

4. Cryogenic freezer

Cryogenic fluids such as liquid nitrogen or carbondioxide or purified chlorinated hydrocarbon (Rl2) is gaining popularity. The use of Rl2 for direct contact of food is not permitted in many countries.

Flow pattern in a liquid nitrogen freezer is shown in Fig. 13.9. The fish on the stainless steel belt initially come into contact with the counter current flow of nitrogen gas at a temperature of about −50°C. As the fish progress through the freezer the temperature of the cooling gas progressively falls to −196°C. At the initial stage of cooling in the gaseous nitrogen, partial freezing of fish takes place and the freezing is completed below the liquid nitrogen spray by the boiling liquid. About 50 per cent of the product heat is extracted at the precooling zone and the remainder heat transfer takes place at the small freezing area below the spray. After leaving the spray zone, the last stage in the freezer is used for the fish temperature to reach equilibrium (equilibration zone) before the fish are discharged.

Freezing is very fast and the freezer size is small. Maintenance is minimum and power consumption is very low. However this method of freezing is more expensive than most others. The flow pattern of a liquid nitrogen freezer is shown in Fig. 13.9.

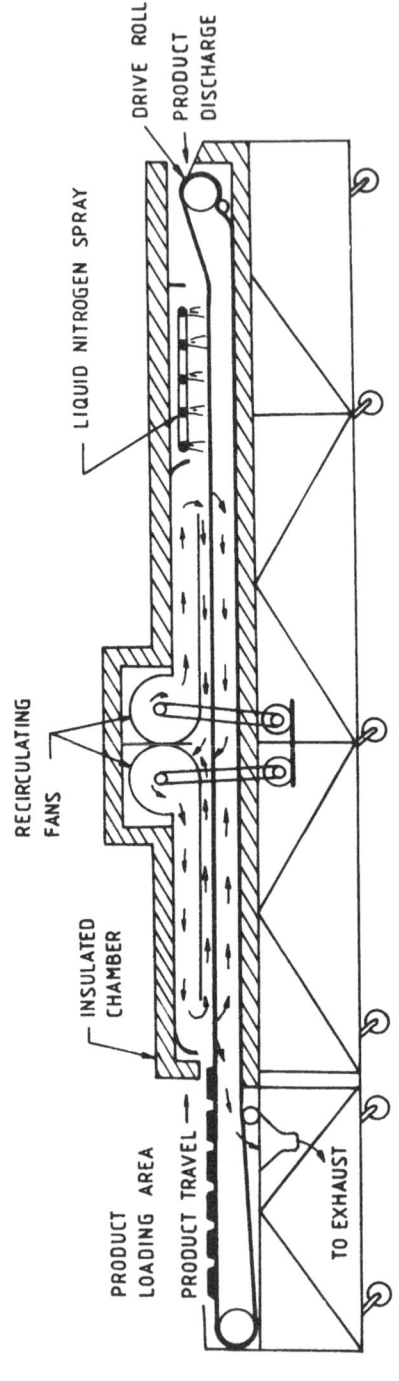

Fig. 13.9. Flow patterns in typical LN$_2$ freezer

13.4.1 INDIVIDUALLY QUICK-FROZEN (IQF) PRODUCTS

Generally quick-frozen fishery products are produced in blocks or regular packages. However individually quick-frozen food materials have advantages in (i) faster heat transfer and hence quicker freezing, (ii) improved processing and packaging efficiencies, and (iii) better consumer preference. Many fishery products such as fish fillets, fish fingers and breaded products have characteristic shapes and size. Packing them in blocks and subsequent recovery from blocks before consumption may cause distortion in shape and size. Besides a housewife prefers them in discreet individual pieces so that the tedious process of chopping or thawing of frozen block will be obviated. Freezing by using continuous conveyors to carry the unpackaged pieces through the freezing atmosphere, generated by cold air or under cryogenic refrigerants, is used for the IQF product. The development of conventional freezers for IQF products has resulted in a multifold increase in freezing capacity.

13.5 Canning of fish

The science and technology of the 'canning' process for preservation of perishables depend on the principle of thermal destruction of mircrobes present in food materials held in hermetically sealed containers. Although identification of microbes as the principle causative agency in spoilage had to await the classical research of Louis Pasteur, another Frenchman, Nicholas Appert, had by virtue of his painstaking work between 1795 and 1810, discovered the art of 'canning' and is generally credited with the discovery of the process.

Canned fish and fishery products accounted for 9,969,200 t of fish and crustaceans in 1983 and constituted 13.1 per cent of the total world fish landing. The share of canned fish and fishery products was however 14.2 per cent in 1979. The declining trend in canned fish production and marketing is continuing due to the preference of consumers for frozen products as against canned ones.

13.5.1 PRINCIPLE OF CANNING

Microbes and enzymes cause spoilage of fish after death and both these agents are effectively controlled by heat. The essential requirements for effective preservation by heat treatment are:

1) The product to be treated in a sealed container with the required amount of heat necessary to inactivate microbes and enzymes.

2) Prevention of subsequent infection with microbes by hermetically sealing the container.

3) The container has to be enduring and any leakage of cans due to corrosion or otherwise must be checked.

4) The product should not react with the material of the container.

The food materials properly sterilised to destroy bacteria and enzymes and protected from further contamination by them, should remain in good condition indefinitely. Under commercial conditions of processing and storage, a life of two years for such treated products may be safely predicted.

Heat-resistant spore-forming bacteria are most important so far as microbiological spoilage of canned fishery products is concerned. The temperature and duration of heating must be such as to ensure the destruction of the most heat-resistant bacterial spores likely to be present in a product to be preserved. Destruction of spores of *Clostridium botulinum*, which are one of the most heat resistant and harmful bacteria encountered in food materials is often used as an indicator of the adequacy of heat treatment. As the chance of survival of bacteria is greatest in the portion of the food which receives the least amount of heat, the temperature profile at the coldest point of the can and the rate of destruction of the target bacteria at this point will determine the time of processing at any specified temperature.

The process time at a specified temperature to which a can is exposed, may be computed graphically (Bigelow *et al.*, 1920). The two sets of information required for the purpose are:

i) time-temperature relation at the point of largest temperature lag in the can during the heating and cooling operations, and

ii) the thermal death time (time, in minutes, to reduce the number of spores from a specified upper to a lower limit of the target organism at the temperature represented by the point in the heating-cooling curve), under the environment present by the food.

The reciprocal of the thermal death time gives the lethal rate at the particular temperature. A lethality curve may be drawn based on the above mentioned two sets of information, by plotting the lethal rate against time of exposure to the particular lethality rates during heating and cooling of the can for processing. The inter-relationship between the heating and cooling and thermal death time curves in shown is Fig. 13.10 (Ball, 1923).

Figure 13.10a shows the temperature at the specified point in the can

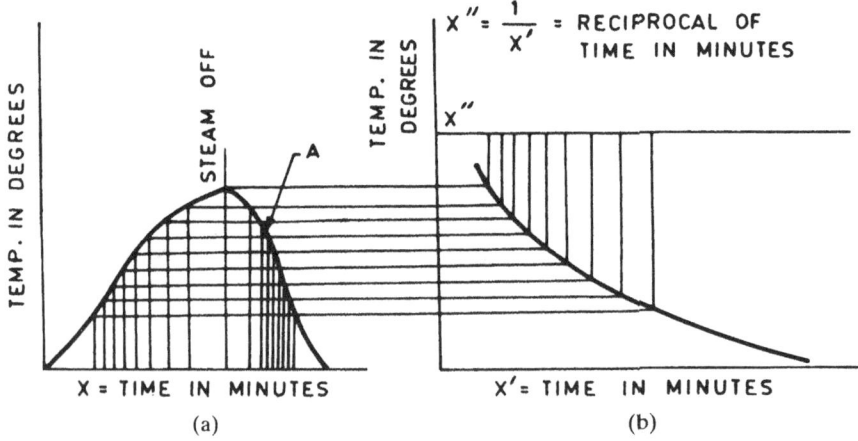

Fig. 13.10. The inter-relationship between the heating, the cooling and the thermal death time curves in the calculation of lethality

during heating and cooling and the typical thermal death curve for bacterial spores is illustrated on Fig. 13.10b. The reciprocal of the thermal death time is the rate of destruction of the spores at any particular temperature. A lethality curve may be drawn with time as in the time temperature curve in Fig. 13.10a with rate of destruction at the temperature corresponding to time as ordinate. A lethality curve is illustrated in Fig. 13.11. When the area under the lethality curve is equal to unity, the process is just sufficient to destroy the spores of the specified organism. The period between the point when steam inside the heating vessel reaches the processing temperature and the point of shutting off the steam designates the process time. The beginning of cooling may be adjusted by drawing lines parallel to the cooling curve to unit value.

The graphical method of plotting a lethality curve and determining the area beneath it is a slow and tedious process and it does not allow the coordinating of various factors which affect processing time. Ball (1923) introduced a direct calculation method based on certain assumptions on the nature of the graphs obtained in the graphical method. Thompson (1919) and Williamson and Adam (1919) mathematically deduced temperature at the center of the cylinder at any time period of heating. They assumed that the time-temperature heating and cooling curves are practically a straight line except for the lag period at the beginning of heating and cooling. Experimentally determined data also show that the heating-cooling curves for most products in a can agree with the theory.

It has also been shown by Bigelow (1912) that experimentally determined thermal death time curves for spores of bacteria are straight lines when plotted on semi-logarithmic paper (Fig. 13.12).

Based on these observations Ball (1923) derived a logarithmic equation for this curve. Deviation from the straight lines occurs at the beginning of the heating and cooling curves. The deviation at the beginning of the heating curve may not be taken into consideration as the temperatures prevailing at this portion of the heating have no significant sterilising value. However, some time is required for bringing the retort (autoclave) to the specified temperature. This is known as the coming up time. While the heating value of the coming up time is not high, it should not be neglected. Conventionally,

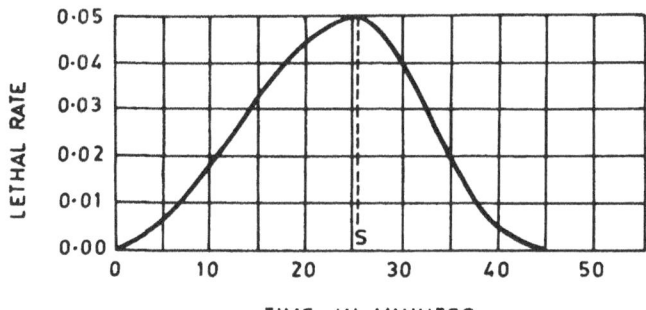

Fig. 13.11. Lethality curves

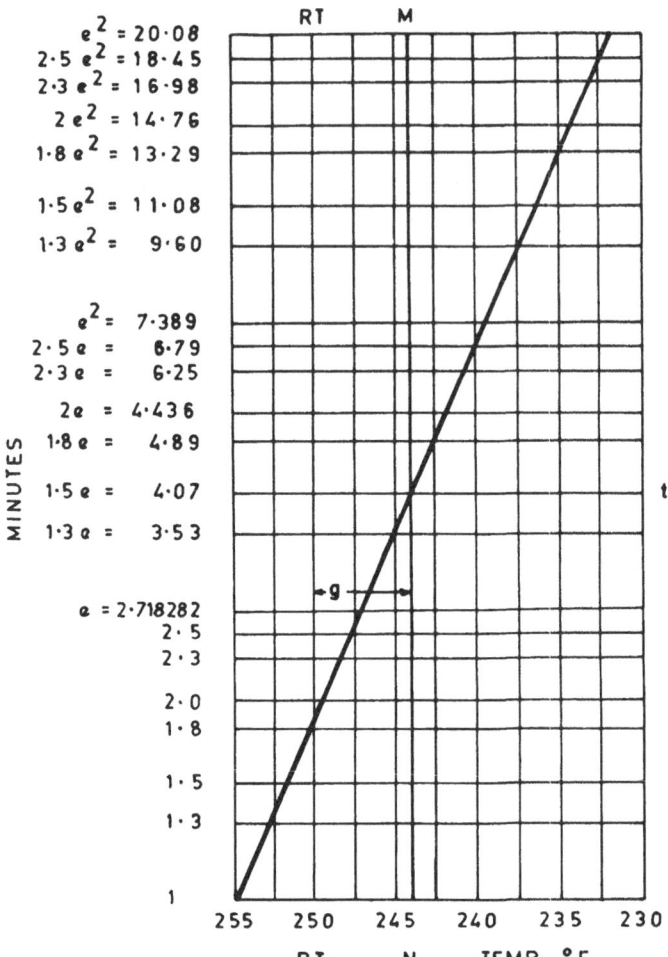

Fig. 13.12. Thermal death time curve

the heating value of the coming up time is taken as 0.42 parts of the coming up time. If the straight line portion of the heating curve is extended to intersect the line representing the zero come up time, the intersection will represent the pseudo-initial temperature. It means that the corrected beginning of the process is not when the retort reaches the stipulated temperature but will be 0.42 multiplied by the coming up time before that point. Ball introduced the factor j to locate the intersection of the heating and the vertical line representing the beginning of the process if no time is consumed to bring the retort to the processing temperatures as:

$$j = \frac{RT - psIT}{RT - IT}$$

Where RT = retort temperature
   ps IT = pseudo-initial temperature
        IT = initial temperature

The difference between retort temperature and the initial temperature is designated as I or $I = RT - IT$. Hence jI represents that part of the intersection of vertical line indicating the beginning of the process (the heating curve being a straight line). This point together with slope of the line of the heat penetration curve will give the position of the straight line. The slope is represented by the symbol fh which is defined as the number of minutes required for the straight line portion of the curve to traverse one log cycle. The length of the process in minutes B_B is arrived at by the equation

$$B_B = fh \ \log \ ji/g$$

Where g is the number of degrees below retort temperature attained by the food before cooling started (Ball, 1923).

13.5.2 Sterilising value of a process
Ball (1923) introduced a symbol F which is defined as time in minutes required to destroy a specified organism at 121.1°C and expressed the sterilising value of the process in terms of F. According to this concept all processes having the same F value will have the same capacity to destroy a given organism in a given product.

The equation to evaluate the sterilising value of a process (F) was developed by Ball (1928) as

$$\frac{250° - RT}{z} = \log \ U/F \qquad\qquad 13.3$$

Temperature 250° is in the Fahrenheit scale, z is the slope of the line on thermal death time curve plotted on a semi-logarithm paper for the giving organism, U is numerically equivalent to the thermal death time of an organism at the retort temperature where thermal death time at 250°F is F.

13.5.3 Container for canned food
Apart from the appreciation of the role of the correct heat treatment in the preservation of a canned food, the other factor which has been largely responsible for the rapid growth of the canned food industry is the introduction of the sanitary tin can. These are open top cans made of tin-coated steel plates. In the fabrication of such a can, chances of contamination of foods by contact with the solder containing lead is reduced to a minimum. This is achieved by closing the side of the container by only external application of the solder while the lids are fixed on the top and bottom of the can body mechanically by a double seaming technique.

To give additional protection against rusting and to prevent reaction with

H_2S with consequent darkening of the inner surfaces of the can, an enamel lining is generally used in cans used for fishery products. The purpose of an enamel lining is mainly to increase the attractiveness of the food and to improve the appearance of the container. The cans for the purpose are produced from enamelled sheets by the same fabrication methods as are employed for sanitary cans made from plain tin plate. For production of enamelled tin plates, the sheets are coated with an enamel compound and baked to provide an insoluble, inert, resinous film on the tin plate.

With a double-seaming machine the sealing of the ends is carried out in two operations. In the first operation a small roller passes around the edge and the can body thereby folding the flanges of the can body and can ends. A pair of rolls in the second operation presses the flange against the can and tightly compresses the folded flanges. The sequence of operation by a double-seaming machine is shown in Figs. 13.13 and 13.14.

Besides tin plated steel, other materials are also used in the fabrication of containers for can foods. Aluminium cans which are generally of rectangular or elliptical cross-section, prepared by stamping out of aluminium sheets, have found wide use in the fish processing industry. A number of other materials such as stainless steel, surface-treated steel etc. have been tried without significant commercial success. Retortable pouches made from laminated plastic or metal films are now being used for a number of heat-processed foods including fish.

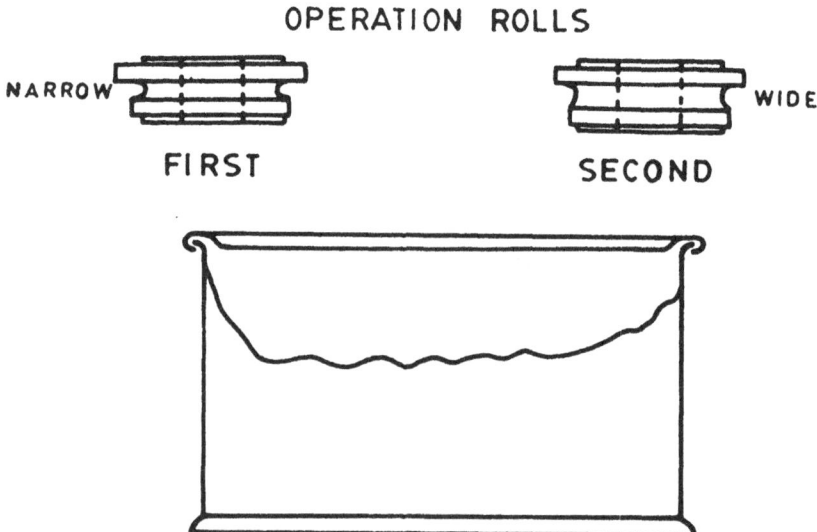

Fig. 13.13. Can sealer rolls (above) and sketch of can below showing different parts (L.G. McKee)

CORRECTLY FORMED FIRST OPERATION SEAMS

TOO TIGHT   TOO LOOSE

SEAMS INCORRECTLY FORMED BY FIRST ROLL

NORMAL   TOO TIGHT

SEAMS FORMED BY SECOND ROLL

Fig. 13.14. Correctly and incorrectly formed can seams (L.G. McKee)

13.5.4 OPERATIONS INVOLVED IN CANNING

Different steps involved in the canning of fishery products vary with species of fish, type of pack and nature of pre-treatment. The following steps are generally common for all such canned products.

13.5.4.1 Selection of raw material

Raw fish for processing must be sound in quality which depends on various factors such as the gear employed for harvesting of fish, post-harvest handling of fish including time and temperature of storage and transport, and implements used in handling. In some cases the state of maturity, particularly in relation to the spawning state is important. Every care has to be taken to prevent damage to the fish by weight load on it or by bruises caused by improper handling.

13.5.4.2 Preparation for the can

The first step in the preparation of fish for the can is dressing and washing of the fish to remove viscera, blood, slime, waste materials, dirt and other extraneous substances, with as little sacrifice of the edible meat as possible keeping in view the type of pack desired. Washing in water may be by agitation or spraying which may or may not be preceded by soaking.

The dressed and washed fish are cut to the size of the can or other sizes as considered necessary and then put in the can or subjected to pretreatments such as pre-cooking, which causes loss of moisture and consequently loss of weight but helps to control the drained weight of the product. Conditions for pre-cooking must be carefully monitored to keep the weight loss to a minimum. The preparations (steps) should be properly synchronized with subsequent operations particularly the one of processing for preservation, as unprocessed material kept for long, specially in the warm atmosphere of the cannery premises, will undergo quick deterioration which will adversely affect its quality.

13.5.4.3 Filling the can

The container is filled with the material to be processed either by hand or by machine. For filling odd size containers or for materials requiring careful handling, hand filling is preferred. The size of the canning unit and relative cost of human labour also determine the choice between mechanical and manual filling. Except in case of 'solid packs' it is customary to add a liquid such as brine (sodium chloride) or oil or a sauce according to the nature of the pack to fill the voids between solid pieces. Liquid helps in better transmission of heat during processing and subsequent cooling besides improving the flavour and reducing disintegration of the solid pieces by providing a cushion.

The cans are filled in such a way that the required empty space termed 'head space' is left between the top level of the food in the can and the inner surface of the upper lid. A head space of over 5 mm should be retained to allow the expansion of the contents in a can during heating so as to enable the creation of the desired amount of vacuum inside the can. The head space also provides space for the small amount of hydrogen generated through electrochemical reactions on the tin plate. However, a larger head space is undesirable because there are greater chances of denting or buckling of the cans. The upper limits for head space are generally prescribed by standards for canned food products in a country's food laws, to prevent any deceptive impressions regarding the amount of food material inside the can.

The filling of the can should be very carefully supervised to prevent the presence of any foreign material inside the can, such as nails and other metal fragments, hair, flies or any other insects that may cause the contents to be declared adulterated and thereby attract legal action apart from adverse consumer reaction.

13.5.4.4 Exhausting and sealing

A good and sound can of a food material should not have any positive pressure inside the can as a positive pressure is generally associated with formation of gas often accompanying microbial growth. Exhausting or removing air from inside the can is desirable for (i) controlling the reaction of food components with oxygen, (ii) preventing corrosion of the container material, and (iii) reducing strain on the can seams due to a pressure difference between the inside and the outside during cooling of the can.

Exhausting may be carried out by (i) raising the temperature of air in the head space and immediately sealing the can, and (ii) vacuum sealing which combines both functions of exhausting and sealing.

Heat exhaustion may be effected by filling hot liquid or a semi-liquid material like fish soup or chowders in cans and sealing the cans immediately. For solid materials, exhausting is carried out by passing open cans with materials in them through a steam heated chamber on a conveyor. Loose can lids are often clinched to the body of the can so that they will hold on to the can and leave sufficient space for air to come out of the can. The temperature inside the exhaust box, the time of holding of the can inside the box and the volume of head space will determine the amount of vacuum created. The exhaust box is however on the way out in the fish canning industry and is being replaced by a steam-blow technique in which air in the head space is purged by blowing a jet of steam over the food in a can while the can is being sealed.

Vacuum sealing machines are favoured in larger fish canning units. These machines are not, however, economical for small scale operations but are very useful in situations where space is limited as in a factory ship.

13.5.5 PROCESSING

Processing or sterilisation is the most important step in the canning preservation of food as it eliminates the chances of microbial deterioration of the canned stuff.

Fish flesh is categorised under 'non-acid' foods and has to be heated to a temperature above 100°C, generally between 115 and 125°C. Heating at temperatures higher than 100°C requires steam to be maintained under pressure. Hence processing is to be carried out in a pressure vessel which is usually termed a 'retort'.

The quality of heat processed foods depends to a large extent on the correct manipulation of the retorts. The colour and flavour of the retort-processed foods should be nearly the same as home-cooked foods. The achieve this, conditions to be observed are (i) the retort is heated rapidly to the holding temperature, (ii) every container inside the retort receives the same heat treatment, and (iii) cans are cooled rapidly after holding at the specified processing temperature and time.

Three distinct stages in heating are recognised (i) the coming up time—the period from the start of the operation to the time when the steam inside the retort reaches specified temperature, (ii) heating time—the period for which the containers are kept at the specified temperature, and (iii) cooling time—the period between shutting of the supply of steam to the retort and the opening of the retort to take out the cans.

During the canning time all the air inside the retort must be swept out so that there is no air 'pocket' in between the cans as an air-steam mixture will cause non-uniform temperature distribution in the retort affecting heat transfer rates to different cans. The bleeders and vents on the retort should be left open as the steam is turned on and bleeding should be kept open all throughout the process to prevent accumulation of air and non-condensable gases in steam in the retort. The agreement between temperature and pressure measuring instruments according to the steam table will indicate the absence of air in the retort. The diagram of a retort is given in Fig. 13.15.

13.5.6 COOLING OF THE PROCESSED CANS
The processed cans should be cooled to near about the room temperature as quickly as practically and economically justified to prevent discolouration, loss of flavour and to have a good control on processing. Processed cans are commonly cooled by water either exposing the cans to water sprays or by conveying the cans in crates through water in tanks. Hot cans may be put in water tanks. Cooling the cans inside the processing retort with water under pressure is also practised to prevent buckling of large diameter cans during cooling due to large pressure differences between the inside and the outside of the can.

13.5.7 LABELLING AND WAREHOUSING
The product may be labelled and packed in cases for shipment immediately after cooling or may be stacked for future shipment. The levels must comply with the regulations of the producing country or those of the importing country.

13.5.7.1 Warehousing
The external appearance of the can may deteriorate unless packed cans are properly stored. The warehouse must be dry, well lighted and reasonably cool. Canned fishery products will resist a fair degree of heat or cold for short periods but continued heat or alternate freezing and thawing are very injurious to quality particularly to texture and flavour. When properly stored, samples of canned fish have been found to be of satisfactory quality for several years.

A flow sheet of a sardine canning plant and the flow plan of a typical shrimp cannery are given on Fig. 13.16 and Fig. 13.17 respectively.

13.6 Dehydration of fish
Although the two terms 'drying' and 'dehydration' are synonymous, they are used for specific processes. While the term 'drying' is used to indicate the process of removal of moisture by exposure to natural currents of air with numidlty determined by climatic conditions, 'dehydration' is carried out in a man-made apparatus when air temperature, current and humidity are controlled as desired.

When the moisture content of fish flesh is brought down to less than 10 per cent, its storage life is extended by several months because at such moisture levels microbial action is arrested.

Drying in natural air current with solar heat is a relatively cheap process

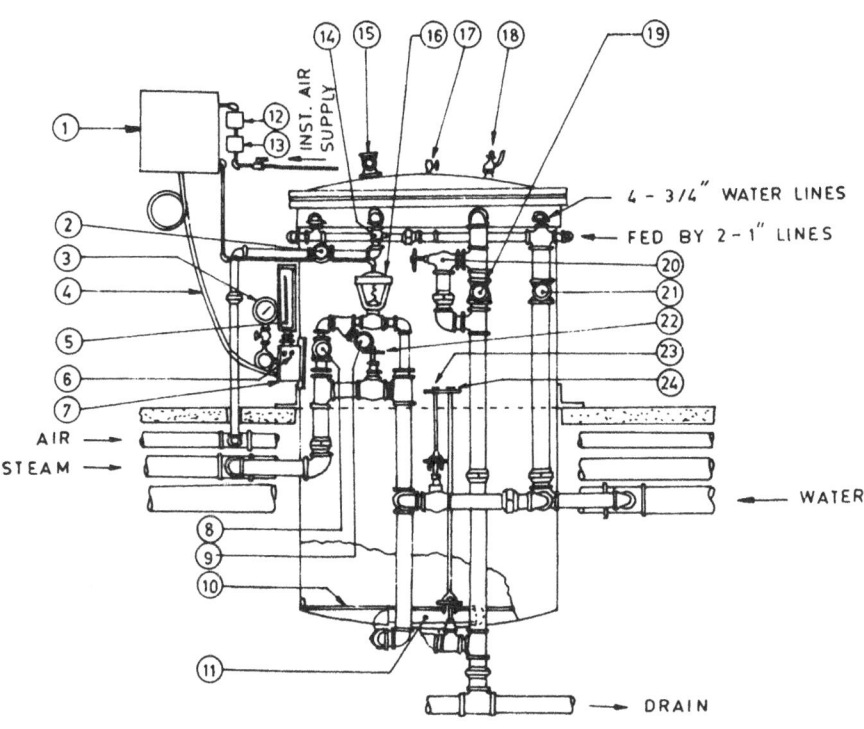

4 - 3/4″ WATER LINES

FED BY 2 - 1″ LINES

AIR

STEAM

WATER

DRAIN

1	TEMPERATURE RECORDER REGULATOR	13	AIR STRAINER
2	COOLING AIR VALVE	14	CHECK VALVE
3	PRESSURE GAUGE	15	BLOW OFF
4	TEMP. BULB	16	AIR OPERATED STEAM CONTROL VALVE
5	THERMOMETER	17	VENT
6	VENT	18	SAFETY VALVE
7	THERMOMETER POCKET	19	OVERFLOW VALVE
8	STEAM VALVE	20	SPRING RELIEF VALVE
9	STRAINER CLEANER VALVE	21	WATER VALVE
10	BAFFLE PLATE	22	BY - PASS VALVE
11	STEAM SPREADER	23	WATER VALVE
12	PRESSURE REDUCER	24	DRAIN VALVE

Fig. 13.15. Diagram of vertical retort equipped for pressure cooling. (Courtesy: American Can Company)

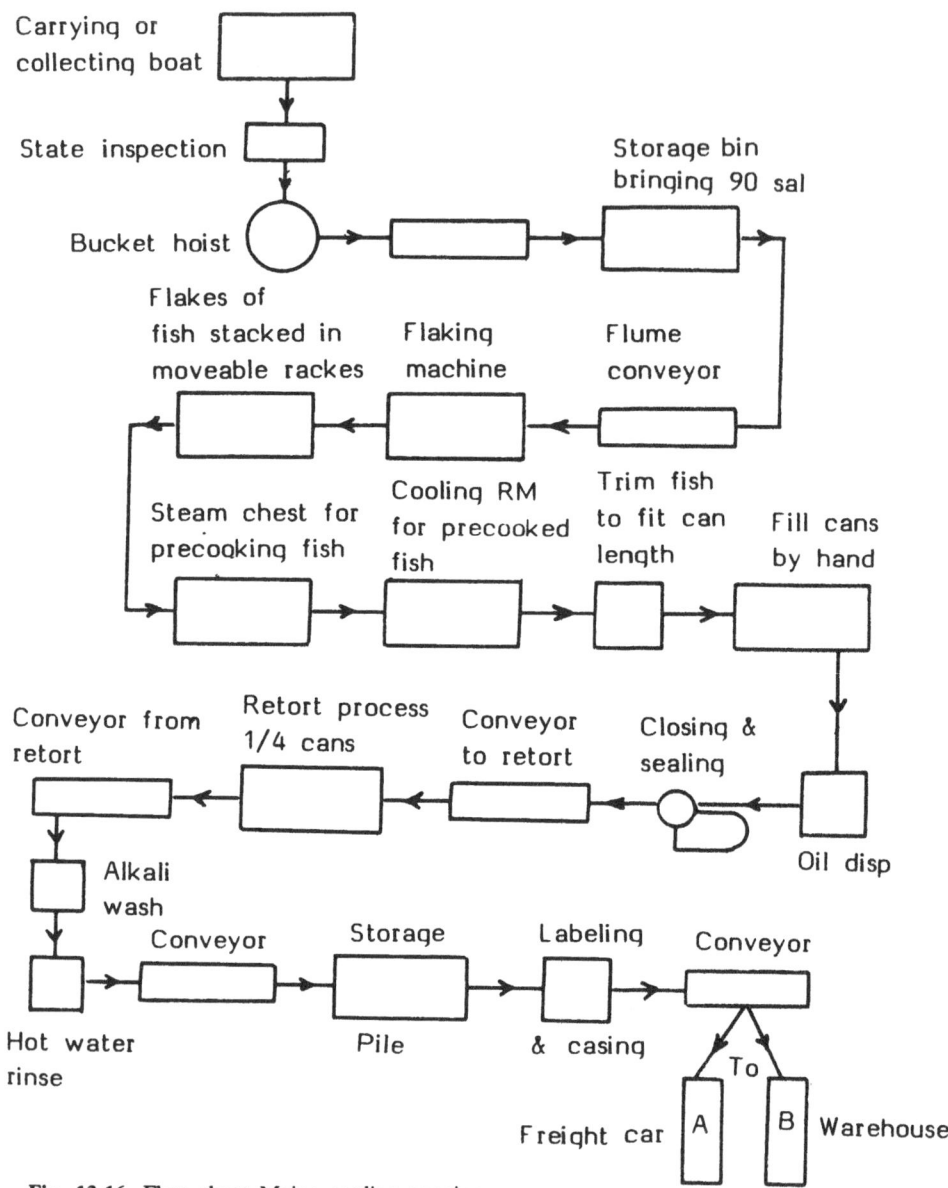

Fig. 13.16. Flow sheet Maine sardine canning

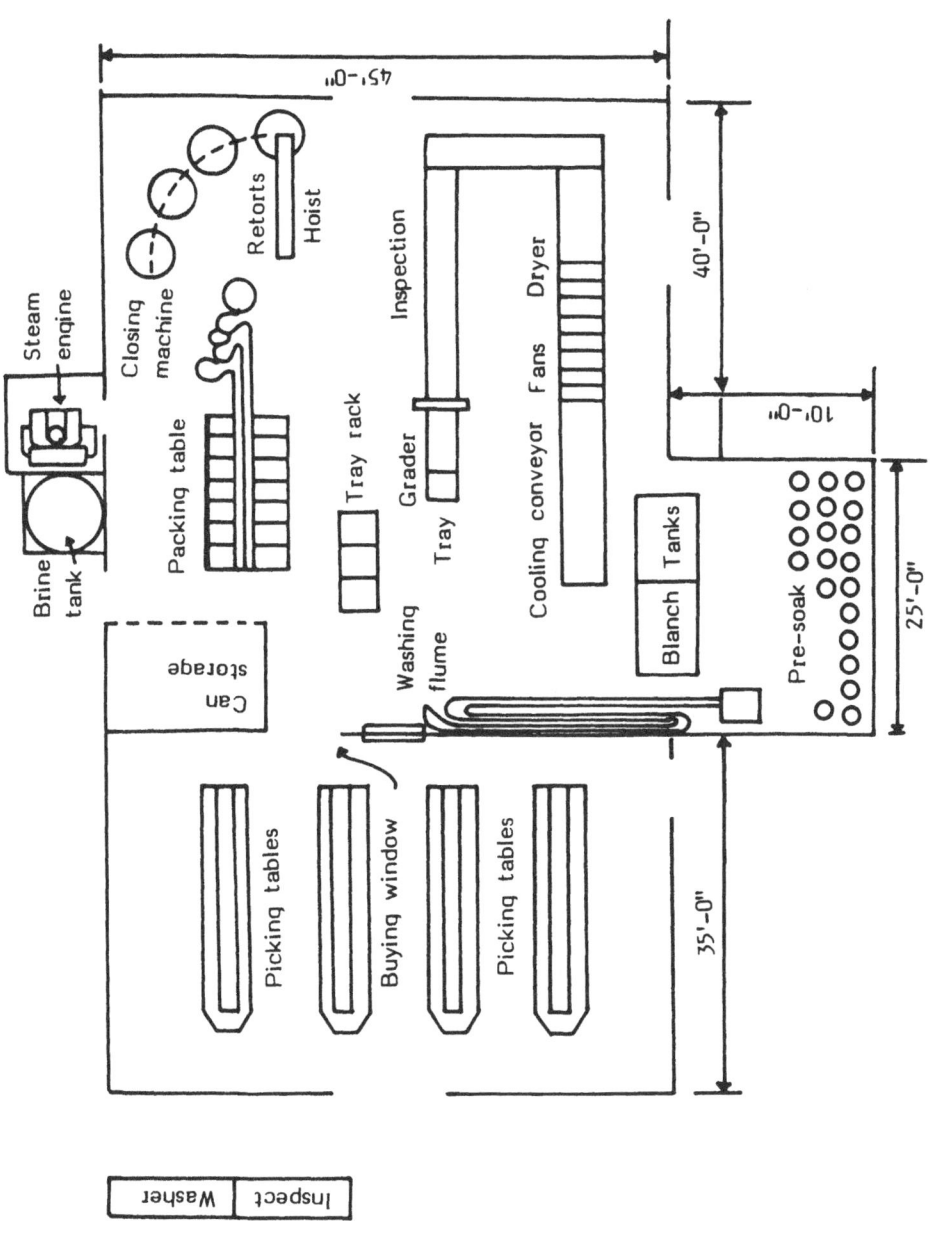

Fig. 13.17. Floor plan of a typical shrimp cannery. (Courtesy: American Can Company)

and is widely used, but it is a lengthy process which may require days and even months to reduce the moisture content to the safe level. Unless the temperature is favourable, as in countries in the Arctic region like Norway, the spoilage of fish will be significant before moisture is brought down sufficiently to stop microbial spoilage. Such a drying process is also affected by atmospheric conditions. Hence the quality of the product is highly variable. Drying in the open exposes the fish to attack by flies and contamination with dust and exposure to unsanitary conditions which often prevail in many of the sea beaches in tropical countries. For hot air drying the temperature of intact fish from temperate waters should not exceed 30°C to avoid break down of the flesh due to denaturation of protein. Fish from tropical water can be dried at a higher temperature round about 50°C (Anon, 1977). Drying at temperatures over 45°C helps in the destruction of fly maggots which often infect dried fish in tropical regions.

Controlled hot air drying is a highly energy consuming process and raises the product cost, often making it not economically viable. Pre-cooking of fish before drying raises the rate of moisture removal. The dehydration of raw fish takes place at a rate approximately one-third that of cooked fish (Stansby 1945). Mincing the cooked fish before drying increases the rate of moisture removal. Pressing cooked fish to expel some of the water in flesh has also been recommended (Young and Sidway, 1943).

Pre-cooking of fish before drying is generally performed at about 100°C for 20 to 30 minutes or under 5 to 10 psig steam for seven to ten minutes. Pre-cooked fish is often minced before dehydration to increase the surface of fish over which drying takes place. Pre-cooked fish can withstand a higher temperature than uncooked fish without loss of quality. Temperatures as high as 85 to 96°C can be used initially, when the moisture in fish is high, without any harm to the quality. But final drying should be at temperatures between 63 and 70°C. Low humidity of air in the drying chamber helps in augmenting the drying rate. Relative humidity of about 20 per cent is desirable. Drying under proper conditions of temperature and humidity of the drying air is accomplished in four to five hours and the moisture content of the dehydrated product is reduced to below 5 per cent. Conventional tunnel dryers or their variations have been used for dehydration of fish. Figure 13.18 shows diagram of three types tunnel dryers.

The storage conditions for dehydrated fish are very important in determining its shelf-life. If the product undergoes changes in storage, this may affect its colour, flavour and appearance. The main factors affecting shelf-life are humidity and temperature of the environment of dehydrated fish. Dried fish also develops a rancid flavour in storage due to the oxidation of lipids. This may happen even in non-oily fish due to the reduction in moisture, as the lipid content of fish relative to protein will be high. Dried fish stored in moisture/vapour-proof containers with nitrogen replacing air and at normal temperature, retains its good quality even after a year's storage (Culting and Reay, 1944).

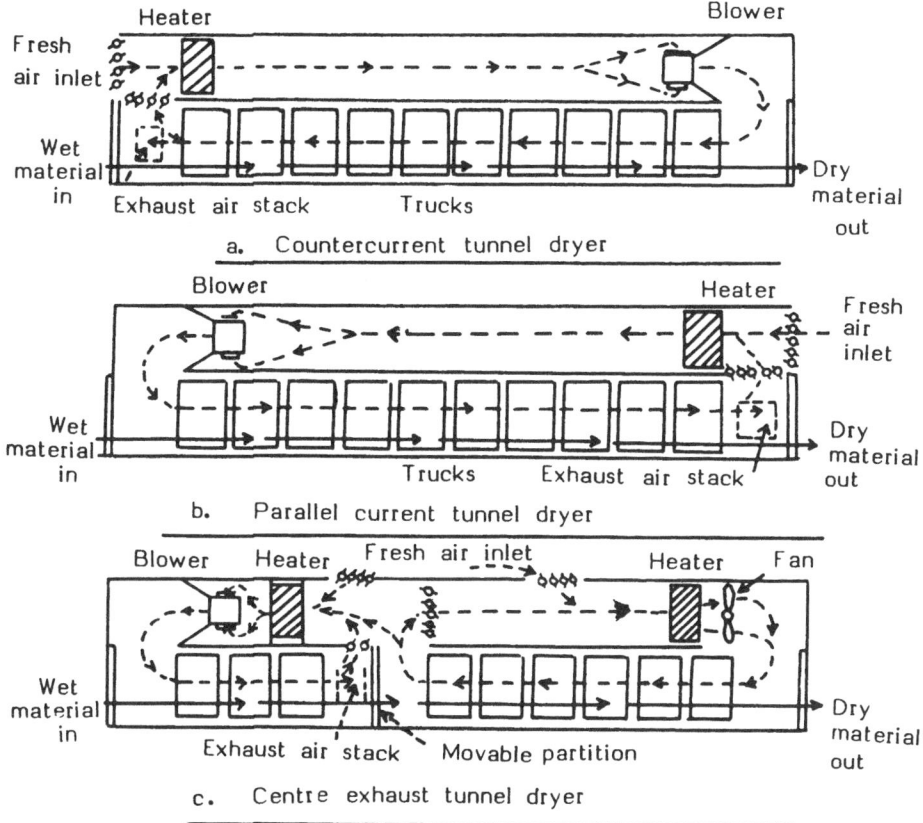

Fig. 13.18. Three types of tunnel driers

Due to the high energy consumption, the dehydration process is, unlike sun drying, rather expensive. In fact cost of production in dehydration is of the same order of magnitude as with canning process. Besides, the requirement of quick drying restricts the thickness or size of the individual pieces. Most of the dehydrated fish is marketed in a granular form which is not popular with consumers who want the identity of the fish to be retained. Hence the production of dehydrated fish in general has not been commercially successful.

Comminuted fish in the form flesh-suspension can be spray dried as in the case of herring. The suspension may also be roller dried provided it does not contain more than 2 per cent of oil (Anon, 1978). Fish meal for human consumption, FPC type B, has been produced in large quantities in Norway from oily fish. It is a relatively cheap protein-rich product with 70 to 75 per cent protein 10 to 15 per cent ash, up to 10 per cent moisture and up to 10 per cent oil. The product has been used mainly in relief feeding (Hansen, 1980).

13.7 Curing, salting and drying of fish

Curing of fish involves smoking, salting, drying and often combinations of these. Pickling and fermentation are also included in curing. Many cured products such as smoked fish are further frozen or canned for the market.

Cured fish including dried and salted products, is next in importance to fresh fish in the disposition of the world's harvested fish (Steinberg, 1980). Figure 13.19 gives the disposition of catch used for food as a percentage of world catch. Figure 13.20 gives the world production of cured fish Asia leads in the production of cured fish with seven million tons of fish used for the purpose in 1976. Production of cured fish has risen steadily in Asia while it is declining in Europe and remaining rather steady in other continents. Dried or salted fish alone constitute 85 per cent of the cured fish. Smoked fish production after remaining nearly stagnant between 1953 and 1963 has shown an upward trend since the mid-sixties.

Two main types of dried fish are commercially produced. (i) stock fish which is dried unsalted cod and related species, and (ii) dried salted fish.

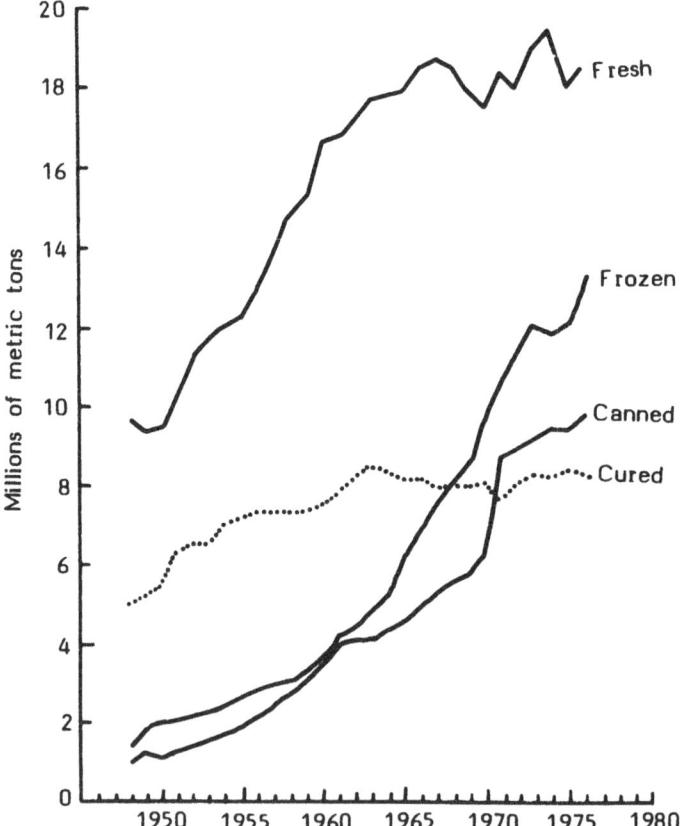

Fig. 13.19. Disposition of the catch used for food (percentage of world catch) (Source: Steinberg, 1980)

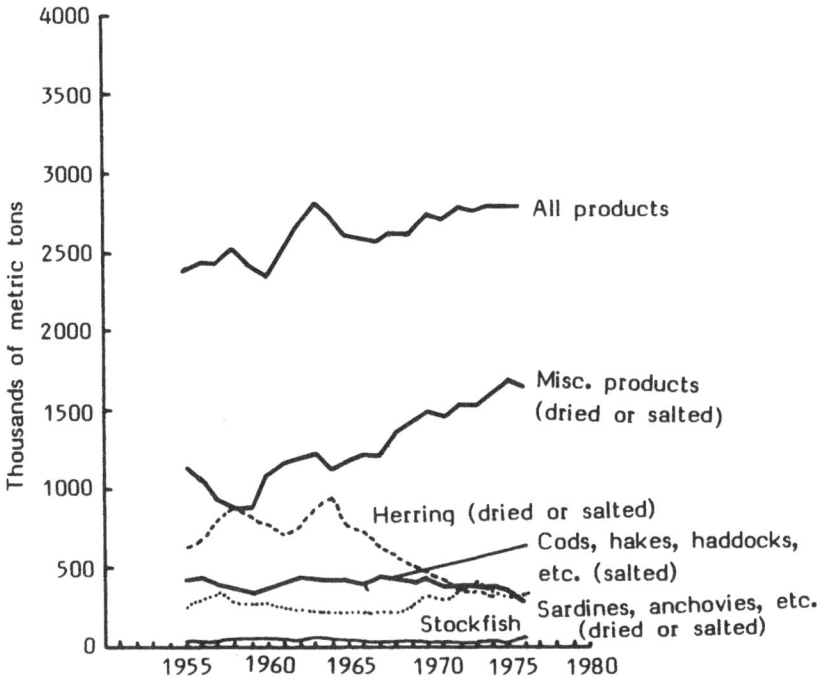

Fig. 13.20. Disposition of the catch, cured products by continents (live weight)
(Source: Steinberg, 1980)

Stock fish production is more or less confined to Norway where the dry air of the Arctic is highly suitable to air-drying of fish. Production is relatively small, about 20 to 50 thousand tons annually.

Sun drying of fish of a number of different species is common in India (where it accounts for 25.9 per cent of landed fish) and in many other countries in Asia, Africa and South America. However Japan and the USSR are the largest producers of miscellaneous dried or salted fish products.

13.7.1 SALTING OF FISH

Salting of fish is guided by the fact that the presence of more than 4 per cent of salt in the solution in the tissues of fish retards both autolytic and bacterial spoilage and when the salt concentration reaches 20 per cent, decomposition is all but stopped unless the product is stored for prolonged periods at elevated temperatures. In a warm climate the rate of decomposition may be so high as to be significant before the salt concentration in tissue fluids rises to a level where such reactions are stopped for all practical purpose. The composition of salt has great influence on the rate of penetration of salt into the tissues of fish and on the physical qualities of the product. Sea salt which is used for salting generally has a high content of calcium, magnesium and sulphates. Salts of calcium and magnesium sulphates retard the rate of penetration of sodium chloride and also affect the colour and firmness of the product. Calcium and magnesium salts give a strong bitter taste to the salted fish. The presence of even 1 per cent of these salts causes a whitening and stiffening of the fish flesh.

Two methods of salting are commercially practised—brine salting and dry salting. Brine salting which is done by putting cleaned fish in concentrated sodium chloride brine is of relatively less importance.

In dry salting, cleaned fish after beheading and splitting when required are packed in water-tight containers with an excess of dry salt. The proportion of salt to fish varies from 10 to 35 per cent by weight of the fish depending on the kind of fish, the weather and the practices followed by the salter. The fish is soon covered with pickle formed with water extracted by salt from the fish flesh. After the fish is completely salted it may be repacked with fresh salt or removed from the pickle and dried (Tressler, 1921). Sun drying of salted fish is commonly practised.

13.7.2 SMOKING OF FISH

Smoking is one of the earliest methods of processing of fish practised by man. The necessity to preserve easily perishable fish and at the same time protect the fish from insect attack which is a common feature with sun-dried fish led men to the discovery of the process of smoking. The specific flavour produced in smoked fish also gained popularity. The main purpose of smoking, as it is practised at present, is to impart to the fish the distinct and desirable flavour and colour. Smoke from burning wood contains a number of chemicals depending on the type of wood used. Although some of the phenols, aldehydes and fatty acids in the smoke may have antiseptic effect on the surface of fish, such effects are only very mild due to the low concentration of these chemicals deposited on fish (Shewan, 1945). Smoke contains some anti-oxidants to fat but salting of fish before smoking will nullify this effect (Stansby and Griffiths, 1943). Deposited phenols from wood smoke which turn a deep brown colour on exposure to oxygen are responsible for the colour of smoked fish (Linton and French, 1945).

Production of smoked fishery products involves four interrelated processes (i) salting, (ii) drying, (iii) heat treatment, and (iv) smoking.

Salting aids in firming the flesh by removing moisture and by denaturation of some proteins. Salt also inhibits the growth of bacteria at certain concentrations and imparts a flavour to the product. Fish may be heavily or lightly salted depending upon the type of product to be prepared.

Drying of the salted fish is necessary for removal of additional moisture thus aiding in its preservation. But reduction of moisture content to 60 to 75 per cent, as in most cases of smoked fish, is not sufficient to prevent bacterial action. The more important effect of drying is in 'pellicle' formation. The pellicle is the glossy firm surface imparted to fish which gives the desired appearance and which also helps in adsorption and development of the delicate smoked flavour.

Drying is usually accomplished in a smoking chamber. If the relative humidity in the chamber is above 75 per cent there will be practically no drying.

Salted fish is exposed to a higher temperature in the smoking chamber. The temperature to which fish will be exposed will depend on the type of smoking chamber used but generally lies between 29.4 and 32.2°C to prevent cooking of the flesh. Fat present in fish and the method of preliminary cure determine the temperature. For fatty fish such as salmon, temperature in the range of 26.7 to 29.4°C will be desirable while for lean fish a higher temperature around 32.2°C may be used. Fish exposed to such temperatures will not be freed from bacteria and will remain highly perishable requiring refrigeration for its preservation.

Hot smoking at a temperature between 48.9 and 82.2° produces commodities like 'kippered' and 'barbecued' items. At this temperature range autolytic enzymes in fish flesh are inactivated and moulds and most of the bacteria are destroyed. The proteins of the fish flesh are denatured. Besides there is significant reduction in the moisture. The product has a better storage life than cold smoked fish.

For smoking purpose, salted fish is exposed to smoke from burning wood in a properly designed chamber. A number of designs of smoke kilns are available. The 'Torry kiln' designed by the Tory Research Station is one such modern smoke kiln. Figure 13.21 gives the diagram of a controlled smoke house.

Hard wood is used for generating smoke. Soft wood is not recommend

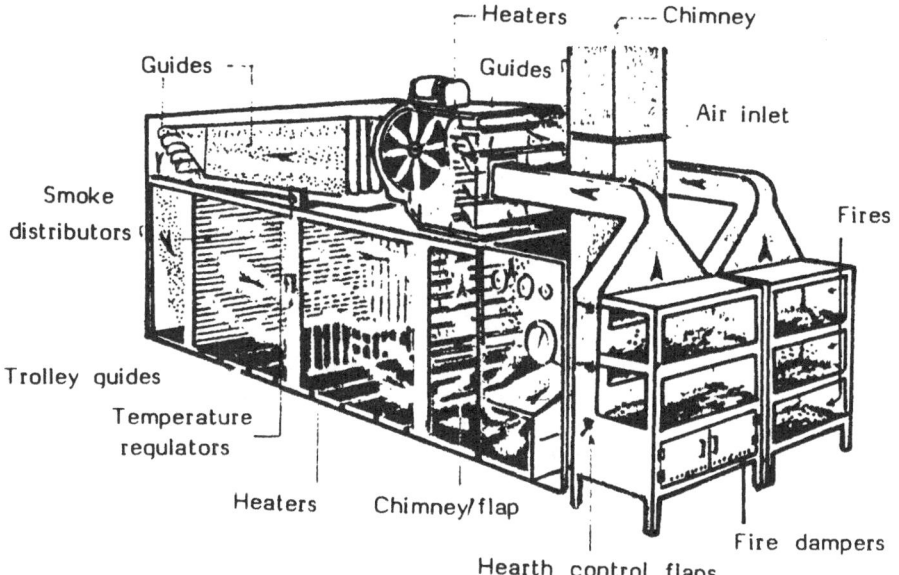

Fig. 13.21. Cutaway diagram of Torry kiln
Source: Fish Handling of Processing, 1965, Eds. Burgess G.H.O., Cutting C.L., Lovern J.A., Waterman J.J., Her Majesty's Stationery Office, Edinburg

because of its resinous nature and tendency to impart an acrid flavour and odour to the smoked product. Two types of smoking 'distilled' and 'blown' are commercially used. Distilled smoke is dense grey smoke produced by incomplete burning of wood. It contains a lot of moisture which reduces the shrinkage of smoked fish but less drying retards suitable pellicle formation. Distilled smoke also contains resinous matter which imparts an acrid smell and flavour to the product. 'Blown' smoke is lighter and produced by complete combustion of wood. As it contains less moisture, more drying helps in the formation of the desirable type of pellicle.

13.8 Fish meal and fish oil

Nearly 30 per cent of the world's fish catch is at present processed into fish meal and oil. Fish meal is used mainly in the preparation of animal feed while oil finds a number of industrial uses such as in the production of soap and glycerol, and leather tanning. The use of a such a large portion of a rich protein source, which fish is, for purposes other than human consumption has rightly been criticised. However 90 per cent of the fish currently being used in production of fish meal and fish oil, are unmarketable in large quantities as human food because they are unpalatable to consumer. Besides many of them are too small in size, spoil readily or turn rancid too quickly for economic storage and subsequent processing.

13.8.1 PRODUCTION OF FISH MEAL AND OIL

In the production of fish meal two categories of fish (i) lean, and (ii) oily fish, are used. A lean fish contains less than 2.5 per cent fat. On the other hand oily fish contains more than 2.5 per cent fat and requires removal of oil to make the fish meal produced from it, acceptable to the market.

The process for production of fish meal consists essentially of the following steps: (i) heating—which coagulates the fish protein, releases the bound water and also ruptures oil cells, (ii) pressing—to remove a large fraction of liquids from the heated mass, (iii) drying—which removes the desired amount of water to ensure the shelf-life of the product, and (iv) grinding of the dried material to the desirable granular form.

Yield and quality of fish meal and fish oil depend to a large extent on the quality of the raw fish; particularly on its freshness. A break down of fish protein and fish oil in raw fish is caused both by bacteria and autolytic enzymes. The prevention of these changes in fish economically, particularly when fish is to be stored for more than about 30 hours, is a challenge to the industry. Chilling to a temperature near about 0°C is an effective method but the economics of the fish meal and oil market do not generally make it a practical proposition. A number of chemicals have been suggested as preservatives but only formaldehyde amongst them is widely used. Formaldehyde however, reacts with lysine and thus affects the nutritive value. However, the use of a small amount, such as 0.05 per cent of the weight of fish, has no deleterious effect on protein quality.

Fish meal and oil is produced mainly by the wet pressing method. Larger raw fish are fed into a steam cooker after pulping. Cooking coagulates the fish protein and thereby liberates the bound water and oil. The cooked mass is drained on a strainer conveyer or a vibrating screen and passed on to a screw press wherefrom press cake and press liquor are obtained.

The press cake is dried in an indirect steam drier or direct fired drier. The dried meal is ground in a hammer mill. The ground meal is the fish meal of common use. The press liquor is passed through a decanter where most of the sludge is removed. Oil is separated from the liquor on a stick-water centrifuge where it is separated into three components; oil, stick water and fine sludge. The sludge from the decanter and the centrifuge is added to the press cake. Oil and stick water are passed on for further processing.

The oil component is poured through an oil separator, to remove sludge impurities and recover the fish oil.

The stick water is generally concentrated in a multi-effect evaporator and the concentrate is added to the press cake before drying.

Fish meal containing residual oil is prone to get heated by autoxidation, depending on the degree of unsaturation of remaining lipids in the meal and on the species of fish. Anti-oxidants are added to reactive fish meal immediately after manufacture to avoid undue heat generation. Anti-oxidants such an ethoxyquinone or BHT (Butylated Hydroxy Toluene) are used.

Most of the fish meal in the market is produced from whole fish and incorporates practically all the chemical constituents in raw fish except the oil. The fat soluble vitamins A and D are removed in the separated oil. However, fish meal still contains about 10 per cent of oil which contributes to the metabolisable energy of animals. Fish meal is also believed to contain an unidentified growth factor.

In fish meal production, particular care is to be taken to avoid its contamination with *Salmonella*. The bacteria are destroyed during the process of cooking and pressing but the temperature reached in drying may not be sufficient to destroy *Salmonella*. It is to be ensured that material does not get re-infected. High standards of sanitation are to be maintained to prevent re-infection.

Commercial fish meal contains 60 to 75 per cent protein, 6 to 14 per cent lipids (ethyl ether extract), 4 to 12 per cent moisture and 6 to 18 per cent ash. The protein in fish meal has a high and well-balanced content of essential amino acids. The damage to protein during processing, particularly drying, is indicated by reduction in the available lysine determined by reaction with flurodinitrobenzene (Sanger's method modified by Carpenter).

Figure 13.22 gives a lay out of a fish meal plant.

The air and vapours emitted from dryer and evaporator have a highly objectionable smell. They should be passed through a scrubber and through a boiler or treated with chlorine before being let out into the atmosphere.

Material balance in fish meal production is given in Fig. 13.23.

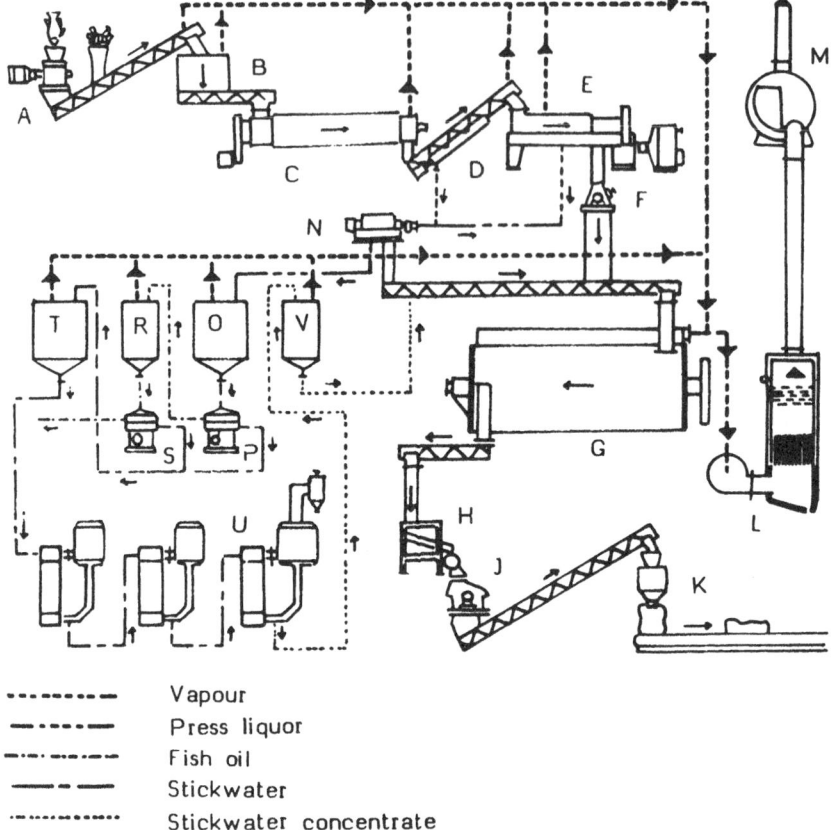

--------   Vapour
— · — · —   Press liquor
—·——·——·—   Fish oil
————— — —   Stickwater
·—··········· Stickwater concentrate

Fig. 13.22. Layout of a fish meal plant

13.9 Chitin and chitosan

Chitin is a linear polymer of anhydro-N-acetyl-D-glucosamine. The monomar units are linked by 1-4, glycosidic bonds as in cellulose. It is insoluble in water and in most of the organic solvents. De-acetylation of chitin gives chitosan, a high molecular weight linear polymer of amino-D-glucose.

Chitin and its derivatives, particularly chitosan, find many industrial applications in paper making, textile printing, and sizing, ion-exchange chromatography, removal of metal ions from industrial effluents, manufacture of pharmaceuticals and cosmetics and as an additive in the food industry. Chitin has also shown to have a growth promoting effect on broiler chicks.

Chitin is the most important organic constituents of exoskeletal materials of animals such as crustacea (shrimp, crab oysters, mulluscs, etc.). The tough and resilient property of chitin is utilised by living organisms as skeletal support and body armour against attack by other animals. The most important economic source of chitin is wastes (head and shell) from the shrimp processing industry. Shells of lobster, crab, squilla and squid pens also provide a good source of chitin.

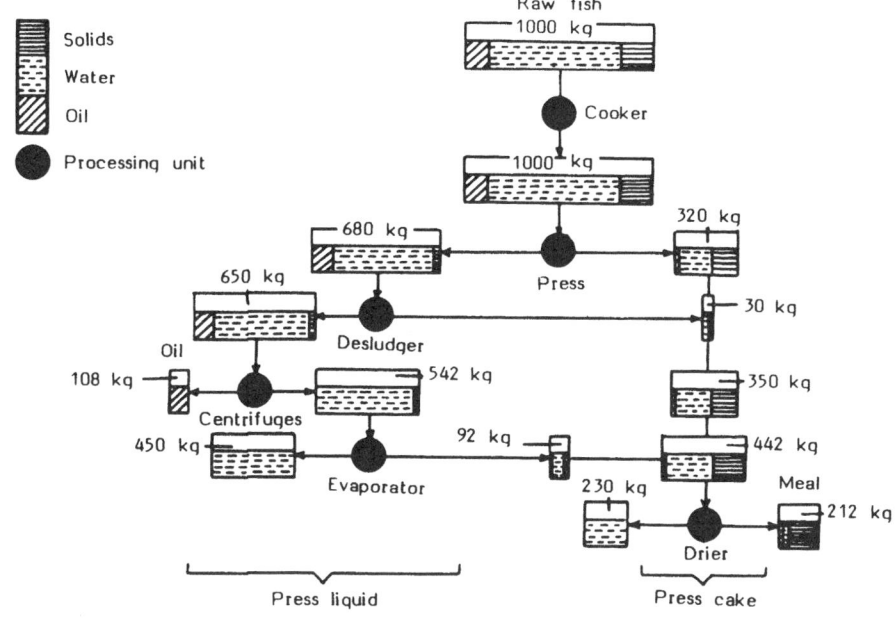

Fig. 13.23. Material balance in fish meal production

Manufacture of purified chitin has been the subject of a number of patent specifications and technical papers. Natural chitin is bound by protein and calcium carbonate, removable to a greater and lower degree by different purification methods. The extracting process involves grinding the raw material, removal of mineral component with dilute hydrochloric acid followed by removal of proteins enzymatically or with dilute NaOH solution. The pink colouration in chitin can be removed by mild oxidation. Chitin can also be purified by decalcifying with ethylene diamine and removal of lipids by various kinds of organic solvents. Removal of the acetyl group with a strong alkali at an elevated temperature readily forms chitosan, its simplest derivative (Anon, 1987). Figure 13.24 gives a flow sheet for production of chitin and chitosan. In the process, prawn waste from processing plants is washed with water mixed in the ratio of two parts of prawn waste to three of alkali solution heated by boiling in 0.5 per cent aqueous sodium hydroxide for 30 minutes. The alkali is drained off and kept separately for recovery of protein. The residual protein in the treated mass is removed with an equal amount of sodium hydroxide solution, draining off the alkali and repeating the process. The residue after draining off the alkali is demineralised by keeping it in hydrochloric acid (1.25N) for one hour. The residue after washing of the acid gives commercially pure chitin. Bleaching may be done if required though it may not be necessary for normal commercial requirement.

Chitosan is produced from chitin manufactured as mentioned by de-acetylation of chitin with 1:1(W/W) aqueous sodium hydroxide for 1.5 to 2.0 hours at 90 to 95°C. The residue after treatment with and draining off of alkali is washed free of alkali, dried and pulverised to the required size (Madhavan *et al.*, 1986).

The properties of chitosan are affected to a large extent by the degree of demineralisation. A progressive increase in the concentration of hydrochloric acid in the treatment liquor increases the degree of demineralisation but an acid concentration above 1.25N HCl affects the viscosity of chitosan. The process described here gives chitosan of low viscosity. For high viscosity chitosan the de-acetylation and demineralisation treatments are suitably modified. Flow diagram for a chitosan plant is given in Fig. 13.24.

13.10 Fish protein concentrate (FPC)

The term "Fish Protein Concentrate" (FPC) applies to all forms of dried fish including fish meal intended for human consumption. Although the process of concentration may just mean the removal of water as in dried fish, further concentration by removal or partial removal of lipids particularly and also other non-protein substances, has given rise to the classification of FPC into different groups on the basis of their protein contents.

Interest in the production of FPC to supplement the protein food supply was created by the presence of large amounts of underutilised or unutilised resources of fish, in the seas surrounding every continent, which could be converted into an acceptable food product. Such a product would help in providing high quality protein as a supplement to lesser quality vegetable proteins which constitute the main source of this important nutrient for a large section of the population particularly in the developing countries. To meet this specific purpose, the product should be such that it can be stored, shipped and marketed at ambient temperatures throughout the world without significant loss in protein quality. The product should also be inexpensive and available in a form suitable for incorporation in to the traditional food preparations of the consumers.

The problems in manufacture of FPC differ according to the extent of removal of lipids from the fish, as it dominates all other factors in determining the taste and flavour of the product. Even the leanest of fish, for instance cod flesh, contains 0.6 per cent of total lipids. If dried to 10 per cent moisture content, the resulting FPC will contain about 2.6 per cent of fat, sufficient to cause an undesirable off-flavour (Lovern, 1965).

Attempts to produce odourless, light coloured, dry powder from underutilised and easily available fish, have been made with varying degrees of success in many countries. These are based on using solvents or their mixtures which would remove fats and also other lipids. Use of chlorinated hydrocarbons as solvents led to the establishment of the first FPC production plant. The process which attracted much interest was developed in the

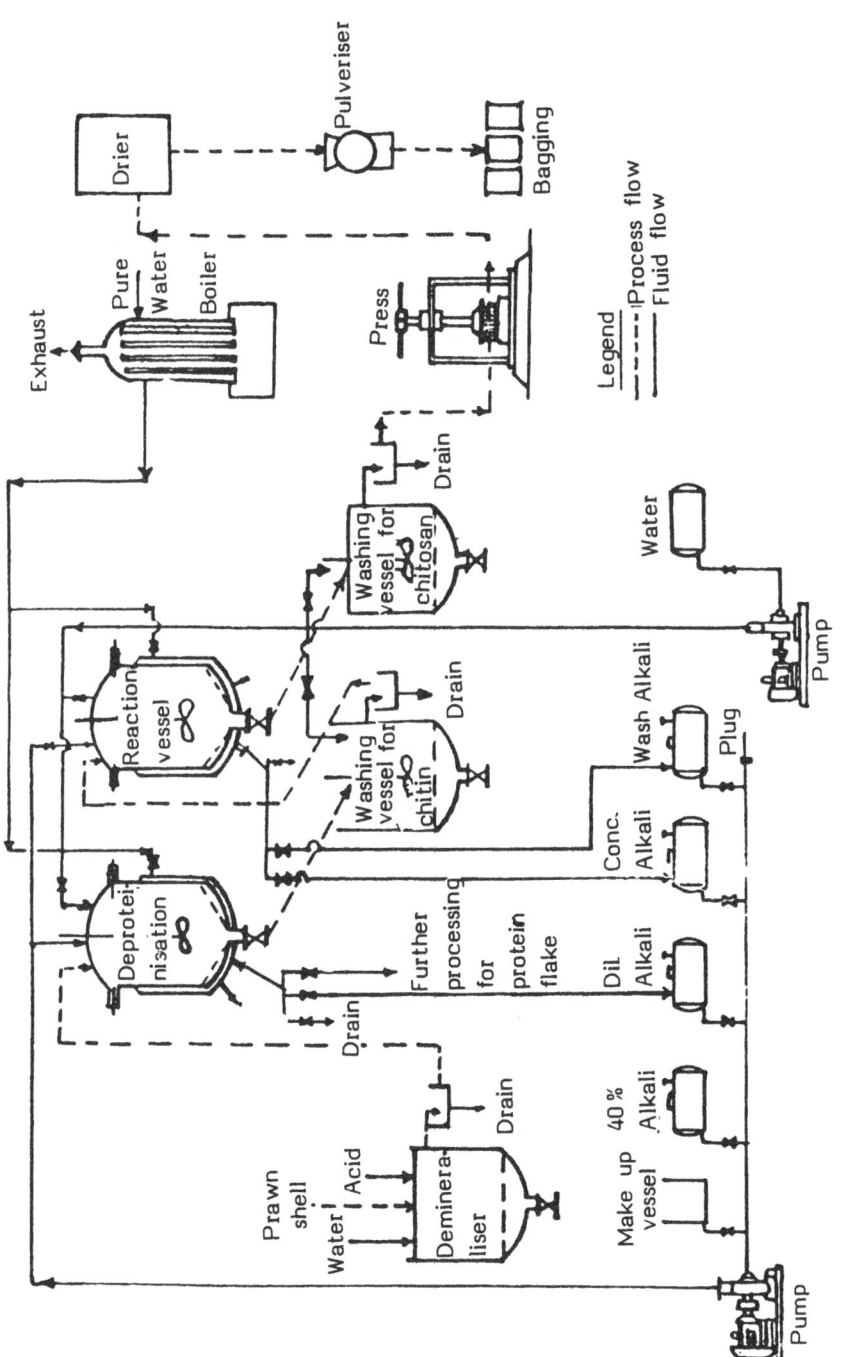

Fig. 13.24. Flow diagram of chitosan plant
Courtesy: Central Institute of Fisheries Technology, Cochin, India

laboratories in the U.S. Bureau of Commercial Fisheries Technological Laboratory. The process is as follows:

The process (Snyder *et al.*, 1976) used the fish Red Hake as the raw material because: (i) it is abundant off the US coast, (ii) it is largely unutilised as a food fish, and (iii) sufficient amount of fish of good quality is available for a considerable period of the year.

Fish to be processed is stored near 0°C. The stored fish is washed, rinsed with fresh water, weighed and ground. The ground fish is fed to the first extractor, which is a jacketed vessel well stirred with solvent. The solvent used is azeotropic isopropyl alcohol containing 87 per cent by weight of alcohol and 13 per cent water.

In the first stage of the extraction process, the dehydration takes place in an unheated vessel where ground fish is mixed with centrifugate Miscella M_2, obtained from stage two, from the previous batch sequence of extraction. The temperature in the first extractor is within the range of 20 to 30°C. The contents of the extractor are fed to a continuous centrifuge from which Miscella M_1 and wet cake S are obtained. M_1 contains the water and lipids from the fish mince and is treated subsequently to recover the solvent and other valuable components. The cake, S_1 is fed to the second extractor which contains Miscella M_3 from the third extractor. Extraction is for 90 minutes at a temperature of 75°C. The contents of the second extractor are centrifuged and the wet cake S_2 is carried to the third extractor containing fresh distilled and purified hot isopropyl alcohol. Extraction is for about 50 minutes. Miscella M_3 is sent to the second extractor for the next batch in the extraction sequence and the wet-cake S is passed on to the desolvenising device, which is a vacuum drum dryer where the solvent is removed from the wet cake. The vapours drawn from the dryer are condensed by using refrigerated brine as the water. The composition of the condensate is essentially an azeotropic mixture of isopropyl alcohol and water.

The dry cake from the dryer is conveyed to the mill to grind FPC to a fine powder and packaged.

The proximate analysis of FPC prepared from is given in Table 13.6.

The amino acid composition of FPC as compared with that of whole egg is given in Table 13.7.

Table 13.6
Proximate analyses of 10 samples of fish protein concentrate prepared by isopropyl alcohol extraction of hake

Component	Mean	Range	Standard error
Crude protein	80.9	78.0–82.3	0.4
Volatiles	7.71	6.25–10.78	0.41
Ash	13.5	12.9–14.4	0.1
Lipids	0.18	0.13–0.22	0.01

(from Snyder *et al.*, 1967)

Table 13.7
Comparison of amino acid composition of fish protein concentrate (FPC) and whole egg

Amino acid	FPC[1]	Whole egg[2]
	g/16 g nitrogen	
Lysine	8.29	6.40
Histidine	1.96	2.40
Arginine	6.83	6.56
Threonine	4.33	4.98
Valine	4.98	7.42
Methionine	3.24	3.14
Cystine	0.86	2.34
Isoleucine	4.37	6.64
Leucine	7.33	8.80
Phenylalanine	4.01	5.78
Tyrosine	3.21	4.30
Tryptophan	0.97	1.65

(from Snyder et al., 1967)

REFERENCES

Alvarez, R.J. and Koburger, J.M. (1978): Microbial development on shrimp as affected by delayed heading: *Proceedings of 3rd Annual Tropical and Sub-Tropical Conference of Americas, New Orleans*, 23–26 April, No. TANU-SG-79.

Anon. (1977): Sun-powered fish drying: *Aust. Fish* **36**(5).

Anon. (1978): *Annual Report for 1977*. Technological Laboratory, Ministry of Fisheries, Lyngby, Denmark.

Anon. (1987): Chitin: *Infofish*, Sept-Oct, 1987 (FAO publication). P. Madhavan, K.G. Ramachandran Nair, T.K. Thankappan, P.V. Prabha and K. Gopakumar. Production of chitin and chitosan, *Extension Note*: Central Institute of Fisheries Technology, Cochin.

Astrom, S. (1971): Freezing equipment influence on weight losses. *Proc. Third International Congress* of Food Science & Technology. Streuat and C.L. Willey eds., Inst. of Food Technologists, Chicago.

Ball, C.O. (1923): Thermal Process Time for canned foods. *Bull. Nat Res. Council*, The National Academy of Science, Washington DC, 7: 37.

Ball, C.O. (1928): Thermal Processing of Canned Food: University of California Publications in Public Health 1:2

Bigelow, W.D., Bohart, G.S., Richardson, A.L. and Ball, C.O. (1920): Heat penetration on processing of canned foods, *NCA Bull. 16-L*.

Bigelow, W.D. (1921): The Logarithmic Nature of Thermal Death Time Curves, *J. Infect. Dis.* **29**(5).

Bose, A.N. (1969): Freezing of tropical fish. In: *Freezing and Irradiation of Fish*, Rudlaf Kreuzer, ed., Fishery News (Books), Ltd., London.

Carpenter, K.J. (1960): The estimation of the available lysine in animal protein foods, *Biochem. J.*, 77.

Cutting, C.L., and Reay, G.A. (1944): The dehydration of fish: *Chem. & Ind.*, **63**(47).

Hansen, P. (1980): Fish Preservation Methods: *Adv. in Fish Sci & Tech.*, Fishing News Book Ltd.

Heen, E. (1981): *Advances in the refrigerated treatment of fish*, International Institute of Refrigeration, Paris.

Horsley, R.W. (1970): The bacterial flora of Atlantic Salmon (*Salmo salar*) in relation to its environment. *J. App. Bact.*, 36.

Laycock, R.A. and Regier, L.W. (1970): Psedomonads and Achromobacters in spoilage of irradiated haddok of different proirradiation quality, *Appl. Microbiol.* **20** (3).

Lekshmy, A., Gobindan, T.K., Mathew, A. and Pillai, V.K. (1962): Studies on frozen storage of prawn. *Indo-Pacific Fish.* Conf.: *Occasional Paper* 63/7.

Linton, C.P. and French, H.V. (1945): Factors affecting disposition of smoke constituents on fish. *J. Biol. Bd., Can.* **6**.

Liston, J. (1980): Microbiology in Fishery Science. *Adv. Fish. Sci. & Tech.*: Fishing Book News Ltd.

Liston, J. and Shewan, J.M. (1958): Bacteria brought into brines of fish. *The Microbiology* of *Fish* and *Meat Curing Brine.* B.P. Eddy, ed., HMSO, London.

Londahl, G. (1981): Refrigerated storage in fisheries. *FAO Fisheries Tech. Paper* No. 214.

Lovern, J.A. (1965): Technological Problems in the production and storage of fish protein concentrate, the solutions to which require more information from fundamental research, In: *The Technology of Fish Utalisation*; contributions from research. R. Kreuzer, ed., Fishing News (Books) Ltdö. London.

Raccach, M. and Baker, P. (1978): Microbial properties of mechanically deboned fish flesh. *J. Food Sci.* **43**.

Raj, H. and Liston, J. (1961): Survival of bacteria of public health importance in frozen sea-foods. *Food Technol.,* **15**.

Raj, H. (1968): Public health bacteriology of proteio foods I Dehydrated and reconstituted batter. In: *Microbiology of Dried Foods*, E.H. Kampelmacher, M. Ingram and D.A.A. Mossel, eds., Grafishe Industrial., Haarlem, the Netherlands.

Reay, G.A. and Shewan, J.M. (1949): The spoilage of fish and its prevention by chilling: *Advances in food Res.,* **11**.

Shewan, J.M. (1945): Some of the principles in the smoke-curing of fish: *Chem. & Ind.* **64**.

Shewan, J.M. (1949): Some bacteriological aspects of handling, processing and distribution of fish, *J. Roy. Sanit. Inst.* **59**.

Shewan, J.M. (1971): The microbiology of fish and fishery products—a progress report—*J. Appl. Bacteriol.,* **38**.

Shewan, J.M. (1977): The bacteriology of fresh and spoiling fish and the biochemical changes induced by bacterial action. *Handling, Processing* and *Marketing* of *Tropical Fish.* Tropical Products Institute, London.

Snyder, D.G., Hammerle, O.A., Brown, N.L. and Knoble, G.M. Jr. (1967): Fish protein concentrate. *Proc. Int. Symp on protein foods and concentrates* CFTRI, Mysore.

Stansby, H.E. and Grififiths, F.P. (1943): Preparation and keeping quality of lightly smoked mackerel: *U.S. Fish & Wildlife Service Res. Rept.* **6**.

Stansby, M.E. (1945): Dehydration of Fishery Products: *U.S. Fish and Wildlife Service, Leaflet* No. 120.

Steinberg, M.A. (1980): *Advances in Fish Science and Technology,* Fishing News Books Ltd.

Thompson, G.E. (1919): Temperature time relations in canned foods during sterilization. *J. Indus. and Engg. Chem.* **11** (7).

Tressler, D.K. (1921): Some considerations concerning the salting of Fish: *U.S. Bureau of Fisheries, Doc.* **884**.

Venkataraman, R. and Sreenivasan, A. (1952): A preliminary investigation of the bacterial flora of the mackerels of the West Coast. *Indian J. Med. Res.* **40**.

Williamson, E.D. and Adams, L.H. (1919): Temperature distribution solids during heating & cooling. *Phys. Rev.* 2nd S. **14** (2).

Young, O.C. and Sidway, E.P. (1943): Dehydration of fish: *Fish. Res. Bd. Can. Prog. Rept., Pacific Coast Station.*, No. 16.

Index